THE GREEN ELEMENT METHOD

THE GREEN ELEMENT METHOD

by

Akpofure E. Taigbenu
Department of Civil and Water Engineering
National University of Science and Technology,
Bulawayo

KLUWER ACADEMIC PUBLISHERS
Boston / Dordrecht / London

Distributors for North, Central and South America:
Kluwer Academic Publishers
101 Philip Drive
Assinippi Park
Norwell, Massachusetts 02061 USA
Telephone (781) 871-6600
Fax (781) 871-6528
E-Mail <kluwer@wkap.com>

Distributors for all other countries:
Kluwer Academic Publishers Group
Distribution Centre
Post Office Box 322
3300 AH Dordrecht, THE NETHERLANDS
Telephone 31 78 6392 392
Fax 31 78 6546 474
E-Mail <services@wkap.nl>

Electronic Services <http://www.wkap.nl>

Library of Congress Cataloging-in-Publication Data

A C.I.P. Catalogue record for this book is available
from the Library of Congress.

Printed on acid-free paper.

Printed in the United States of America

To
Foluke
— my best friend and wife

TABLE OF CONTENTS

PREFACE xiii

ACKNOWLEDGMENTS xv

CHAPTER 1: PRELIMINARIES 1

- 1.1 Introduction 1
- 1.2 Motivation for the Green Element Method 4
- 1.3 Elementary Matrix Algebra 9
- 1.4 References 12

CHAPTER 2: LINEAR LAPLACE/POISSON EQUATION 15

- 2.1 Introduction 15
- 2.2 Derivation of Linear Laplace/Poisson Equation 15
- 2.3 Boundary Conditions 17
- 2.4 Integral Equations 18
- 2.5 Conventional Procedure of Implementing the Boundary Element Method 26
- 2.6 Solution Procedure of the Green Element Method 27
  - 2.6.1 Choice of the Value of k 29
  - 2.6.2 Assembling the Global Matrix 30
- 2.7 Comparison of Boundary Element and Green Element Formulations 33
- 2.8 Evaluation of the Line Integral of the Recharge Term 34
- 2.9 Numerical Examples 39
- 2.10 Remarks 47
- 2.11 Exercises 48
- 2.12 References 49

CHAPTER 3:	**NONLINEAR LAPLACE/POISSON EQUATION**	51
3.1	Introduction	51
3.2	Derivation of Differential Equation (Heat Conduction Example)	51
3.3	Integral Representation	53
3.4	First Green Element Model (GESNH-1)	54
3.5	Interpolation	56
3.6	One-Dimensional Elements	58
	3.6.1 Linear Elements	58
	3.6.2 Quadratic Elements	61
3.7	Second Green Elements Model (GESNH-2)	62
3.8	Third Green Element Model (GESNH-3)	67
3.9	Examples on Linear Heterogeneous Laplace/Poisson Problems	69
	3.9.1 Implementation of GESNH-1 on a 3-Element Computational Region	71
	3.9.2 Implementation of GESNH-2 on a 3-element Computational Region	72
	3.9.3 Implementation of GESNH-3 on a 3-Element Computational Region	73
3.10	Nonlinear Solution Strategies	77
	3.10.1 Linear Iteration or Picard scheme	77
	3.10.2 The Newton-Raphson Solution Algorithm	79
3.11	Applications of Picard and Newton-Raphson Algorithms in Green Element Calculations	82
3.12	Example 2 — Nonlinear Example	84
3.13	Applications to Structural Analysis of Beams	86
	3.13.1 Example 3	88
	3.13.2 Example 4	91
3.14	Remarks	94
3.15	Exercises	94
3.16	References	96
CHAPTER 4:	**HELMHOLTZ EQUATION**	97
4.1	Introduction	97
4.2	Derivation of Helmholtz Equation	97
4.3	Two Green Element Models	99
	4.3.1 Model 1	99
	4.3.2 Model 2	101
4.4	Numerical Experiments to Establish Suitable Range of Values for the Penalty Parameter	108
4.5	Hand Computations	110
4.6	Computer Simulations of the Three Helmholtz Examples	112

4.7	Remarks	113
4.8	Exercises	114
CHAPTER 5:	**TRANSIENT DIFFUSION**	117
5.1	Introduction	117
5.2	Derivation of the Transient Diffusion Equation	117
5.3	Boundary and Initial Conditions	120
5.4	Quasi-Steady Green Element Model (QSGE)	120
5.5	Transient Green Element Model (TGE)	124
5.6	Stability Characteristics of GEM for the Diffusion Equation	131
5.7	Numerical Examples of Transient Linear Diffusion	140
5.8	Nonlinear Diffusion	142
5.9	Numerical Example on Transient Nonlinear Diffusion	146
5.10	Remarks	149
5.11	Exercises	150
5.12	References	151
CHAPTER 6:	**TRANSPORT EQUATION**	153
6.1	Introduction	153
6.2	Model 1: Quasi-Steady Green Element Model (QSGE)	154
6.3	Mode 2 (TGE Model)	158
6.4	Model 3 (ADGE Model)	161
6.5	Stability Characteristics of **GEM** for the Transport Equation	170
6.6	Numerical Examples	181
6.7	Remarks	191
6.8	Exercises	193
6.9	References	194
CHAPTER 7	**BURGERS EQUATION**	195
7.1	Introduction	195
7.2	Model 1 for the Burgers Equation	197
7.3	Model 2 for the Burgers Equation	200
7.4	Model 3 for the Burgers Equation	202
7.5	Numerical Examples	204
7.6	Numerical Experiments and Results	206
7.7	Remarks	214
7.8	Exercises	215
7.9	References	215
CHAPTER 8	**UNSATURATED FLOW (RICHARDS EQUATION)**	217
8.1	Introduction	217
8.2	Derivation of the Unsaturated Flow Equation	219
8.3	Green Element Formulation	222

8.4	Numerical Examples	225
8.5	References	230
CHAPTER 9	**HIGHER-ORDER ELEMENTS**	231
9.1	Introduction	231
9.2	Green Element Equations with Hermitian Interpolation Polynomials	232
	9.2.1 Hermitian **GEM** for the Transport Equation	233
	9.2.2 Hermitian **GEM** for Burgers Equation	235
	9.2.3 Hermitian **GEM** for the Unsaturated Flow Equation	237
9.3	Numerical Stability Characteristics of Hermitian **GEM** for the Transport Equation	240
9.4	Numerical Calculations	243
	9.4.1 Examples of Contaminant Transport	243
	9.4.2 Examples of Momentum Transport (Burgers Equation)	244
	9.4.3 Examples of Unsaturated Flow	247
9.5	References	250
CHAPTER 10	**STEADY TWO-DIMENSIONAL PROBLEMS**	251
10.1	Introduction	251
10.2	Steady Second-Order Differential Equation	251
10.3	Boundary Conditions	252
10.4	Integral Equations	253
10.5	Green Element Models	255
10.6	Polygonal Elements and their Interpolation Functions	260
	10.6.1 Bilinear Rectangular Element	260
	10.6.2 Linear Triangular Element	261
10.7	Boundary Integrations	263
10.8	Domain Integrations	265
10.9	Numerical Examples	271
10.10	Remarks	288
10.11	References	288
CHAPTER 11	**UNSTEADY TWO-DIMENSIONAL PROBLEMS**	289
11.1	Introduction	289
11.2	Flow Equation	289
11.3	Green Element Modeling	290
	11.3.1 Quasi-Steady Models	290
	11.3.2 Transient Model	295
11.4	Numerical Examples	299
11.5	Remarks	311
11.6	References	313

CHAPTER 12	**FURTHER CONSIDERATIONS**	315
12.1	Introduction	315
12.2	Quadratic Rectangular and Triangular Elements	315
	12.2.1 Quadratic Rectangular Element	316
	12.2.2 Quadratic Triangular Elements	317
12.3	Domain Integrations For Quadratic Elements	318
12.4	Simulation Of Example With Quadratic Elements	324
12.5	Isoparametric Elements	324
12.6	Three-Dimensional Problems	329
12.7	References	331
APPENDIX A		333
A.1	Capabilities of the Program GEMLN1D	333
A.2	User Manual of the Executable Program GEMLN1D	333
A.3	Hint to Solution to some of the Exercises	334
APPENDIX B		335
B.1	Element Matrices for Linear Rectangular Elements	335
B.2	Element Matrices for Linear Triangular Elements	337
APPENDIX C		339
C.1	Element Matrices for Quadratic Rectangular Elements	339
C.2	Element Matrices for Quadratic Triangular Elements	344
AUTHOR INDEX		349
SUBJECT INDEX		351

PREFACE

Most texts on computational methods are borne out of research activities at postgraduate study programs, and this is no exception. After being introduced to the *boundary element method* (**BEM**) (then referred to as the *boundary integral equation method* (**BIEM**)) in 1981 by Prof. Jim Liggett of Cornell University, a number of graduate students and myself under his supervision took active interest in the development of the theory and its application to a wide range of engineering problems. We certainly achieved some amount of success. A personal desire to have a deeper understanding and appreciation of computational methods prompted one to take related courses in finite deference method, and to undertake a self-instructed study of variational and finite element methods. These exposures were not only quite instructive but fruitful, and may have provided the motivation for the current research on the *Green element method* (**GEM**) — a name coined by Prof. Liggett in 1987 during my visit as Professor to the School of Civil & Environmental Engineering, Cornell University.

The main objectives of this text are to serve as an instructional material to senior undergraduate and first year graduate students undertaking a course in computational methods, and as a resource material for research scientists, applied mathematicians, numerical analysts, and engineers who may wish to take these ideas to other frontiers and applications. To achieve the first objective, the first six chapters of the text have been written in such a way that it can serve as an instruction text to students or as self-teaching material for those who wish to study at their pace. To this end, calculations of some of the numerical examples have been done by hand so that the student can have a good feel for the method, and exercises have been designed at the end of some of the chapters with which to practice. Some of these exercises, which are presented in data files, can be run with the executable program GEMLN1D that can be accessed either at: www.nust.ac.zw/aetaigbenu/gem/GEMLN1D or at: www.lafetech.com/gem/GEMLN1D.

Most of the theoretical aspects of the Green element method are discussed in chapter 2 for 1-D. problems, and chapter 10 for 2-D. problems. Thus, both chapters are essential in our understanding of other chapters of the text. The 1-D. linear Laplace differential equation is used in chapter 2 to highlight the differences between **BEM** and

GEM, and the ensuing discussions on the strength and weaknesses of both methods provide the basis for extending the Green element approach to more challenging problems in which nonlinearity and heterogeneity are encountered. As stated in the text, the theoretical framework of **GEM** is founded essentially on the boundary element theory, but it is its implementation that differs from the conventional approach of global coupling of nodal information. In **GEM**, as in **FEM**, only local coupling of nodal information is done. Our achievements by this approach are: (1) the boundary element theory is made more versatile in the sense that a good number of engineering problems are for the first time effectively addressed by the theory, and (2) computational difficulties from applying the boundary element theory to problems with heterogeneity and nonlinearity have essentially been overcome. While the Green element method could be viewed as another weighted residual approach in which the free-space Green's functions are the weighting or test functions, no attempt is made in this text to prove this. Our approach throughout the text is to present in a systematic manner the additional computational gains achieved by implementing the boundary element theory in an element-by-element manner. A more rigorous mathematical treatment of the subject which ties together these concepts and related computational methodologies is outside the scope of this text, and will be left to mathematicians.

Many developments and refinements to the basic concept of **GEM** described here are expected in the coming years. Thus, this text may only serve as a reference for those developments. We may conjecture that those developments may proceed along the line of applying the method to problems more challenging than the ones addressed here and to problems in three spatial dimensions.

In the chapters following that in which we discussed the basic concept of the Green element method for the 1-D. linear Laplace operator (chapter 2), the method is applied to heterogeneous, nonlinear, and transient problems. Chapter 3 addresses nonlinear 1-D. Laplace/Poisson problems in heterogeneous media, and demonstrates the extension of the concept to the beam deflection problem. Chapter 4 treats the Helmholtz equation, while chapter 5 applies **GEM** to the transient diffusion equation. The linear and nonlinear diffusion-advection equations are solved by **GEM** in chapters 6 and 7, while chapter 8 addresses the transient nonlinear Richards equation. The numerical stability characteristics of **GEM** for diffusion and advection problems are evaluated in chapters 5 and 6. Hermitian interpolation functions are incorporated into **GEM** in chapter 9 to demonstrate the advantages that may be derived from the use of higher-order elements in **GE** applications. Chapters 10, 11, and 12 examine steady and unsteady problems in 2-D. spatial dimensions. Numerical examples are provided to demonstrate the capabilities of **GEM** for 2-D. problems.

Akpofure E. Taigbenu, Ph.D

ACKNOWLEDGMENTS

It may be difficult not to be somewhat sentimental here. There are a number of individuals who helped to hatch and shape most of the thoughts which have culminated in the preparation of this text, and many others too numerous to mention who were of inspiration and encouragement before and during the preparation of this book. Notably among these is Prof. J. Liggett of Cornell University who was and remains my teacher of hydraulics and computational methods. A significant portion of the thoughts here came from him during my graduate days, and had it not been for distance and communication difficulties, he should have coauthored this text. Outside the rigors of academic pursuits, his wife Carole made their home in Ithaca, New York a second home for me. Those precious moments at dinner table remain ever fresh in my memories.

My thanks go to Prof. O.O. Onyejekwe of the University of Durban-Westville, South Africa (previously of the National University of Science and Technology (NUST), Zimbabwe) whose invitation to spend my sabbatical leave at NUST in 1994 availed us the opportunity to carry out collaborative work on the Green element method. His constant persuasion for a formal presentation of the concepts of the method certainly spurred the writing of this text.

I must not fail to mention the support that I enjoyed from former colleagues at the University of Benin — Prof. D.N. Nwokoye, Dr. B.U. Anyata, N. Aniekwu, and others — who, for nearly 12 years of association, made the department a conducive place for teaching and research. Very special thanks go to Walter Osedeme who drafted some of the drawings, and to my numerous graduate students of the Department of Civil Engineering, University of Benin who allowed themselves to be used for trial test runs with the manuscript. Their positive response helped in ensuring the completion of the manuscript. The finishes on this work were done at the Department of Civil & Water Engineering, National University of Science & Technology, and I am grateful to the institution for the excellent facilities with which these finishes were carried out.

I also ascribe thanks to my Lord Jesus Christ who provided the inner strength during all stages of the research that culminated in this text.

Permissions by Elsevier Science to use materials in: *Advances in Engineering Software*, Vol. 29, No. 2, 113-118, 1998; *Applied Mathematical Modelling*, Vol. 22, 687-702, 1998; Vol. 19, 675-684, 1995; *Engineering Analysis with Boundary Elements,* Vol. 22, No. 2, 161-165, 1998, by John Wiley & Sons in: *International Journal for Numerical Methods in Fluids*, Vol. 25, No. 4, 437-454, 1997; Vol. 24, 563-578, 1997, *International Journal for Numerical Methods in Engineering*, Vol. 38, . 2241-2263, 1995, and by the Water Research Commission in: *Water South Africa* Vol. 24, No. 4, 303-307, 1998 are gratefully acknowledged.

1 PRELIMINARIES

1.1 Introduction

We may not be properly poised to talk about the Green element method (**GEM**) without making reference to the boundary element method (**BEM**) since both methods are founded on the same singular integral theory. The theoretical developments in boundary element (**BE**) circles can be traced to the eighteenth century when related theories of ideal fluid flow (potential flow) and of integral transforms were established, but it was not until the 1960's that flurries of research activities on **BEM** intensified because of its acclaimed advantage of being a *boundary-only* method. Because the earlier boundary element applications were directed at elliptic boundary-value problems, the *free-space Green's function* of the differential operator, which is $ln r$ in 2-D. spatial dimensions and $1/r$ in 3-D. spatial dimensions, is amenable to Green's identity which transforms the differential equation into an integral one that can essentially be implemented on the boundary (along a line for 2-D. and on a surface for 3-D.) of the computational domain. In the classical approach, a system of discrete equations is obtained from the integral equation by subdividing the boundary into segments over which distributions are prescribed for the primary variable and the flux (Jaswon [1], Jaswon and Ponter [2], Symm [3], Liggett [4], Liu and Liggett [5], Fairweather *et al.* [6], Rizzo [7], and others). These research thrusts were considered successful when viewed against the background that contemporary methods, like the finite element method (**FEM**), solved similar problems by discretizing the entire computational domain. This led to claims by most investigators that **BEM** was more superior than existing computational methods in terms of accuracy and computational efficiency. It is not in doubt that the integral replication of the differential equation provided by the boundary element theory evolves quite naturally, making use of the response function (free-space Green's function) to a unit instantaneous input. It is only at the computational stage that the distribution of the dependent variable is approximated by some interpolating polynomial function. It is, thus, expected that for problems where the unit response function can be obtained, **BEM** achieves second-order accuracy. The claim that **BEM** is superior to **FEM** in computational efficiency is not, in most cases, arrived at after actual CPU comparisons are carried, but it is alluded to on the basis of the boundary-only character of the method.

Efforts were made at applying **BEM** to steady potential flows in heterogeneous media. Clements [8], Cheng [9], and Lafe [10,11], derived the free-space Green functions for certain known spatial distributions of the medium property and then used the classical boundary element technique to provide solutions to idealized 1-D. problems. These attempts were quite restrictive because the models applied only to the few distributions of the medium properties that were considered. To achieve a more general approach Lafe and Cheng [12] employed a perturbation approach in conjunction with **BEM** which had the major drawback of being iterative and exhibiting convergence difficulties for rapidly varying medium property. Even with their perturbation approach, it was still not a strictly boundary method as the domain had to be discretized. Furthermore, an iterative scheme seemed inconsistent with the linear nature of the problem.

Extension of the boundary element formulation to unsteady parabolic problems was attended with mixed results. One of the earlier formulation of **BEM** for the diffusion equation employed the *Laplace transform (**LT**) method* which has the main attraction of transforming the parabolic equation into an elliptic type which is readily amenable to **BEM**. Following this approach are Rizzo and Shippy [13] who applied it to heat conduction problems in solids, Liggett and Liu [14] who solved unsteady confined flow problems in aquifers, Taigbenu *et al.* [15] who applied the method to seawater intrusion into coastal aquifers, and Cheng *et al.* [16,17] who applied the method to problems of soil consolidation, and wave-solid interaction. It is widely recognized that the **LT** method has a major drawback that requires that the general behavior of the solution be known *apriori* so that this behavior can be represented in a numerical inversion scheme which inverts the solution from the transformed plane into the physical (time) plane. Since a general numerical inversion procedure is difficult to achieve, the **LT** method can only be applied on an *ad hoc* basis. Other boundary element formulations have been proposed for unsteady parabolic problems, and although these schemes eliminate the major drawback of the **LT** method, they have their peculiar advantages and demerits. A direct Green function approach is expensive to implement because the Green function is complicated so that quite a considerable amount of computing resource is expended in evaluating the boundary integrals. This approach has, however, been followed by Chang *et al.* [18], Shaw [19], Liggett and Liu [14], Brebbia and Wrobel [20], Banerjee and Butterfield [21], Wu [22], Taigbenu and Liggett [23], among others. To enhance the computational efficiency of the direct Green function approach, another numerical formulation was proposed by Taigbenu [24], Taigbenu and Liggett [25,26], and Liggett and Taigbenu [27] which derives its fundamental solution from the terms of the equation with the highest derivatives while the other terms are treated as forcing terms. Apart from this formulation being more computationally efficient because of its simpler kernel which makes the evaluation of the integral equations much easier, it is of more general application than the direct Green function method. However, because this formulation involves domain discretization and integration, it may not be considered an attractive method, as the much desired feature of a strictly boundary method is sacrificed.

To retain the boundary-only character of **BEM** for parabolic problems, a technique identified as dual reciprocity **BEM (DR-BEM)** was proposed by Brebbia and Nardini [28] in the context of elasto-dynamic problems. This technique is later extended to transient heat conduction problems by Wrobel *et al.* [29]. The success of **DR-BEM** is strongly dependent on the choice of certain predefined secondary transformation or coordinate functions which have to be harmonic functions. The **DR-BEM** is reported to be quite efficient, and retains the second-order accuracy usually associated with the boundary element method. As an update on the work of Taigbenu and Liggett [25], Aral and Tang [30] proposed a technique which they called a secondary reduction **BEM (SR-BEM)** for the diffusion and nonlinear Boussinesq equations. It uses a secondary interpolation technique for the temporal derivative that limits the evaluation of the domain integration to a one-time process. They reported improved efficiency with their formulation.

Application of the boundary element theory to the diffusion-advection equation has been quite limited because of the unique nature of this equation that behaves as a parabolic equation when advection is absent and as a hyperbolic equation when diffusion is absent. The equation is more difficult to handle when advection dominates, and for such a case most numerical schemes produce infamous spurious oscillations and/or numerical diffusion because of their inability to accurately reproduce the sharp gradients of the concentration. All boundary element formulations so far proposed for the time-dependent transport problem involve domain integrations and discretizations. Using a direct Green function approach, Brebbia and Skerget [31] solved both steady and transient transport problems but restricted themselves to problems with small values of Peclet number. Following the approach that uses the fundamental solution to the part of the governing equation which constitutes the Laplacian, numerical solutions, which covered practically all values of Peclet number, compared better than the Galerkin finite element solutions with linear elements [24,26,27].

The application of **BEM** to nonlinear problems has received the greatest setback for the method. Here, most researchers on boundary element technology agree that **FEM** has overwhelming advantages. Except where gross simplifications of the nonlinear differential operator are effected **BEM** may not be applied.

In reappraising the boundary-only character of **BEM**, it is observed that character is attainable only for very idealized and sometimes hypothesized flow problems. Even for elliptic problems, **BEM** fails to be a strictly boundary method when an external forcing term which is not a harmonic function is present, and many such practical flow problems abound. In practical **BEM** applications, therefore, there is usually the need to discretize the domain in order to account for terms that cannot be incorporated into boundary integrals.

We may point out that it appears researchers in boundary element technology have been too inclined to achieving a boundary-only character of the method to the extent that in many cases computational efficiency is sacrificed. That leads one to ask: Does achieving a strictly boundary formulation always guarantee attaining an efficient numerical scheme? By the very manner that conventional **BEM** is implemented, it can be readily shown that **BEM** cannot produce an efficient numerical scheme for the kinds of problems that are of practical interest to engineers. Such problems, in general, are large and entail having as many as 10^6 or more number of nodes on the boundary. The computational procedure of **BEM** requires that the solution at one node must directly involve every node on the boundary. In other words, as many as N^2 elemental integrations have to be evaluated in order to determine the solutions at N boundary nodes in the computational domain. This computational procedure has three important implications: (1) Enormous amount of computing resource is dedicated to evaluating the elemental integrals when N is large; (2) The resultant coefficient matrix must be a square and fully populated one; and (3) Medium properties have to be homogeneous throughout the computational domain. In contrast, an element-by-element approach in the **FE** sense provides the solution at a source node using information from only nodes with which the source node shares common elements. In other words, the approach defines a *zone of influence* for each source node which embraces only neighboring nodes. The implications of this numerical procedure are: (1) Fewer elemental integrations are evaluated per source node; (2) The resultant coefficient matrix is sparse and banded; and (3) Medium properties need only be uniform within each element but can vary from element to element [32].

1.2 Motivation for the Green Element Method

We have gone through the foregoing discussions on some of the developments in boundary element circles so that we may have an idea of the stage to which we have attained by the conventional approach in the implementation of the boundary element theory. It seems that if we may apply the theory to more robust, practically-based, and nonlinear problems, we may have to break faith with the boundary-only feature of **BEM**. That is exactly what we have done by adopting the Green element approach [32 - 36]. The Green element method is based on the boundary element theory, and as such the second-order accuracy commonly associated with **BEM** is retained. However, **GEM** implements the theory in an element-by-element manner so that, in contrast to the implementation procedure of **BEM** in which solution information are coupled for all nodes in the computational domain, **GEM** couples information of nodes that share common elements. The advantages of proceeding along these lines are discussed below.

Firstly, by this approach domain integrations are more easily carried out in comparison to the conventional **BE** approach. One reason why it is desirable to avoid domain integrations in **BE** formulations is because it is most often the case that the source node is outside the element to which belongs the field point and over which integration has to be performed (see Figure 1.1a). That situation is more difficult to handle than with the Green element method where the source and field nodes always belong to the same element (see Figure. 1.1b). In the latter case, domain integrations over regular polygonal elements can be handled analytically for most Green's functions.

Secondly, because **GEM** like **FEM** requires only *local* support of neighboring nodes, the global coefficient matrix arising from solving and assembling the discretized integral equations is sparse and banded. The sparseness of the global coefficient matrix arises from the fact that column entries at a row which corresponds to a source point come only from neighboring nodes with which the source node shares the same elements. Although it is argued that the dense matrix of **BEM** has row dimension that is smaller than that of **GEM** (which is only true for some elliptic problems for which **BEM** retains a strict boundary-only character), the tall but slender matrix of **GEM** is still easier to invert in terms of computing time and storage than the short but robust (dense) matrix of **BEM**, especially when dealing with large regional problems.

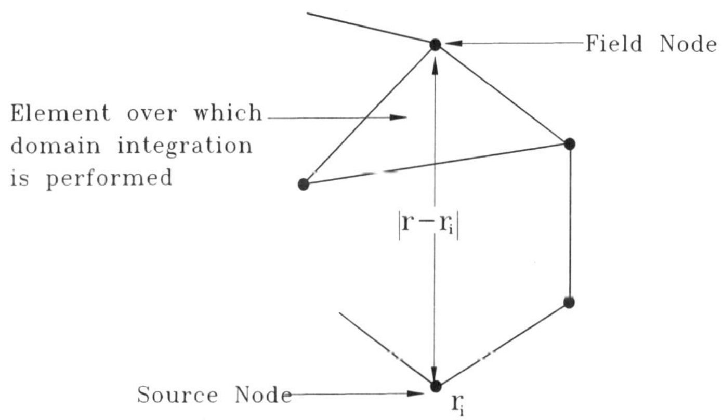

Figure 1.1a Typical situation of domain integration in **BEM**.

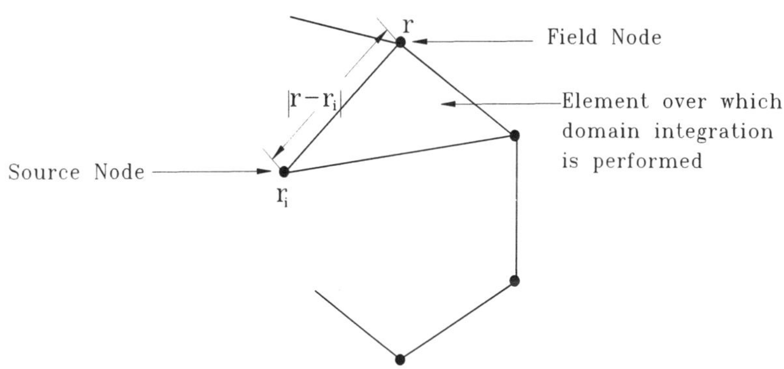

Figure 1.1b Typical situation of domain integration in **GEM**.

Thirdly, **GEM** is in general expected to evaluate a larger number of elemental integrals than **BEM**. We employ an example of potential flow in a rectangular domain to illustrate this point. The example is governed by $\nabla^2 h = 0$, and its integral representation based on **BEM** and **GEM** is

$$-\lambda h(r_i) + \int_{\Gamma} [h\nabla G(r,r_i) - G(r,r_i)\nabla h].\boldsymbol{n}\, ds = 0 \tag{1.1}$$

in which λ is the nodal angle, $G(r,r_i) = \ln(r - r_i)$ is the unit response function, Γ is the boundary of the domain Λ, and $\boldsymbol{n}$ is the unit outward pointing normal vector. Based on the implementation procedure of **BEM**, the solution at any point requires evaluation of elemental integrals equal to the total number of boundary nodes, that is, 24. Thus to obtain the solutions at all 24 boundary nodes, 24^2 or 576 elemental integrals ought to be evaluated (Figure 1.2a). For the Green element formulation, the solution at a source node involves evaluating the integrals along the sides of the rectangular elements to which it belongs (Figure 1.2b). Thus, as shown in Figures 1.3a through 1.3c, the solution at a source node located at a corner demands evaluation of 4 elemental integrals, that on a side 6 elemental integrals, and that at an interior demands that 8 elemental integrals be evaluated. Since there are altogether 4 corner nodes, 20 side nodes, and 21 interior nodes in the computational domain, it means that the Green element simulation requires that a total of 304 elemental integrals be evaluated (Table 1.1) or 52.8% of those required by the **BE** simulation. Here we have only used $N = 24$ boundary elements, and achieved savings of 47.2% on the number of elemental integrals to evaluate. For a large value of N, more savings can be expected with **GEM**. In addition to these savings in computing cost, a more comprehensive information on the solution is obtained from the **GE** simulation because of the information from the internal nodes.

Table 1.1 Book keeping of number of elemental integrals by **GEM** for simple illustrative potential flow problem

Node	Number of nodes	No. of elemental integral per node	Total number of elemental integrals
Corner	4	4	16
Side	20	6	120
Internal	21	8	168
Total	45		304

Fourthly, the Green element method is more amenable to accommodating medium heterogeneities into its computational procedure than **BEM**. This advantage of **GEM** is illustrated by the problem governed by the nonlinear Boussinesq equation which describes a heat conduction problem in which the heat conduction coefficient varies linearly with the temperature. The governing partial differential equation is given as

$$\nabla.(h\nabla h) - \partial h/\partial t = 0 \tag{1.2a}$$

Even if the equation is linearized to

$$\bar{h}\nabla^2 h = \partial h/\partial t \tag{1.2b}$$

(where $\bar{h}$ is some spatial average value of h) so that the free-space Green function can be derived, **BEM** requires that $\bar{h}$ be an average quantity of h throughout the flow region so that problem becomes one of linear heat conduction, whereas **GEM** only requires that $\bar{h}$ be an average quantity within an element, and be allowed to vary from element to element, so that the medium heterogeneity inherent in the original differential equation is still represented to some extent [33,34,36]. Thus, **GEM** treats the problem as a quasi-linear one which is a better replication of the original differential equation.

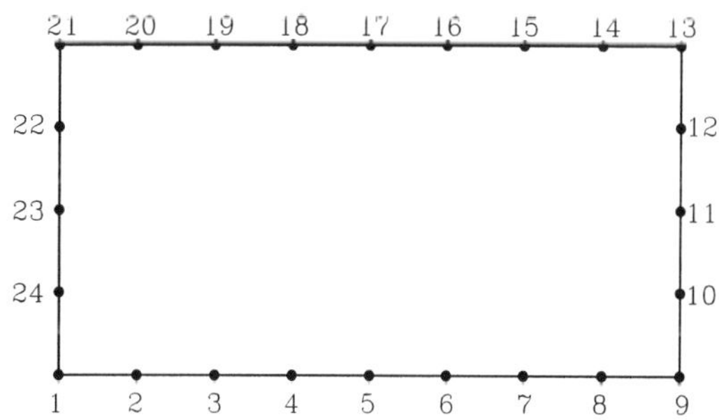

(a)

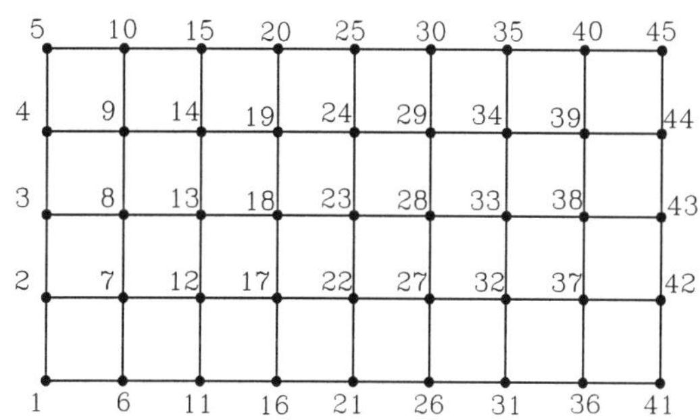

(b)

Figure 1.2 Discretization of domain for potential example — **(a) BEM, (b) GEM.**

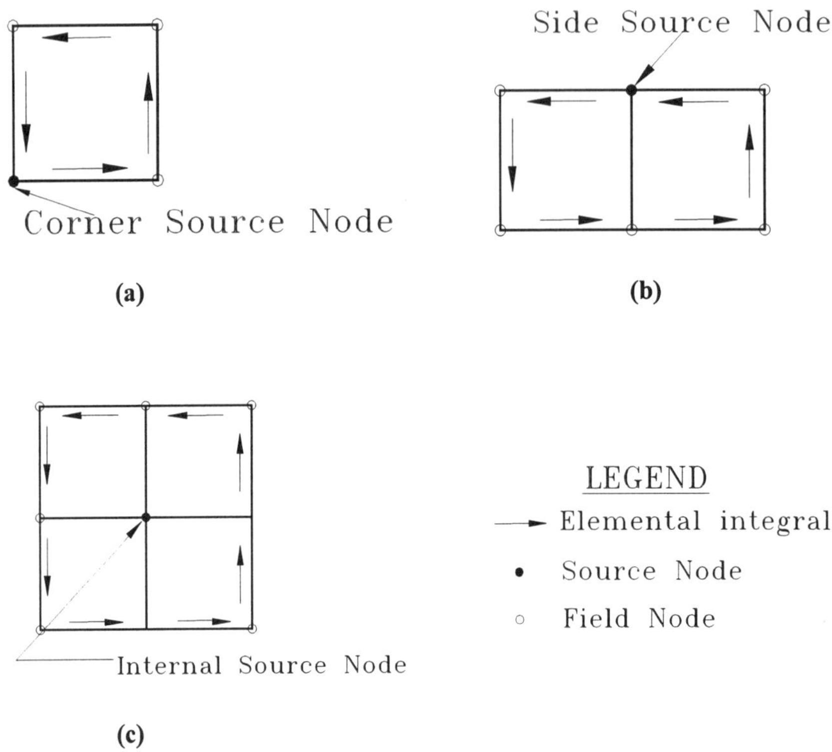

Figure 1.3 Boundary integrations for different locations of source nodes — **(a)** corner node, **(b)** side node, **(c)** internal node.

It may be worth mentioning that one of the strengths of **BEM** is with handling problems with singularities that arise from the influence of sources/sinks. They usually pose some computational difficulties to other computational methods. **FEM** like other computational methods requires that elements in the neighborhood of such singularities be small enough to accommodate the rapid change of the dependent variable over small distances. In contrast, because the free-space Green function of **BEM** is the contribution from a source/sink of unit magnitude, sources/sinks are more readily accommodated. This computational advantage of **BEM** is also exploited by **GEM** since both methods are based on the same theory.

It is with the background of the foregoing computational issues that there is a motivation for the Green element approach. A number of these issues will again be visited in greater detail in the text.

1.3 Elementary Matrix Algebra

Through out this text matrix manipulations constitute a significant part of our computations. To enhance our understanding of matrix techniques, we shall summarize the important matrix notations, properties and their operations.

As definition, a *matrix* is defined as a rectangular array of numbers or elements, and it is referred to by the number of rows and columns in the array. A matrix, given by

$$\boldsymbol{A} = A_{ij} = \begin{bmatrix} 7 & 1 & -1 & 8 \\ 0 & 5 & -8 & 1 \\ 12 & 3 & -4 & 5 \end{bmatrix}, \quad i=1,3;\ j=1,4 \tag{1.3}$$

is called a 3 (number of rows) by 4 (number of columns) matrix (pronounced as "3 by 4 matrix"). When a matrix is denoted as A_{ij}, the first subscript stands for the row, and the second for the column. Thus, $A_{14} = 8$. As it is in eq. (1.3), it is not necessary to indicate that the row index i covers the range between 1 and 3, and that the column index j runs between 1 and 4 since that is quite obvious from the arrangement of the elements of the matrix. However, in certain instances it is necessary to specify the row and column dimensions.

In this text we shall deal mainly with *square* matrices and matrices with column dimension of unity or with row dimension of unity. A square matrix is one in which the row and column dimensions are equal. A matrix with column dimension of unity is called a *column vector*. An example is

$$V = \{V\} = \begin{Bmatrix} 6 \\ -8 \\ 0 \\ 3 \end{Bmatrix} \tag{1.4}$$

A *row vector* is a matrix whose row dimension is unity. There are matrices with special configurations which have special names. A matrix in which all the non-zero entries or elements are along the main diagonal and above it is an *upper triangular matrix*. An example is

$$\boldsymbol{A} = \begin{bmatrix} a_{11} & a_{12} & a_{13} & a_{14} \\ & a_{22} & a_{23} & a_{24} \\ & & a_{33} & a_{34} \\ & & & a_{44} \end{bmatrix} \tag{1.5}$$

While a *lower triangular matrix* is one whose elements above the diagonal are all zero. A *diagonal matrix* is a matrix in which all elements are zero except those along the main diagonal. A diagonal matrix whose diagonal elements are all unity is the *identity matrix* or *unit matrix*. The identity matrix is given as

$$\boldsymbol{I} = \begin{bmatrix} 1 & & & & \\ & 1 & & & \\ & & 1 & & \\ & & & 1 & \\ & & & & 1 \end{bmatrix} \tag{1.6}$$

A matrix is said to *banded* if the elements outside the band about the main diagonal are all zero. An example is

$$\boldsymbol{A} = \begin{bmatrix} a_{11} & a_{12} & a_{13} & & & \\ a_{21} & a_{22} & a_{23} & a_{24} & & \\ a_{31} & a_{32} & a_{33} & a_{34} & a_{35} & \\ & a_{42} & a_{43} & a_{44} & a_{45} & a_{46} \\ & & a_{53} & a_{54} & a_{55} & a_{56} \\ & & & a_{64} & a_{65} & a_{66} \end{bmatrix} \tag{1.7}$$

which is said to have a half bandwidth of 2. A special banded matrix with half bandwidth of 1 is called a *tridiagonal matrix*.

We focus next on some basic matrix operations and properties. The addition of two matrices can be carried out only if the two matrices are of the same size, and the outcome is a new matrix whose elements are the sums of corresponding elements of the two matrices. That is

$$S_{ij} = A_{ij} + B_{ij} \tag{1.8}$$

The commutative property holds for addition,

$$\boldsymbol{A} + \boldsymbol{B} = \boldsymbol{B} + \boldsymbol{A} \tag{1.9}$$

and for subtraction

$$\boldsymbol{A} - \boldsymbol{B} = -\boldsymbol{B} + \boldsymbol{A} \tag{1.10}$$

The multiplication operation of two matrices is only valid when the number of columns of the premultiplying matrix is equal to the number of rows of the postmultiplying matrix. The resulting matrix has row dimension of the premultiplying matrix and the column dimension of the postmultiplying matrix. Adopting the *Einstein summation convention*, the product of two matrices is given by

$$C_{ij} = A_{ik}B_{kj} \tag{1.11}$$

(The Einstein summation convention states that when the same index is repeated once in a term, then summation is performed on that index.) The identity matrix has the following properties for square matrices

$$\boldsymbol{I}\boldsymbol{A} = \boldsymbol{A}\boldsymbol{I} = \boldsymbol{A} \tag{1.12a}$$

and

$$\boldsymbol{A}\boldsymbol{A}^{-1} = \boldsymbol{I} \tag{1.12b}$$

where $\boldsymbol{A}^{-1}$ is the *inverse* of the matrix $\boldsymbol{A}$. We define the *transpose* of a matrix as one obtained from interchanging the rows and columns of the matrix. The transpose of a matrix $\boldsymbol{A}$, denoted as $\boldsymbol{A}^T$, is given by

$$\boldsymbol{A} = A_{ij}; \quad \boldsymbol{A}^T = A_{ji} \tag{1.13}$$

A square matrix $\boldsymbol{A}$ is said to be orthogonal if the transpose of the matrix is equivalent to its inverse.

$$\boldsymbol{A}^T = \boldsymbol{A}^{-1} \tag{1.14}$$

A matrix is said to be *symmetric* if its transpose is the same as the original matrix.

$$\boldsymbol{A}^T = \boldsymbol{A} \tag{1.15}$$

1.4 References

1. Jaswon, M.A., "Integral Equation Methods in Potential Theory, I", *Proc. Roy. Soc. Ser.*, **A 275**, pp. 23-32, 1963.

2. Jaswon, M.A. and A.R. Ponter, "An Integral Equation Solution of the Torsion Problem", *Proc. Roy. Soc. Ser.*, **A 275**, pp. 237-246. 1963.

3. Symm, G.T., "Integral Equation Methods in Potential Theory, II", *Proc. Roy. Soc. Ser.*, **A 275**, pp. 33-46, 1963.

4. Liggett, J.A., "Location of Free Surface in Porous Media", *J. Hyd Div., ASCE,* **103** (HYA), pp. 353-365, 1977.

5. Liu, P. L-F. and J.A. Liggett, "An Efficient Numerical Method of Two-Dimensional Steady Groundwater Problems", *Water Resources Res.*, **14**(3), pp. 385-390, 1978.

6. Fairweather, G., F.J. Rizzo, D.J. Shippy and Y.S. Wu, "On the Numerical Solution of Two-Dimensional Potential Problems by an Improved Boundary Integral Equation Method", *J.Comp. Phy.*, **31**, pp. 96-112, 1979.

7. Rizzo, F.J., "An Integral Equation approach to Boundary Value Problems of Classical Elastostatics", Quart. Appl. Math (**25**), 1967.

8. Clements, D.L., "A Boundary Integral Equation Method for the Numerical Solution of a Second-order Elliptic Equation with Variable Coefficients", *J. Austrl. Math. Soc.*, **22B**, pp. 218-228, 1980.

9. Cheng, A. H-D., "Darcy's flow with variable permeability - A Boundary Integral Solution", *Water Resources Res.*, **20** (7), pp. 980-984, 1984.

10. Lafe, O.E., "On Boundary Integral Formulation for the Direct and Inverse Problems in Heterogeneous and Deformable Aquifers", Paper presented at *Proceedings of the International Atomic Agency/UNESCO Seminar on Applications of Isotope and Nuclear Techniques in Hydrology of Arid and Semi Arid Lands*, Adana, Turkey, 1985a.

11. Lafe, O.E., "A Boundary Integral Formulation for the Direct and Inverse Problem in Heterogeneous and Compressible Porous Media", Paper presented at *Proceedings, NATO Advanced Study Institute on the Fundamentals of Transport Phenomena in Porous Media*, Newark, Del., 1985b.

12. Lafe, O.E. and A. H-D. Cheng, "A Perturbation Boundary Element Code for Steady State groundwater Flow in Heterogeneous Aquifers", *Water Resources Res.*, **23**(6), pp. 1079-1084, 1987.

13. Rizzo, F.J. and D.J. Shippy, "A Method of Solution for Certain Problems of Transient Heat Conduction", *AIAA*, **8**(11), pp. 2004-2009, 1970.

14. Liggett, J.A. and P.L-F. Liu, "Unsteady Flow in Confined Aquifers: A comparison of Two Boundary Integral Methods", *Water Resources Res.*, **15**(4), pp. 861-9886, 1979.

15. Taigbenu, A.E., A.H-D. Cheng and J.A. Liggett, "Boundary Integral Solution to Seawater Intrusion into Coastal Aquifers", *Water Resources Res.*, **20**(8), pp.1150-1158, 1984.

16. Cheng, A. H-D and J.A. Liggett, "Boundary Integral Equation Methods for Linear Porous-Elasticity with Applications to Soil Consolidation", *Int. J. Numer. Meth. Engrg.*, **20**, pp. 255-278, 1984.

17. Cheng, A. H-D and P. L-F Liu, "Seepage Forces on a Pipeline buried in a Poroelastic seabed under wave Loading", *Appl. Ocean Res.*, **8**, pp. 22-32, 1986.

18. Chang, Y.P., C.S. Kang and D.J. Chen, "The Use of Fundamental Green's Functions for the Solution of Problems of Heat Conduction in Anisotropic Media", *Int. J. Heat Mass Transfer*, **16**, pp. 1905-1918, 1973.

19. Shaw, R.P., "An Integral Equation Approach to Diffusion", *Int. J. Heat Mass Transfer*, **17**, pp. 693-699, 1974.

20. Brebbia, C.A. and L.C. Wrobel, "The Boundary Element Method for steady State and Transient Heat Conduction", *Numerical Methods in Thermal Problems*, ed. by R.W. Lewis and K. Morgan, Prineridge Press, Swansea, Wales, 1979.

21. Banerjee, P.K. and R. Butterfield, *Boundary Element Methods in Engineering Science*, McGraw-Hill, London, U.K., 1981.

22. Wu, J.C., "Fundamental Solutions and Numerical Methods for Flow Problems", *Int. J. Numer. Meth. Fluids*, 4, pp. 185-201, 1984.

23. Taigbenu, A.E. and J.A. Liggett, "Boundary Element Calculations for Diffusion Equation", *J. Engrg. Mechanics, ASCE*, **111**(3), pp. 311-328, 1985.

24. Taigbenu, A.E., "A new Boundary Element Formulation applied to unsteady aquifer problems", *Ph.D Thesis*, Cornell University, 1985.

25. Taigbenu, A.E. and J.A. Liggett, "An Integral Formulation Applied to the Diffusion and Boussinesq Equations", *Int. J. Numer. Meth. Engrg.*, **23**, pp. 1057-1079, 1986a.

26. Taigbenu, A.E. and J.A. Liggett, "An Integral Solution for the Diffusion-Advection Equation", *Water Resources Res.*, **22**(8) pp.1237-1246, 1986b.

27. Liggett J.A. and A.E. Taigbenu, "Calculation of Diffusion, Advection-Diffusion and Boussinesq Flow by Integral Methods", *Proc. VI Int. Conf. on the Finite Element Method in Water Resources*, Lisbon, Portugal; pp 723-733, 1986.

28. Brebbia, C.A. and D. Nardini, "Dynamic analysis in Solid Mechanics by an alternative Boundary Element Procedure", *Int. J. Soil Dynam. Earthquate Engrg.*, **2**, pp. 228-233, 1983.

29. Wrobel, L.C., and C.A. Brebbia, and D. Nardini, "The Dual Reciprocity Boundary Element Formulation for Transient Heat Conduction", *Proc. VI Int. Conf. on Finite Elements in Water Resources*, pp. 801-812, 1986.

30. Aral, M.M. and Y Tang, "A New Boundary Element Formulation for Time-Dependent Confined and Unconfined Aquifer Problems", *Water Resources Res.*, **24**(6), pp. 831-842, 1988.

31. Brebbia, C.A. and P. Skerget, "Diffusion-Advection Problems using Boundary Elements", *Proc. 5th Int. Conf. on Finite Elements in Water Resources*, Burlington Vermont, USA, pp. 747-768, 1984.

32. Taigbenu, A.E., "The Green Element Method," *Int. J. for Numerical Methods in Engineering*, **38**, pp. 2241-2263, 1995.

33. Taigbenu, A.E., and O.O. Onyejekwe, "Green Element simulations of the Transient Unsaturated Flow Equation," *Applied Mathematical Modelling*, **19**, pp. 675-684, 1995.

34. Taigbenu, A.E., "Green Element Solutions of the 1-D. Steady Groundwater flow in Heterogeneous Aquifers", *Ife Journal of Technology*, **5**(2), pp. 29-43, 1996.

35. Taigbenu,, A.E. and O.O. Onyejekwe, "A Mixed Green element formulation for transient Burgers' equation", *Int. J. Num. Methods in Fluids*, **24**, pp. 563-578, 1997.

2 LINEAR LAPLACE/POISSON EQUATION

2.1 Introduction

A good starting point to derive and apply the Green element method is to use a simple second-order ordinary differential equation, and the Laplace/Poisson equation, which is encountered in many engineering applications, serves that purpose. It is expressed as

$$\frac{d}{dx}[K(x,h)\frac{dh}{dx}] = f(x) \tag{2.1}$$

As presented in eq. (2.1), the differential equation is said to be *nonlinear* when the parameter K is a function of the variable h which is herein referred to as the *primary* or *dependent variable*, otherwise it is *linear*. If K is only a function of the independent variable x then the flow medium is considered *heterogeneous*, but if it is a constant then the flow medium is *homogeneous*. The actual definitions of each of the parameters, variables, and terms in eq. (2.1) depend on the physical problem being addressed. In general, the differential equation arises from examining in a differential element a combination of: (1) the statement of mass balance (continuity or mass conservation), (2) momentum or energy balance, and (3) constitutive relations between the parameter K and the other variables. We shall in this chapter only address the linear form of eq. (2.1) in a domain that is homogeneous. That is

$$K\frac{d^2h}{dx^2} = f(x) \tag{2.2}$$

2.2 Derivation of the Linear Laplace/Poisson Equation

An example of flow in a groundwater or porous medium is used to derive eq. (2.2). Figure 2.1a shows a confined aquifer in which the distribution of the *piezometric head h* is depth-averaged and varies only in the x-direction, and which receives distributed recharge through a leaky strip and a well, both of which are represented as a known function $N(x)$. Considering a *control volume* of length Δx (unit length is assumed in the other orthogonal direction on the horizontal plane), the statement of mass conservation under steady conditions is that mass flux into the element must be equal to mass flux out of the element (see Figure 2.1b). That is

$$q(x) + N(x)\Delta x \;=\; q(x+\Delta x) \;=\; q(x) + \Delta q \tag{2.3a}$$

or

$$\frac{\Delta q}{\Delta x} \;=\; N(x) \tag{2.3b}$$

where q [L^2/T] is the *specific discharge* or volumetric discharge per unit width of aquifer, and it has been assumed in eq. (2.3a) that the density of the fluid is constant. If we consider an element of infinitesimal length, then the above equation becomes

$$\frac{dq}{dx} \;=\; N(x) \tag{2.4}$$

The simplified form of the momentum equation, which essentially neglects the effects of acceleration or convective forces, is the *Darcy law* which states that the specific discharge is proportional to the gradient of the piezometric head.

$$q \;=\; -Kb\frac{dh}{dx} \tag{2.5}$$

in which K is the *hydraulic conductivity* which is taken to be a constant because the medium is assumed homogeneous, and b is the thickness of the confined medium. Substituting eq. (2.5) into eq. (2.4) yields

$$K\frac{d^2h}{dx^2} \;=\; -\frac{N(x)}{b} \;=\; f(x) \tag{2.6}$$

which is the differential equation (2.2) that we started with. Eq. (2.6) can be referred to as the linear one-dimensional form of the *Poisson equation*, and when $f(x) = 0$ then it is the linear 1-D. form of the *Laplace equation.* In heat conduction problems where the temperature distribution due to heat transfer is of interest, the equivalent of the Darcy law is the *Fourier law*, K is the *heat conduction coefficient*, $N(x)$ is a distribution of external heat source applied along the length of the bar, and h is the temperature. There are other engineering applications in which equivalent interpretations can be given to the variables and the parameter in eq. (2.1). The reader may wish to refer to Table 2.2 in Reddy [1] for a summary of the fields of engineering where eq. (2.1) is encountered.

Although, later in this text eq. (2.1), in the form presented (heterogeneous and nonlinear form), will be solved by the Green element method, its simplification to

(2.6) is to allow for a systematic presentation of the numerical formulation. Eq. (2.6) is one of the simplest second-order differential equations to which analytical treatment can readily be achieved. It offers the starting point for developing the Green element method.

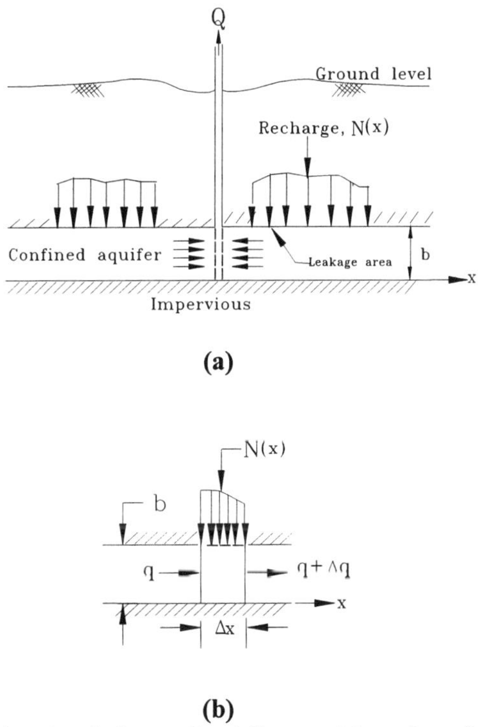

Figure 2.1 Definition sketch for confined flow — **(a)** section of confined aquifer, **(b)** infinitesimal element.

2.3 Boundary Conditions

The differential equation usually applies to a region or domain which is either finite or infinite. We denote the location of the starting point of the flow region as x_0 and that of the terminal point as x_L, and the flow length as $L = x_L - x_0$. At those two points, some conditions must be imposed on h and/or dh/dx (usually derivatives of the primary variable up to an order less than the highest order of the differential equation) so that a unique solution of the differential equation can be obtained. Without these conditions, there are infinitely many distributions of h that can satisfy the differential equation. The boundary conditions are, in general, of three types. The *1st-type* boundary condition, also called the *Dirichlet* or *essential condition* (use of this latter terminology is more appropriate in variational calculus), specifies the value of the primary variable h at a boundary point. For example,

$$h(x=x_0) = h_0 \quad \text{and} \quad h(x=x_L) = h_L \tag{2.7}$$

The *2nd-type* condition is also referred to as the *Neumann* or *natural* or *flux condition*, and it specifies the flux (fluid flux or heat flux) across an area. This condition is expressed as

$$-K\frac{dh}{dx}\bigg]_{x=x_0} = q_0 \quad \text{and} \quad -K\frac{dh}{dx}\bigg]_{x=x_L} = q_L \tag{2.8}$$

A *3rd-type* condition, called the *Cauchy condition*, prescribes a value for the linear combination of h and the flux at a boundary. That is,

$$\left[a_0 h + b_0 K\frac{dh}{dx}\right]_{x=x_0} = v_0, \text{ and } \left[a_L h + b_L K\frac{dh}{dx}\right]_{x=x_L} = v_L \tag{2.9}$$

in which a_0, b_0, a_L, and b_L are known coefficients. In subsequent chapters when we shall address transient problems, it will become necessary to provide an additional condition on h or its derivatives up to an order less than that of the differential equation at the initial time.

2.4 Integral Equations

In numerical computations, it is usually easier to work with an integral equation than a differential one. One reason for this can be deduced from introductory concepts of integration and differentiation. Integration is essentially a summation process, whereas differentiation is a difference procedure. The integral of a function $f(x)$ between two limits a and b is the limiting case of the summation of *elemental areas* or *panels* as shown in Figure 2.2a. That is

$$\int_a^b f(x)\,dx = \lim_{\Delta x \to 0} \sum_{i=1}^{M} f(x_i)\Delta x \tag{2.10}$$

in which $\Delta x = (b-a)/M$, and M is the number of subdomains or panels into which the interval is divided. On the other hand, the definition of the derivative of a function $f(x)$ is a difference expression of the form (Figure 2.2b)

$$\frac{df(x)}{dx} = \lim_{\Delta x \to 0} \frac{f(x+\Delta x) - f(x)}{\Delta x} \tag{2.11}$$

It is easier to implement eq. (2.10) on the computer to a higher level of accuracy without encountering overflow problems than eq. (2.11), and when a discontinuity in $f(x)$ is encountered (point c) it poses computational difficulties to the derivative at that point but none to the integration process. A major thrust of most numerical schemes is to reduce the order of derivatives of the primary variable in the original differential equation, and that usually results in an integral equation. Integral equations arise more naturally when the principles of mass and momentum or energy conservation are derived by considering a *finite control volume* instead of an infinitesimal one. Although such equations have the limitation of not providing information on the variation of the flow variables at every point, the relative ease with which they are solved makes them quite valuable, especially in engineering design. Integral equations are, in general, classified into two types — *Fredholm* and *Volterra* types. A Fredholm integral equation is one in which the limits of integration are fixed, while the Volterra integral equation is one in which at least one of the limits of integration is a function of the independent variable. Each of these two integral equations is again classified into two types depending on whether the dependent variable and/or its derivatives are/is expressed explicitly or implicitly in the integral equation. It is of the *first kind* when the dependent variable and/or its derivatives are/is expressed explicitly, and of the *second kind* when the dependent variable is expressed implicitly, being expressed in terms of an integral whose integrand also comprises the dependent variable. Examples of each of these kinds of integral equations are:

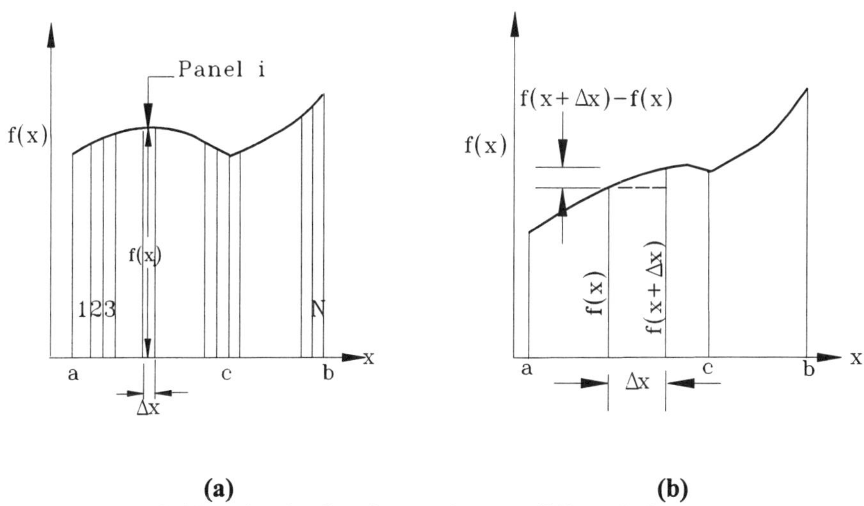

Figure 2.2 Definition sketch of **(a)** integration, **(b)** differentiation

Fredholm equation of the first kind

$$F(x) = \int_a^b G(x,s)h(s)ds \tag{2.12}$$

$$F(x) = \int_a^b G(x,s)[\alpha_1 h(s) + \alpha_2 \frac{dh}{ds}]ds \tag{2.13}$$

Fredholm equation of the second kind

$$\lambda h(x) = F(x) + \int_a^b G(x,s)h(s)ds \tag{2.14}$$

$$\lambda h(x) = F(x) + \int_a^b G(x,s)\alpha_1 \frac{dh}{ds} ds \tag{2.15}$$

Volterra equation of the first kind

$$F(x) = \int_a^x G(x,s)h(s)ds \tag{2.16}$$

$$F(x) = \int_a^{g(x)} G(x,s)[\alpha_1 h(s) + \alpha_2 \frac{dh}{ds}]ds \tag{2.17}$$

Volterra equation of the second kind

$$\lambda h(x) = F(x) + \int_a^x G(x,s)h(s)ds \tag{2.18}$$

$$\lambda h(x) = F(x) + \int_a^{g(x)} G(x,s)\alpha_1 \frac{dh}{ds} ds \tag{2.19}$$

In all the above examples, h is the primary or dependent variable, λ, α_1 and α_2 are constants, $g(x)$ and $f(x)$ are known functions, and $G(x,s)$ is usually called the *kernel function.*

Our goal in boundary element methodology is to achieve an integral representation of eq. (2.6) in which the order of the derivatives of the dependent variable are lower than those of the original differential equation. A more elegant way of achieving the integral representation of eq. (2.6) by the boundary element theory enroute the 'weak' statement of the weighted residual approach has been presented by Brebbia [2] and Gibson [3]. The major achievement of that approach has been to attempt to highlight the similarities and dissimilarities of the boundary element theory and the various versions of the weighted residual approach. For two-dimensional problems, a more direct approach to derive the integral equations of Laplace/Poisson equations has been presented by Liggett and Liu [4]. We shall proceed here along the lines of Liggett and Liu [4] for the one-dimensional problem. That approach appears more appealing. Our starting point is to proposed an appropriate complementary differential equation to that of eq. (2.6). The choice of that equation depends, among other factors, on being able to obtain its solution. Our complementary differential equation is

$$\frac{d^2G}{dx^2} = \delta(x - x_i) \quad \text{on} \quad -\infty \leq x \leq \infty \tag{2.20}$$

The solution to eq. (2.20) is not only useful in deriving the integral representation of eq. (2.6) but it also completely characterizes the response of the system governed by the differential equation. $G = G(x,x_i)$ is referred to by a number of names which are closely associated to the mathematical intuition of the physical process being model. It is called the *free-space Green's function*, or *fundamental solution*, or the *unit response function,* or simply the *kernel.* Before embarking on attaching a physical meaning to $G(x,x_i)$ and eq. (2.20), we may need to have an understanding of the function $\delta(x-x_i)$, referred to as the *Dirac delta* or the *spike function.* It is a function which is not difficult to describe from a mathematical view point, but one that seems to lend itself to different physical interpretations. We may graphically conceive it as the limiting case of the normal density function with a standard deviation σ approaching zero, and modal probability becoming infinite at the mean x_i (see Figure 2.3). Some of the properties of the normal density function — its symmetry about the mean and a unit area under the distribution — do also apply to the Dirac delta function. These representations of this function are given as:

$$\delta(x - x_i) = \begin{cases} \infty & x = x_i \\ 0 & x \neq x_i \end{cases} \tag{2.21}$$

$$\int_{-\infty}^{\infty} \delta(x - x_i) dx = \int_{x_i-\varepsilon}^{x_i+\varepsilon} \delta(x - x_i)\, dx = 1 \tag{2.22}$$

$$\int_{-\infty}^{\infty} f(x)\, \delta(x - x_i)\, dx = \int_{x_i-\varepsilon}^{x_i+\varepsilon} f(x)\, \delta(x - x_i)\, dx = f(x_i) \tag{2.23}$$

where ε is any small positive quantity. One difficulty commonly associated with the Dirac delta function which, in two- and three-dimensions, is more difficult to resolve arises when the point x_i is on the boundary. In one dimension, it involves evaluating $\int^{\infty} f(x)\,\delta(x-x_i)dx$. The symmetric shape of the Dirac delta function provides the clue. From Figure 2.3, it is observed that one half of the area under the distribution lies to the right of x_i, while the other half of the area lies to its left, so that the following relations hold:

$$\int_{x_i}^{\infty} \delta(x-x_i)dx = \int_{x_i}^{x_i+\varepsilon} \delta(x-x_i)\,dx = \frac{1}{2} \tag{2.24a}$$

$$\int_{-\infty}^{x_i} \delta(x-x_i)dx = \int_{x_i-\varepsilon}^{x_i} \delta(x-x_i)\,dx = \frac{1}{2} \tag{2.24b}$$

$$\int_{x_i}^{\infty} f(x)\,\delta(x-x_i)\,dx = \int_{x_i}^{x_i+\varepsilon} f(x)\,\delta(x-x_i)\,dx = \frac{f(x_i)}{2} \tag{2.24c}$$

In the literature, it is usual to refer to the point x_i as the *source point* and x as the *field point*. With these definitions of the Dirac delta function, the physical meaning of $G(x,x_i)$, therefore, is the response of a system governed by eq. (2.20) when it is subjected to a unit instantaneous input.

The only additional information required now before embarking on deriving the integral representation of eq. (2.6) is the statement of one of Green's identities. For any two functions, U and V, which are at least twice differentiable with respect to x, Green's identity states

$$\int_{x_0}^{x_L} \left(U\frac{d^2V}{dx^2} - V\frac{d^2U}{dx^2} \right) dx = U\frac{dV}{dx}\bigg|_{x_0}^{x_L} - V\frac{dU}{dx}\bigg|_{x_0}^{x_L} \tag{2.25}$$

The proof of this identity is easily achieved by *integration by parts*

$$\int_{x_0}^{x_L} U\frac{d^2V}{dx^2}dx = U\frac{dV}{dx}\bigg|_{x_0}^{x_L} - \int_{x_0}^{x_L} \frac{dU}{dx}\frac{dV}{dx}dx \tag{2.26a}$$

$$\int_{x_0}^{x_L} V\frac{d^2U}{dx^2}dx = V\frac{dU}{dx}\bigg|_{x_0}^{x_L} - \int_{x_0}^{x_L} \frac{dU}{dx}\frac{dV}{dx}dx \tag{2.26b}$$

Subtracting eq. (2.26b) from eq. (2.26a) yields the result of eq. (2.25). Setting U to h and V to G, and substituting eqs. (2.6) and (2.20) into the Green identity of eq. (2.25) yields

$$\int_{x_0}^{x_L} \left(h\frac{d^2G}{dx^2} - G\frac{d^2h}{dx^2} \right) dx = h\frac{dG}{dx}\bigg|_{x_0}^{x_L} - G\frac{dh}{dx}\bigg|_{x_0}^{x_L} \tag{2.27}$$

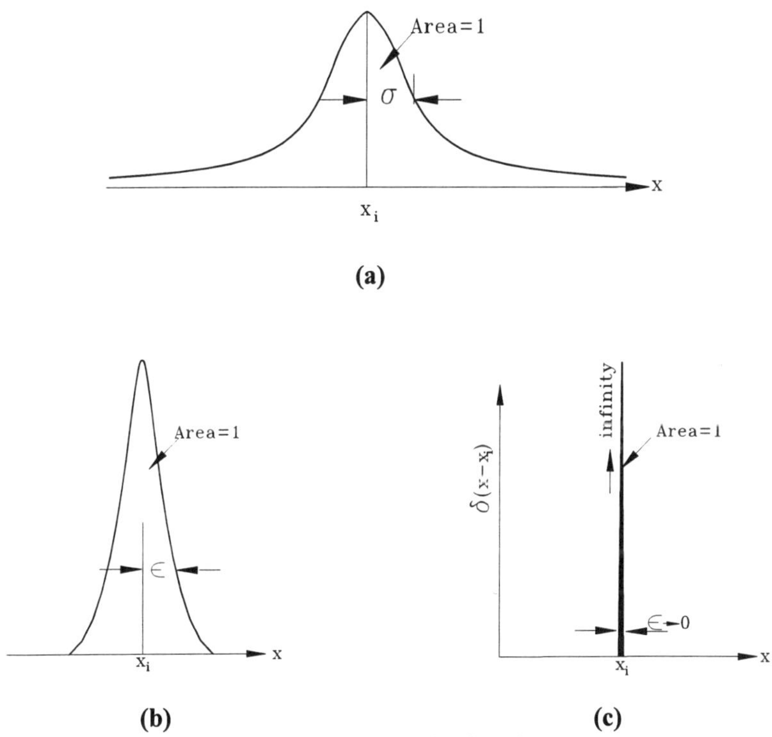

Figure 2.3 Definition sketches of the Dirac-delta function

We can now introduce into eq. (2.27) the quantity on the right hand side of eq. (2.6) instead of d^2h/dx^2 and that on the right hand side of eq. (2.20) instead of d^2G/dx^2. That results in

$$\int_{x_0}^{x_L} \left(h\delta(x-x_i) - G\frac{f(x)}{K} \right) dx = h\frac{dG}{dx}\bigg|_{x_0}^{x_L} - G\frac{dh}{dx}\bigg|_{x_0}^{x_L} \tag{2.28}$$

Using the definition for the Dirac delta function from eq. (2.23), eq. (2.28) becomes

$$-\lambda h_i + h\frac{dG}{dx}\Big|_{x_0}^{x_L} - G\frac{dh}{dx}\Big|_{x_0}^{x_L} + \int_{x_0}^{x_L} G\frac{f(x)}{K}\,dx = 0 \tag{2.29}$$

The parameter λ takes into account the location of the source point *vis-a-vis* the properties of the Dirac delta function earlier stated. $\lambda = 1$ when x_i is within the domain $[x_0, x_L]$, and $\lambda = 0.5$ when x_i is at the end points x_0 and x_L. Eq. (2.29) is the integral representation of eq. (2.6) based on the boundary element theory. It is also from this same equation that the Green element method takes its root. The integral equation is a degenerate form of the Fredholm integral equation of the second kind. The next stage is the numerical implementation of eq. (2.29), and there lies the difference between **GEM** and what we may refer to as the conventional implementation procedure of **BEM**. The implementation procedures of both methods are explained in the subsequent sections in order to provide a basis for the current motivation in favor of **GEM**.

In one dimension, the free-space Green's function or the solution to eq. (2.20) is given as

$$G(x,x_i) = \frac{|x - x_i| + k}{2} \tag{2.30}$$

in which k is an arbitrary constant. The first derivative of G with respect to x is

$$\frac{dG(x,x_i)}{dx} = \frac{1}{2}[H(x-x_i) - H(x_i-x)] \tag{2.31}$$

in which H is the *Heaviside function* that has the following representation

$$H(x-x_i) = \begin{cases} 1, & x>x_i \\ 0, & x<x_i \end{cases} \tag{2.32}$$

Figures 2.4a to 2.4c show graphical representations of G, dG/dx and H. Introducing eqs. (2.30) and (2.31) into the integral equation (2.29) yields

$$-2\lambda h_i + [H(x_L - x_i) - H(x_i - x_L)]h(x_L) - [H(x_0 - x_i) - H(x_i - x_0)]h(x_0)$$

$$-(|x_L - x_i| + k)\frac{dh}{dx}\bigg]_{x=x_L} + (|x_0 - x_i| + k)\frac{dh}{dx}\bigg]_{x=x_0} + \int_{x_0}^{x_L}(|x - x_i| + k)\frac{f(x)}{K}dx = 0$$

or

$$\boxed{\begin{aligned}&-2\lambda h_i + [H(x_L - x_i) - H(x_i - x_L)]h_L - [H(x_0 - x_i) - H(x_i - x_0)]h_0 -\\&(|x_L - x_i| + k)\varphi_L + (|x_0 - x_i| + k)\varphi_0 +\\&\int_{x_0}^{x_i}(x_i - x + k)\frac{f(x)}{K}dx + \int_{x_i}^{x_L}(x - x_i + k)\frac{f(x)}{K}dx = 0\end{aligned}} \tag{2.33}$$

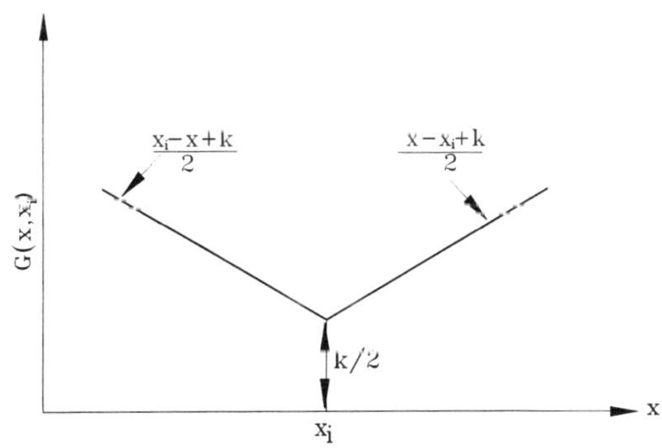

(a)

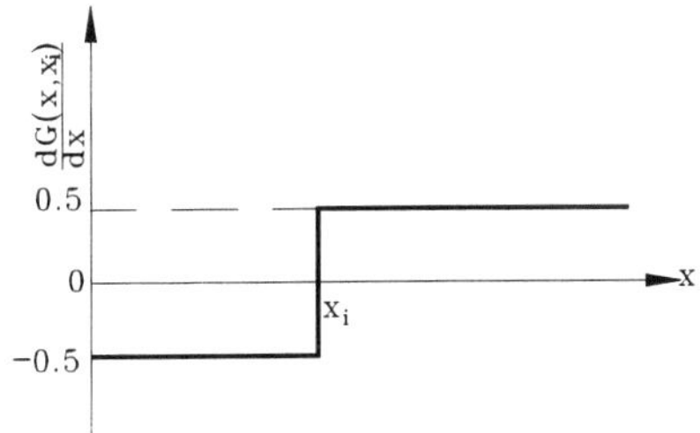

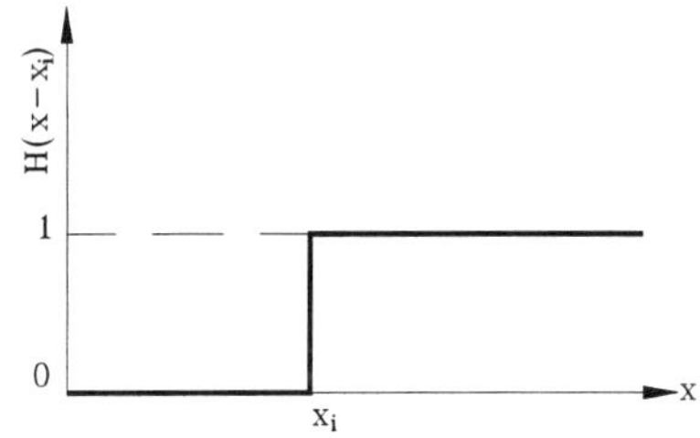

(b) **(c)**

Figure 2.4 Representations of **(a)** $G(x,x_i)$, **(b)** $dG(x,x_i)/dx$, **(c)** $H(x - x_i)$

in which

$$h_L \equiv h(x_L),\ h_0 \equiv h(x_0),\ \varphi_L \equiv \frac{dh(x=x_L)}{dx} \text{ and } \varphi_0 \equiv \frac{dh(x=x_0)}{dx} \tag{2.34}$$

It is necessary to extend the definition of the Heaviside function so that eq. (2.33) can be evaluated.

$$H(x_L-x_i) - H(x_i-x_L) = \begin{cases} 1, & x_i<x_L \\ 0, & x_i=x_L \end{cases} \tag{2.35a}$$

and

$$H(x_0-x_i) - H(x_i-x_0) = \begin{cases} -1, & x_i>x_0 \\ 0, & x_i=x_0 \end{cases} \tag{2.35b}$$

2.5 Conventional Procedure of Implementing the Boundary Element Method

The numerical implementation of eq. (2.33) by the conventional method uses $k = 0$ so that the solution at any point x_i is obtained by introducing the specified boundary conditions into the integral equation. Following this approach, the complete information on the boundary data are obtained by setting $x_i = x_0$ which gives

$$\boxed{-h_0 + h_L - L\varphi_L + \frac{1}{K}\int_{x_0}^{x_L} f(x)(x-x_0)dx = 0,} \tag{2.36a}$$

and setting $x_i = x_L$ gives

$$\boxed{h_0 - h_L + L\varphi_0 + \frac{1}{K}\int_{x_0}^{x_L} f(x)(x_L-x)dx = 0} \tag{2.36b}$$

Whichever information of the solution is not specified at the end points, eqs. (2.36a) and (2.36b) supply them. With the information at the end nodes completely evaluated, eq. (2.33) can then be used to explicitly evaluate the primary variable at any source point x_i. Additional information about the solution is the first derivative of the dependent variable, which can also be obtained from eq. (2.33). It is obtained by taking the derivative of h at the source point x_i with respect to x_i, that is

$$2\lambda \frac{dh(x_i)}{dx_i} = 2\lambda \varphi_i = \varphi_L + \varphi_0 + \frac{1}{K}\int_{x_0}^{x_L} f(x)[H(x_i - x) - H(x - x_i)]dx \tag{2.37}$$

A summary of the steps of the solution procedure of **BEM** is outlined below:

1. Solve eqs. (2.36a) and (2.36b) to obtain the solution data that are not specified at the end points.
2. Use eq. (2.33) to evaluate the primary variable h at any other point between x_0 and x_L using the information supplied in step 1.
3. Use eq. (2.37) to evaluate the first derivative of h, that is φ, at any other point between x_0 and x_L.

As you may have observed, the solution procedure of **BEM** is essentially an explicit one which relies solely on the complete solution information at the end nodes of the computational domain. We have intentionally refrained from commenting on how the line integral in eqs. (2.33), (2.36a), (2.36b), and (2.37) can be evaluated. Since the integral is that of a known function, the evaluation of the line integral should pose no serious difficulties. We shall reserve comments on the evaluation of that integral after describing the **GEM** solution procedure.

2.6 Solution Procedure of the Green Element Method

The Green element implementation procedure follows that of the *finite element method* (**FEM**). The solution domain between x_0 and x_L is divided into subdomains or elements which do not necessarily have to be of equal sizes. Let the solution domain be divided into M elements so that a typical element is represented as in Figure 2.5. The integral representation of the original differential equation which is given by eq. (2.33) can now be expressed as a summation of the integral representation in each element. That is

$$\sum_{e=1}^{M} -2\lambda^{(e)} h_i^{(e)} + [H(x_2^{(e)} - x_i^{(e)}) - H(x_i^{(e)} - x_2^{(e)})]h_2^{(e)} - [H(x_1^{(e)} - x_i^{(e)}) - H(x_i^{(e)} - x_1^{(e)})]h_1^{(e)} - (|x_2^{(e)} - x_i^{(e)}| + k)\varphi_2^{(e)} + (|x_1^{(e)} - x_i^{(e)}| + k)\varphi_1^{(e)} + \frac{1}{K}\int_{x_1^{(e)}}^{x_2^{(e)}} (|x - x_i^{(e)}| + k) f(x)dx = 0, \quad i = 1,2 \tag{2.38}$$

in which $x_1^{(e)}$ and $x_2^{(e)}$ are the coordinates of the end points of each element and

$l^{(e)}=x_2^{(e)}-x_1^{(e)}$. It should be pointed out that in going from eq. (2.33) to eq. (2.38), no approximation has been made at the integral equation level, except imposing the condition on h and φ that they be piecewise continuous. What we have actually done is to express the integration procedure between the limits x_0 and x_L as a sum of integrals within a number of subdomains or elements or panels. Eq. (2.38) can be said to be simply a *discretized version* of eq. (2.33). The way eq. (2.38) is expressed means there are four unknown quantities for the typical element — $h_1^{(e)}$, $\varphi_1^{(e)}$, $h_2^{(e)}$, and $\varphi_2^{(e)}$, which implies 2 degrees of freedom at each node. Eq. (2.38) can be split into two equations by evaluating it with $x_i^{(e)} = x_1^{(e)}$ and $x_i^{(e)} = x_2^{(e)}$. They are

$$\sum_{e=1}^{M} -h_1^{(e)}+h_2^{(e)}+k\varphi_1^{(e)}-(k+l^{(e)})\varphi_2^{(e)}+\frac{1}{K}\int_{x_1^{(e)}}^{x_2^{(e)}} f(x)(x-x_1^{(e)}+k)dx = 0 \tag{2.39a}$$

and

$$\sum_{e=1}^{M} h_1^{(e)}-h_2^{(e)}+(k+l^{(e)})\varphi_1^{(e)}-k\varphi_2^{(e)}+\frac{1}{K}\int_{x_1^{(e)}}^{x_2^{(e)}} f(x)(x_2^{(e)}-x+k)dx = 0 \tag{2.39b}$$

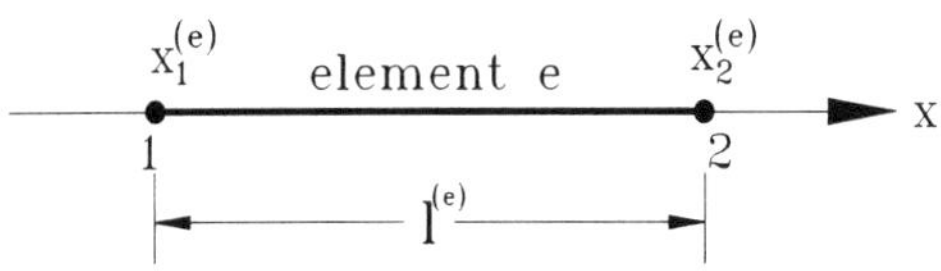

Figure 2.5 Definition sketch of a 1-D linear element.

Our knowledge of matrix algebra and indicia notation enables us to combine (2.39a) and (2.39b) into a matrix equation of the form

$$\sum_{e=1}^{M}\begin{bmatrix}-1 & 1\\ 1 & -1\end{bmatrix}\begin{Bmatrix}h_1^{(e)}\\ h_2^{(e)}\end{Bmatrix}+\begin{bmatrix}k & -(k+l^{(e)})\\ (k+l^{(e)}) & -k\end{bmatrix}\begin{Bmatrix}\varphi_1^{(e)}\\ \varphi_2^{(e)}\end{Bmatrix}+\begin{Bmatrix}F_1^{(e)}\\ F_2^{(e)}\end{Bmatrix}=\begin{Bmatrix}0\\ 0\end{Bmatrix} \tag{2.40}$$

or

$$\boxed{\sum_{e=1}^{M} R_{ij}^{(e)}h_j^{(e)} + L_{ij}^{(e)}\varphi_j^{(e)} + F_i^{(e)} = 0, \quad i,j=1,2} \tag{2.41}$$

in which

$$R_{ij}^{(e)} = \begin{bmatrix} -1 & 1 \\ 1 & -1 \end{bmatrix} = (-1)^{i+j-1} \tag{2.42a}$$

$$L_{ij}^{(e)} = \begin{bmatrix} k & -(k+l^{(e)}) \\ (k+l^{(e)}) & -k \end{bmatrix} \tag{2.42b}$$

$$F_1^{(e)} = \frac{1}{K}\int_{x_1^{(e)}}^{x_2^{(e)}} f(x)(x-x_1^{(e)}+k)dx \tag{2.42c}$$

$$F_2^{(e)} = \frac{1}{K}\int_{x_1^{(e)}}^{x_2^{(e)}} f(x)(x_2^{(e)}-x+k)dx \tag{2.42d}$$

It is observed that the matrix $R_{ij}^{(e)}$ is independent of the length of the element. Eq. (2.41) is the discretized form of the original differential equation by the Green element method. It shall be referred to as the system of *element equations*. The construction of the element equations for a differential operator is the heart of **GEM**. Those equations provide a complete information of the solution at every node in the solution domain. The formulation like the one just described may be said to be a *mixed formulation* because it calculates not only the primary variable, but also the flux at each node.

2.6.1 *Choice of the Value of k*

The value of the arbitrary constant k in the system of discrete element equations (2.41) has to be chosen in a prudent manner so that the numerical scheme produces the solutions at the nodes, and furthermore that those solutions do not depend on the value of k employed (uniqueness). The value of k cannot be set to zero because that makes the diagonal elements of the element matrix $L_{ij}^{(e)}$ have zero value, and thereby singular. We are now left to choose the value of k from the set of positive and negative real or integer numbers, except zero. One reasonable choice for k is the longest length of spatial element used in the discretization of the computational domain which we shall denote as $\tilde{l}$. That is

$$k \equiv \tilde{l} = \underset{\{e=1,2,\ldots,M\}}{Max}\left[l^{(e)}\right] \tag{2.43a}$$

If the computational domain is uniformly discretized, then

$$k = \tilde{l} = l \tag{2.43b}$$

where l is the length of each element. With this choice of the value of k, the expressions for the element matrix $L_{ij}^{(e)}$ and the vectors $F_1^{(e)}$ and $F_2^{(e)}$ become

$$L_{ij}^{(e)} = \begin{bmatrix} \tilde{l} & -(\tilde{l}+l^{(e)}) \\ (\tilde{l}+l^{(e)}) & -\tilde{l} \end{bmatrix} \tag{2.44a}$$

$$F_1^{(e)} = \frac{1}{K}\int_{x_1^{(e)}}^{x_2^{(e)}} f(x)(x-x_1^{(e)}+\tilde{l})dx \tag{2.44b}$$

$$F_2^{(e)} = \frac{1}{K}\int_{x_1^{(e)}}^{x_2^{(e)}} f(x)(x_2^{(e)}-x+\tilde{l})dx \tag{2.44c}$$

For a computational domain that is uniformly discretized

$$L_{ij}^{(e)} = L_{ij} = l\begin{bmatrix} 1 & -2 \\ 2 & -1 \end{bmatrix} \tag{2.44d}$$

2.6.2 *Assembling the Global Matrix*

Since M line segments or elements are used to discretize the flow domain, it implies that there are a total of $N = M+1$ nodes and $2N$ unknown quantities to compute. But the Green element formulation generates two equations per element or $2M=2(N-1)$ total number of equations which falls short of the total number of unknown quantities by two. The values of these two extra unknown quantities are supplied by the imposed boundary conditions. Eq. (2.41) has been derived for a typical element e and as such it applies to every element used in subdividing the flow domain. Assembly of the discrete element equations is done taking cognisance of the continuity requirements at nodes shared by two elements. Using Figure 2.6 to illustrate these requirements at the common node shared by elements e and f, we have

$$h_2^{(e)} = h_1^{(f)} \tag{2.45a}$$

and

$$K^{(e)}\varphi_2^{(e)} = K^{(f)}\varphi_1^{(f)} \tag{2.45a}$$

in which $K^{(e)}$ and $K^{(f)}$ are the hydraulic conductivities of elements e and f, respectively. However, since we are dealing with a homogeneous medium $K^{(e)} = K^{(f)}$.

To demonstrate the procedure of assembling the global equations from the element equations, we use an example whose flow domain is discretized into 5 elements (Figure 2.7). We assume that the primary variable h is specified at node 1, that is $h(x=0) = h_1$, and the first derivative of h is specified at node 6, that is $\varphi(x=L) = \varphi_6 = -q_L/K$. Each element is considered separately and renumbered as in Figure 2.7. The *source node numbering* is circled to distinguish it from the *global node numbering*. The global matrix equation from implementing eq. (2.41) for each of the 5 elements is

$$\begin{bmatrix} L_{11}^{(1)} & R_{12}^{(1)} & L_{12}^{(1)} & 0 & 0 & 0 & 0 & 0 & 0 & 0 \\ L_{21}^{(1)} & R_{22}^{(1)} & L_{22}^{(1)} & 0 & 0 & 0 & 0 & 0 & 0 & 0 \\ 0 & R_{11}^{(2)} & L_{11}^{(2)} & R_{12}^{(2)} & L_{12}^{(2)} & 0 & 0 & 0 & 0 & 0 \\ 0 & R_{21}^{(2)} & L_{21}^{(2)} & R_{22}^{(2)} & L_{22}^{(2)} & 0 & 0 & 0 & 0 & 0 \\ 0 & 0 & 0 & R_{11}^{(3)} & L_{11}^{(3)} & R_{12}^{(3)} & L_{12}^{(3)} & 0 & 0 & 0 \\ 0 & 0 & 0 & R_{21}^{(3)} & L_{21}^{(3)} & R_{22}^{(3)} & L_{22}^{(3)} & 0 & 0 & 0 \\ 0 & 0 & 0 & 0 & 0 & R_{11}^{(4)} & L_{11}^{(4)} & R_{12}^{(4)} & L_{12}^{(4)} & 0 \\ 0 & 0 & 0 & 0 & 0 & R_{21}^{(4)} & L_{21}^{(4)} & R_{22}^{(4)} & L_{22}^{(4)} & 0 \\ 0 & 0 & 0 & 0 & 0 & 0 & 0 & R_{11}^{(5)} & L_{11}^{(5)} & R_{12}^{(5)} \\ 0 & 0 & 0 & 0 & 0 & 0 & 0 & R_{21}^{(5)} & L_{21}^{(5)} & R_{22}^{(5)} \end{bmatrix} \begin{Bmatrix} \varphi_1 \\ h_2 \\ \varphi_2 \\ h_3 \\ \varphi_3 \\ h_4 \\ \varphi_4 \\ h_5 \\ \varphi_5 \\ h_6 \end{Bmatrix} = \begin{Bmatrix} -R_{11}^{(1)}h_1 - F_1^{(1)} \\ -R_{21}^{(1)}h_1 - F_2^{(1)} \\ -F_1^{(2)} \\ -F_2^{(2)} \\ -F_1^{(3)} \\ -F_2^{(3)} \\ -F_1^{(4)} \\ -F_2^{(4)} \\ -L_{12}^{(5)}\varphi_6 - F_1^{(5)} \\ -L_{22}^{(5)}\varphi_6 - F_2^{(5)} \end{Bmatrix}$$

or

$$[A]\begin{Bmatrix} h \\ \varphi \end{Bmatrix} = \{S\} \tag{2.46}$$

The matrix $[A]$ is referred to as the *global coefficient matrix*. For a general case, the global coefficient matrix has a half bandwidth of 2 and a row dimension of $2M$. Its general shape is given in Figure 2.8. Because of its banded nature its elements should be stored in a different fashion to conserve computer memory storage. It is advisable to have the elements of $[A]$ stored as:

Diagonal elements

⇓

$$[A] = \begin{bmatrix} 0 & 0 & L_{11}^{(1)} & R_{12}^{(1)} & L_{12}^{(1)} \\ 0 & L_{21}^{(1)} & R_{22}^{(1)} & L_{22}^{(1)} & 0 \\ 0 & R_{11}^{(2)} & L_{11}^{(2)} & R_{12}^{(2)} & L_{12}^{(2)} \\ R_{21}^{(2)} & L_{21}^{(2)} & R_{22}^{(2)} & L_{22}^{(2)} & 0 \\ 0 & R_{11}^{(3)} & L_{11}^{(3)} & R_{12}^{(3)} & L_{12}^{(3)} \\ R_{21}^{(3)} & L_{21}^{(3)} & R_{22}^{(3)} & L_{22}^{(3)} & 0 \\ 0 & R_{11}^{(4)} & L_{11}^{(4)} & R_{12}^{(4)} & L_{12}^{(4)} \\ R_{21}^{(4)} & L_{21}^{(4)} & R_{22}^{(4)} & L_{22}^{(4)} & 0 \\ 0 & R_{11}^{(5)} & L_{11}^{(5)} & R_{12}^{(5)} & 0 \\ R_{21}^{(5)} & L_{21}^{(5)} & R_{22}^{(5)} & 0 & 0 \end{bmatrix} \tag{2.47}$$

With this arrangement, the third column contains the diagonal elements. Because the global coefficient matrix is banded, it can be more efficiently decomposed by a banded matrix solver. The vector of unknowns $\{h, \ \varphi\}^T$ is a mixed vector of unknowns, and the right-hand-side vector S is a vector of known quantities obtained from the specified boundary conditions, and point and distributed recharge inputs. Having derived the boundary element and Green element formulations for the 1-D. Poisson equation, we are now in a position to make some general comments on both formulations.

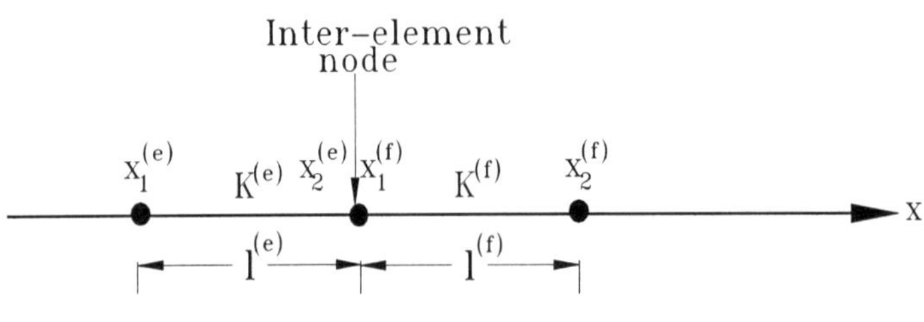

Figure 2.6 Conditions at an inter-element node.

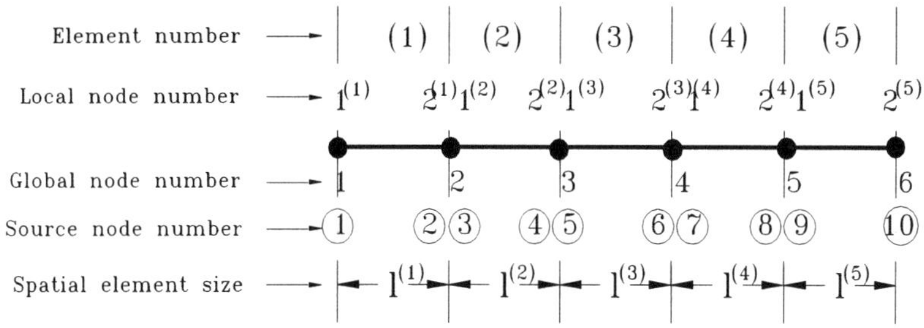

Figure 2.7 GEM discretization of a domain into 5 uniform elements.

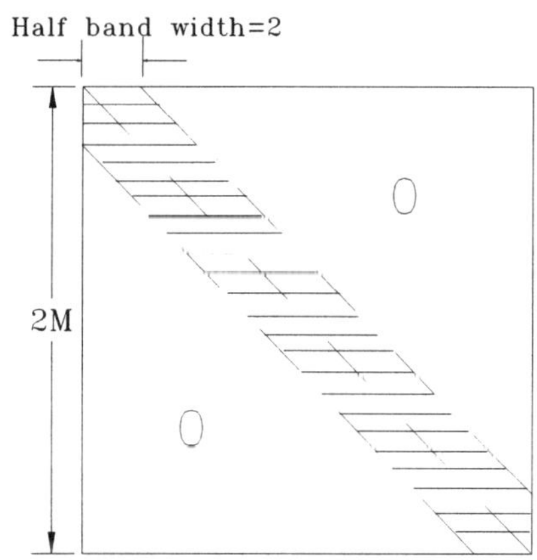

Figure 2.8 Shape of the global coefficient matrix of the 1-D. **GE** formulation.

2.7 Comparison of Boundary Element and Green Element Formulations

We shall restrict our comparison of the **BE** and the **GE** formulations to the 1-D. steady second-order differential equation (2.6). The following observations can be made.

1. The two formulations would produce identical solutions because no approximations were made on h or φ. In fact, if the integral of the recharge term is done exactly, both numerical solutions will reproduce the exact solution, when one can be found.
2. The Green element formulation produces global solutions h and φ at every node, whether or not those solutions are desired. In contrast, **BEM** gives the particular solution only at a point of interest. This certainly is an advantage of the **BE** formulation over the **GE** formulation.
3. The procedure involved in the **GE** formulation requires the solution of a system

of linear simultaneous equations. This is not required in the boundary element formulation which, after calculating the unspecified boundary values, evaluates either h or φ at any point in an explicit manner from eqs. (2.33) and (2.37). This is another advantage of **BEM** over **GEM**.

4. The only advantage that **GEM** has over **BEM** in its formulation of this linear 1-D. Poisson equation is in handling the line integral of the recharge term. Because **GEM** segmentizes the solution domain, it is able to readily accommodate more general variation of the recharge input function. For **BEM** to exhibit similar capability it has to discretize the region, which it attempts to avoid in the first place.

From the foregoing observations, it is obvious that the boundary element formulation for this 1-D. Poisson equation is superior to that of the Green element method. One may then ask, "why border with the Green element method?" It may be stated that if our concerns do not go beyond solving linear homogeneous problems, there may not be the need to be involved with the Green element method nor with other numerical methods like the finite element method or the *finite difference method* (**FDM**). The boundary element method is certainly most ideal for such problems, when their fundamental solutions exist. Unfortunately the real world is *diverse* (heterogeneous) and *complex* (nonlinear). When the hydraulic conductivity parameter K is dependent on the spatial location (heterogeneous flow) or on the primary variable (nonlinear) as in flow in unsaturated media or heat conduction in a nonlinear material, then the story of **BEM** becomes a totally different one. You may have observed that most texts on boundary element method do not go beyond considering linear homogeneous problems, and the reasons will become obvious as we proceed with this text. But before we consider nonlinear heterogeneous problems, it may be necessary to consider the treatment of the line integral of the recharge term, and attempt a few examples in homogeneous media.

2.8 Evaluation of the Line Integral of the Recharge Term

The line integral encountered in **BE** and **GE** formulations of the 1-D. Poisson equation represents the contribution of external recharge to the flow. Recharge usually consists of concentrated or point and distributed sources/sinks. For computational purposes they are treated differently, and as such the recharge term can be considered as a superposition of concentrated and distributed sources/sinks.

$$f(x) \quad = \quad f_p(x) \quad + \quad f_d(x) \tag{2.48}$$

Total Recharge	=	Point or Concentrated sources/sinks	+	Distributed sources/sinks

Point or concentrated sources/sinks are represented mathematically using the earlier stated concept of the Dirac delta function

$$f_p(x) = \sum_{i=1}^{N_p} Q_i \delta(x - x_i) \tag{2.49}$$

in which Q_i is the strength of the i-th source or sink located at x_i, and N_p is the total number of sources/sinks.

In the boundary element formulation, the line integrals that have to be evaluated are extracted from eqs. (2.33) and (2.37) and restated below

$$F_i = \frac{1}{K}\int_{x_0}^{x_L} f(x)(x-x_i)[H(x-x_i) - H(x_i-x)]\,dx \tag{2.50}$$

and

$$G_i = \frac{1}{K}\int_{x_0}^{x_L} f(x)[H(x_i-x) - H(x-x_i)]\,dx \tag{2.51}$$

If $f(x)$ represents only concentrated sources/sinks, the integral of (2.50) yields

$$F_i = \frac{1}{K}\sum_{j=1}^{N_p} Q_j(x_j-x_i)[H(x_j-x_i) - H(x_i-x_j)] \tag{2.52}$$

and similarly the line integral of eq. (2.51)

$$G_i = \frac{1}{K}\sum_{j=1}^{N_p} Q_j[H(x_i-x_j) - H(x_j-x_i)] \tag{2.53}$$

For distributed sources/sinks, Gaussian quadrature is most appropriate in evaluating the line integral. Without providing a rigorous theoretical framework of Gaussian quadrature, the integral of a function $f(x)$ between two limits a and b is approximated by

$$\int_a^b f(x)dx = \frac{(b-a)}{2}\int_{-1}^{1} f(\xi)d\xi \approx \frac{(b-a)}{2}\sum_{m=1}^{N_g} w_m f(x_m) \tag{2.54}$$

in which w_m and x_m are the weights and sampling points of the Gauss formula, and N_g is the number of Gauss points. Numerical integration is based on exact integration of a linear combination of known basis functional vectors which have been employed to approximate the integrand. In Gauss quadrature, these basis functions are orthogonal functions whose natural interval is [-1,1] and have the property

$$\int_{-1}^{1} \psi_m(\xi)\psi_n(\xi)d\xi = \delta_{mn} \tag{2.55}$$

where $\psi_m(\xi)$ are the orthogonal functions, and δ_{mn} is the *Kronecker delta* or *identity matrix* which is a second ranked tensor with the property

$$\delta_{mn} = \begin{cases} 1, & m=n \\ 0, & m \neq n \end{cases} \tag{2.56}$$

When the number of orthogonal functions $\psi_m(\xi)$, $m=1,2,...,N_g$ completely span the vector space of the integrand $f(x)$ then Gaussian quadrature is exact. Evaluating these orthogonal functions at their zeros optimizes their functional approximation of the integrand and thus the optimal evaluation of the integral. The zeros of the orthogonal functions are the sampling points x_m of eq. (2.54). By a simple transformation, the interval $[a,b]$ is mapped to the natural interval of the orthogonal functions $[-1,1]$ by the relationship:

$$x_m = \frac{1}{2}[a+b+(b-a)\xi_m] \tag{2.57}$$

There are different Gauss quadrature methods which employ the general formula of eq. (2.54) or a variation of it, and the different methods depend on the kind of orthogonal functions used. A particular set of orthogonal functions may minimize the error of approximating the integrand within the limits of integration than others. The *Legendre* orthogonal polynomials do seem to approximate most regular functions in finite intervals quite reasonably and, thus, are the most popular Gaussian integration formulas. The quadrature based on the Legendre polynomials is referred to as the *Gauss-Legendre quadrature.* We have provided a list of sampling points and weights of the Gauss-Legendre quadrature in Table 2.1 for values of m from 2 to 10. A more comprehensive listing of Gauss weights and sampling points for different orthogonal polynomials can be obtained from Abramowitz and Stegun [5].

In boundary element calculations, the use of two to four Gauss points should produce acceptable level of accuracy for most recharge distributions that can be encountered. The line integral of eq. (2.50) is approximated by Gauss-Legendre formula

$$\begin{aligned} F_i &= \frac{1}{K}\left[\int_{x_0}^{x_i} f(x)(x_i-x)dx + \int_{x_i}^{x_L} f(x)(x-x_i)dx\right] \\ F_i &\approx \frac{1}{K}\left[\frac{(x_i-x_0)}{2}\sum_{j=1}^{N_g^{(1)}} w_j f(x_j)(x_i-x_j) + \frac{(x_L-x_i)}{2}\sum_{j=1}^{N_g^{(2)}} w_j f(x_j)(x_j-x_i)\right] \end{aligned} \tag{2.58}$$

while that of eq. (2.51) is approximated by

$$G_i = \frac{1}{K}\left[\int_{x_0}^{x_i} f(x)dx - \int_{x_i}^{x_L} f(x)\,dx\right]$$

$$G_i \approx \frac{1}{K}\left[\frac{(x_i - x_0)}{2}\sum_{j=1}^{N_g^{(1)}} w_j f(x_j) \;-\; \frac{(x_L - x_i)}{2}\sum_{j=1}^{N_g^{(2)}} w_j f(x_j)\right] \tag{2.59}$$

It should be noted that we have had to split the line integration into two parts ranging from x_0 to x_i, and from x_i to x_L in order to account for the discontinuity of the free-space Green's function and the Heaviside function at x_i. This is the usual approach when domain integrations are encountered in boundary element calculations. It should be pointed out as well that the number of Gauss points for each of the two intervals of integration does not have to be the same, and that is why $N_g^{(1)}$ Gauss points are used for the interval between x_0 and x_i, and $N_g^{(2)}$ for the second interval between x_i and x_L.

Table 2.1 Weights and sampling points of the Gauss-Legendre Quadrature

$\pm\xi$	N_G	w_i	$\pm\xi$	N_G	w
0.57735026919	2	1	0.861136311594053	4	0.347854845137454
0.774596669241483	3	0.555555555555556	0.339981043584856		0.652145154862546
0.000000000000000		0.888888888888889			
0.906179845938664	5	0.236926885056189	0.932469514203152	6	0.171324492379170
0.538469310105683		0.478628670499366	0.661209386466265		0.360761573048139
0.000000000000000		0.568888888888889	0.238619186083197		0.467913934572691
0.949107912342759	7	0.129484966168870	0.960289856497536	8	0.101228536290376
0.741531185599394		0.279705391489277	0.796666477413627		0.222381034453374
0.405845151377397		0.381830050505119	0.525532409916329		0.313706645877887
0.000000000000000		0.417959183673469	0.183434642495650		0.362683783378362
0.968160239507626	9	0.081274388361574	0.973906528517172	10	0.066671344308688
0.836031107326636		0.180648160694857	0.865063366688985		0.149451349150581
0.613371432700590		0.260610696402935	0.679409568299024		0.219086362515982
0.324253423403809		0.312347077040003	0.433395394129247		0.269266719309996
0.000000000000000		0.330239355001260	0.148874338981631		0.295524224714753

In Green element calculations, line integrals or domain integrations are easier to handle. These integrals are again re-stated

$$F_1^{(e)} = \frac{1}{K}\int_{x_1^{(e)}}^{x_2^{(e)}} f(x)(x-x_1^{(e)}+\tilde{l})dx \tag{2.42c}$$

$$F_2^{(e)} = \frac{1}{K}\int_{x_1^{(e)}}^{x_2^{(e)}} f(x)(x_2^{(e)}-x+\tilde{l})dx \tag{2.42d}$$

A closer examination of the above integrals shows that the source point and the field point are always in the same element, and that makes it easier in evaluating the line integrals. Whereas boundary element technology experts do always shy from domain integrations, they pose no difficulties in Green element calculations because the source and field nodes always share the same element. In most situations, these domain integrals are evaluated exactly. Thus, the use of quadrature technique is rarely attempted in Green element calculations because they offer no added advantage.

For concentrated or point sources/sinks, $F_1^{(e)}$ and $F_2^{(e)}$ are non-zero only when those sources/sinks are within the element. Their contributions become

$$F_1^{(e)} = \frac{1}{K}\int_{x_1^{(e)}}^{x_2^{(e)}}\sum_{j=1}^{N_p^{(e)}} Q_j\delta(x-x_j)(x-x_1^{(e)}+\tilde{l})dx = \frac{1}{K}\sum_{j=1}^{N_p^{(e)}} Q_j(x_j-x_1^{(e)}+\tilde{l}) \tag{2.60}$$

and

$$F_2^{(e)} = \frac{1}{K}\int_{x_1^{(e)}}^{x_2^{(e)}}\sum_{j=1}^{N_p^{(e)}} Q_j\delta(x-x_j)(x_2^{(e)}-x+\tilde{l})dx = \frac{1}{K}\sum_{j=1}^{N_p^{(e)}} Q_j(x_2^{(e)}-x_j+\tilde{l}) \tag{2.61}$$

in which $N_p^{(e)}$ is the number of concentrated sources/sinks within the e-th element.

For distributed sources/sinks, Gaussian quadrature approximates the line integrals by the following formula

$$F_1^{(e)} \approx \frac{l^{(e)}}{2K}\sum_{j=1}^{N_g} w_j\, f(x_j)(x_j-x_1^{(e)}+\tilde{l}) \tag{2.62}$$

and

$$F_2^{(e)} \approx \frac{l^{(e)}}{2K}\sum_{j=1}^{N_g} w_j\, f(x_j)(x_2^{(e)}-x_j+\tilde{l}) \tag{2.63}$$

In many practical situations the expression for $f(x)$ is not given, but only its values at a number of points. In that case a functional interpolation is carried out to obtain a functional representation of the recharge term. We shall address this situation after discussing the theory of interpolation in chapter 3. However, in situations where the analytical expression for $f(x)$ is given, then the integrals denoted by $F_1^{(e)}$ and $F_2^{(e)}$ of eqs. (2.42c) and (2.42d) may be evaluated analytically. If we define a local coordinate ζ whose origin is at $x_1^{(e)}$ and has a value of unity at $x_2^{(e)}$, that is $\zeta = (x - x_1^{(e)})/l^{(e)}$, then the line integrals of eqs. (2.42c) and (2.42d) become

$$F_1^{(e)} = \frac{l^{(e)}}{K}\int_0^1 f(x_1^{(e)} + l^{(e)}\zeta)(\tilde{l} + l^{(e)}\zeta)d\zeta \tag{2.64a}$$

$$F_2^{(e)} = \frac{l^{(e)}}{K}\int_0^1 f(x_1^{(e)} + l^{(e)}\zeta)[\tilde{l} + l^{(e)}(1 - \zeta)]d\zeta \tag{2.64b}$$

The analytical expressions of those integrals in eqs. (2.64a) and (2.64b) when $f(x)$ takes on some simple functional expressions are given in Table 2.2

2.9 Numerical Examples

Four examples of linear 1-D. Laplace/Poisson problems are used to assess the capabilities of the Green element model that had been developed in the previous sections. In the first three examples the Green element solutions are obtained by hand calculations, while that of the fourth is obtained using the executable computer program **GEMLN1D**. These hand calculations are done because, as shown by the experiences of most practitioners in the field of computational methods, they enhance comprehension of the computational aspects of the numerical scheme.

Example 1

Our first example is given by

$$\frac{d^2h}{dx^2} = 0; \quad h(x=0) = 1, \text{ and } \frac{dh(x=2)}{dx} \equiv \varphi(x=2) = 2 \tag{2.65}$$

The reader can easily verify that the exact solution is $h = 1 + 2x$. We shall discretize the computational domain over which the differential equation is valid into 3 equal elements, and that means that the spatial element size is uniform, that is . The Green element equations for this problem are obtained from eq. (2.41)

Table 2.2 Expressions for line integrals for simple functional expressions of recharge input

$f(x)$	$K.F_1^{(e)}$
a_0	$a_0 l^{(e)}(\tilde{l} + \frac{l^{(e)}}{2})$
$a_0 x$	$a_0 l^{(e)}\left[(\tilde{l} + l^{(e)})(x_1^{(e)} + \frac{l^{(e)}}{2}) - \frac{l^{(e)3}}{3} - \frac{x_1^{(e)} l^{(e)}}{2}\right]$
$a_0 x^2$	$a_0 l^{(e)}\left[x_1^{(e)2}\tilde{l} + \frac{x_1^{(e)}(2\tilde{l} + x_1^{(e)})l^{(e)}}{2} + \frac{(\tilde{l} + 2x_1^{(e)})l^{(e)2}}{3} + \frac{l^{(e)3}}{4}\right]$
$e^{a_0 x}$	$\frac{1}{a_0^2}\left\{(1 - a_0\tilde{l})e^{a_0 x_1^{(e)}} + [a_0(\tilde{l} + l^{(e)}) - 1]e^{a_0(x_1^{(e)} + l^{(e)})}\right\}$
$\sin(a_0 x)$	$\frac{1}{a_0^2}\left[a_0\tilde{l}\cos(a_0 x_1^{(e)}) - \sin(a_0 x_1^{(e)}) - .a_0(\tilde{l} + l^{(e)})\cos[a_0(x_1^{(e)} + l^{(e)})] + \sin[a_0(x_1^{(e)} + l^{(e)})]\right]$
$\cos(a_0 x)$	$\frac{1}{a_0^2}\left[-\cos(a_0 x_1^{(e)}) - a_0\tilde{l}\sin(a_0 x_1^{(e)}) - .a_0(\tilde{l} + l^{(e)})\sin[a_0(x_1^{(e)} + l^{(e)})] + \cos[a_0(x_1^{(e)} + l^{(e)})]\right]$

$f(x)$	$K.F_2^{(e)}$
a_0	$a_0 l^{(e)}(\tilde{l} + \frac{l^{(e)}}{2})$
$a_0 x$	$a_0 l^{(e)}\left[x_1^{(e)}\tilde{l} + \frac{(\tilde{l} + x_1^{(e)})l^{(e)}}{2} + \frac{l^{(e)2}}{3}\right]$
$a_0 x^2$	$a_0 l^{(e)}\left[x_1^{(e)2}(\tilde{l} + l^{(e)}) + \frac{l^{(e)2}(\tilde{l} - 2x_1^{(e)} + l^{(e)})}{3} + \frac{x_1^{(e)} l^{(e)}(2\tilde{l} - x_1^{(e)} + 2l^{(e)})}{2} - \frac{l^{(e)3}}{4}\right]$
$e^{a_0 x}$	$\frac{1}{a_0^2}\left\{-[1 + a_0(\tilde{l} + l^{(e)})]e^{a_0 x_1^{(e)}} + (1 + a_0\tilde{l})e^{a_0(x_1^{(e)} + l^{(e)})}\right\}$
$\sin(a_0 x)$	$\frac{1}{a_0^2}\left[\cos(a_0 x_1^{(e)}) + a_0(\tilde{l} + l^{(e)})\sin(a_0 x_1^{(e)}) - \cos[a_0(x_1^{(e)} + l^{(e)})] + a_0\sin[a_0(x_1^{(e)} + l^{(e)})]\right]$
$\cos(a_0 x)$	$\frac{1}{a_0^2}\left[a_0(\tilde{l} + l^{(e)})\cos(a_0 x_1^{(e)}) + \sin(a_0 x_1^{(e)}) - a_0\tilde{l}\cos[a_0(x_1^{(e)} + l^{(e)})] + \sin[a_0(x_1^{(e)} + l^{(e)})]\right]$

$$\sum_{e=1}^{M} R_{ij}^{(e)} h_j^{(e)} + L_{ij}^{(e)} \varphi_j^{(e)} = 0, \qquad i,j=1,2$$

where, for this example,

$$R_{ij}^{(e)} = R_{ij} = (-1)^{i+j-1} \text{ and } L_{ij}^{(e)} = L_{ij} = \frac{2}{3}\begin{bmatrix} 1 & -2 \\ 2 & -1 \end{bmatrix}$$

The element equations are assembled for the three elements, while the specified boundary conditions are applied to give

$$\begin{bmatrix} 2/3 & 1 & -4/3 & 0 & 0 & 0 \\ 4/3 & -1 & -2/3 & 0 & 0 & 0 \\ 0 & -1 & 2/3 & 1 & -4/3 & 0 \\ 0 & 1 & 4/3 & -1 & -2/3 & 0 \\ 0 & 0 & 0 & -1 & 2/3 & 1 \\ 0 & 0 & 0 & 1 & 4/3 & -1 \end{bmatrix} \begin{Bmatrix} \varphi_1 \\ h_2 \\ \varphi_2 \\ h_3 \\ \varphi_3 \\ h_4 \end{Bmatrix} = \begin{Bmatrix} 1 \\ -1 \\ 0 \\ 0 \\ 8/3 \\ 4/3 \end{Bmatrix}$$

The solution to the above matrix equation, obtained by a matrix solver, is: $h_2 = h(2/3) = 2.33$, $h_3 = h(4/3) = 3.67$, $h_4 = h(2) = 5$, $\varphi_1 = \varphi(0) = 2$, $\varphi_2 = \varphi(2/3) = 2$, and $\varphi_3 = \varphi(4/3) = 2$. The **GEM** and exact solutions are identically the same.

Example 2

The second example is a linear Poisson equation given by

$$\frac{d^2h}{dx^2} = N_0 \sin(\frac{\pi}{L}x), \quad h(x=0) = 0 \text{ and } h(x=L) = 0 \tag{2.66}$$

The exact solution is

$$h = -\frac{N_0 L^2}{\pi^2} \sin(\frac{\pi}{L}x) \tag{2.67}$$

The numerical calculations are done with $L = 1$. We employ a 4-element uniform spatial discretization of the computational domain so that $l^{(e)} = \tilde{l} = l = 0.25$. The Green element equations for this example are eq. (2.41) and re-stated below

$$\sum_{e=1}^{M} R_{ij}^{(e)} h_j^{(e)} + L_{ij}^{(e)} \varphi_j^{(e)} + F_i^{(e)} = 0, \qquad i,j=1,2$$

where, for this example,

$$R_{ij}^{(e)} = R_{ij} = (-1)^{i+j-1}, \quad L_{ij}^{(e)} = L_{ij} = \frac{1}{4}\begin{bmatrix} 1 & -2 \\ 2 & -1 \end{bmatrix}.$$

$F_1^{(e)}$, and $F_2^{(e)}$ are obtained from Table 2.2

$$F_1^{(1)} = \frac{N_0}{\pi^2}[\frac{\pi}{4}\cos(0) - \sin(0) - \frac{\pi}{2}\cos(\frac{\pi}{4}) + \sin(\frac{\pi}{4})] = 0.3818\frac{N_0}{\pi^2}$$

$$F_2^{(1)} = \frac{N_0}{\pi^2}[\frac{\pi}{2}\cos(0) + \sin(0) - \frac{\pi}{4}\cos(\frac{\pi}{4}) - \sin(\frac{\pi}{4})] = 0.3083\frac{N_0}{\pi^2}$$

$$F_1^{(2)} = \frac{N_0}{\pi^2}[\frac{\pi}{4}\cos(\frac{\pi}{4}) - \sin(\frac{\pi}{4}) - \frac{\pi}{2}\cos(\frac{\pi}{2}) + \sin(\frac{\pi}{2})] = 0.8483\frac{N_0}{\pi^2}$$

$$F_2^{(2)} = \frac{N_0}{\pi^2}[\frac{\pi}{2}\cos(\frac{\pi}{4}) + \sin(\frac{\pi}{4}) - \frac{\pi}{4}\cos(\frac{\pi}{2}) - \sin(\frac{\pi}{2})] = 0.8178\frac{N_0}{\pi^2}$$

$$F_1^{(3)} = \frac{N_0}{\pi^2}[\frac{\pi}{4}\cos(\frac{\pi}{2}) - \sin(\frac{\pi}{2}) - \frac{\pi}{2}\cos(\frac{3\pi}{4}) + \sin(\frac{3\pi}{4})] = 0.8178\frac{N_0}{\pi^2}$$

$$F_2^{(3)} = \frac{N_0}{\pi^2}[\frac{\pi}{2}\cos(\frac{\pi}{2}) + \sin(\frac{\pi}{2}) - \frac{\pi}{4}\cos(\frac{3\pi}{4}) - \sin(\frac{3\pi}{4})] = 0.8483\frac{N_0}{\pi^2}$$

$$F_1^{(4)} = \frac{N_0}{\pi^2}[\frac{\pi}{4}\cos(\frac{3\pi}{4}) - \sin(\frac{3\pi}{4}) - \frac{\pi}{2}\cos(\pi) + \sin(\pi)] = 0.3083\frac{N_0}{\pi^2}$$

$$F_2^{(4)} = \frac{N_0}{\pi^2}[\frac{\pi}{2}\cos(\frac{3\pi}{4}) + \sin(\frac{3\pi}{4}) - \frac{\pi}{4}\cos(\pi) - \sin(\pi)] = 0.3818\frac{N_0}{\pi^2}$$

From the information above, the global matrix equation can now be assembled. It should be noted that the right-hand-side vector comprises solely the quantities $F_1^{(e)}$ and $F_2^{(e)}$.

$$\begin{bmatrix} 1/4 & 1 & -1/2 & 0 & 0 & 0 & 0 & 0 \\ 1/2 & -1 & -1/4 & 0 & 0 & 0 & 0 & 0 \\ 0 & -1 & 1/4 & 1 & -1/2 & 0 & 0 & 0 \\ 0 & 1 & 1/2 & -1 & -1/4 & 0 & 0 & 0 \\ 0 & 0 & 0 & -1 & 1/4 & 1 & -1/2 & 0 \\ 0 & 0 & 0 & 1 & 1/2 & -1 & -1/4 & 0 \\ 0 & 0 & 0 & 0 & 0 & -1 & 1/4 & -1/2 \\ 0 & 0 & 0 & 0 & 0 & 1 & 1/2 & -1/4 \end{bmatrix} \begin{Bmatrix} \varphi_1 \\ h_2 \\ \varphi_2 \\ h_3 \\ \varphi_3 \\ h_4 \\ \varphi_4 \\ \varphi_5 \end{Bmatrix} = -\frac{N_0}{\pi^2}\begin{Bmatrix} 0.3818 \\ 0.3083 \\ 0.8483 \\ 0.8178 \\ 0.8178 \\ 0.8483 \\ 0.3083 \\ 0.3818 \end{Bmatrix}$$

The solution to the global matrix equation is presented in Table 2.3. We also evaluated the line integral with the sinusoidal external distributed recharge using a 2-point Gauss-Legendre quadrature. The natural sampling points are $\xi_1 = -1/\sqrt{3}$ and $\xi_2 = 1/\sqrt{3}$, while the Gauss weights are $w_1 = w_2 = 1$. Within any element with end nodes $x_1^{(e)}$ and $x_2^{(e)}$, the corresponding locations of the sampling points are obtained from the transformation given by eq. (2.57), that is

$$x_m = x_1^{(e)} + \frac{l^{(e)}}{2}(1 + \xi_m) = x_1^{(e)} + \frac{1}{8}(1 + \xi_m)$$

With this, the right-hand-side quantities $F_1^{(e)}$ and $F_2^{(e)}$ become

$$F_1^{(1)} = \frac{N_0}{8\times 24}\left[(9-\sqrt{3})\sin\frac{\pi}{24}(3-\sqrt{3}) + (9+\sqrt{3})\sin\frac{\pi}{24}(3+\sqrt{3})\right] = 0.3820\frac{N_0}{\pi^2}$$

$$F_2^{(1)} = \frac{N_0}{8\times 24}\left[(9+\sqrt{3})\sin\frac{\pi}{24}(3-\sqrt{3}) + (9-\sqrt{3})\sin\frac{\pi}{24}(3+\sqrt{3})\right] = 0.3080\frac{N_0}{\pi^2}$$

$$F_1^{(2)} = \frac{N_0}{8\times 24}\left[(9-\sqrt{3})\sin\frac{\pi}{24}(9-\sqrt{3}) + (9+\sqrt{3})\sin\frac{\pi}{24}(9+\sqrt{3})\right] = 0.8483\frac{N_0}{\pi^2}$$

$$F_2^{(2)} = \frac{N_0}{8\times 24}\left[(9+\sqrt{3})\sin\frac{\pi}{24}(9-\sqrt{3}) + (9-\sqrt{3})\sin\frac{\pi}{24}(9+\sqrt{3})\right] = 0.8176\frac{N_0}{\pi^2}$$

$$F_1^{(3)} = \frac{N_0}{8\times 24}\left[(9-\sqrt{3})\sin\frac{\pi}{24}(15-\sqrt{3}) + (9+\sqrt{3})\sin\frac{\pi}{24}(15+\sqrt{3})\right] = 0.8176\frac{N_0}{\pi^2}$$

$$F_2^{(3)} = \frac{N_0}{8\times 24}\left[(9+\sqrt{3})\sin\frac{\pi}{24}(15-\sqrt{3}) + (9-\sqrt{3})\sin\frac{\pi}{24}(15+\sqrt{3})\right] = 0.8483\frac{N_0}{\pi^2}$$

$$F_1^{(4)} = \frac{N_0}{8\times 24}\left[(9-\sqrt{3})\sin\frac{\pi}{24}(21-\sqrt{3}) + (9+\sqrt{3})\sin\frac{\pi}{24}(21+\sqrt{3})\right] = 0.3080\frac{N_0}{\pi^2}$$

$$F_2^{(4)} = \frac{N_0}{8\times 24}\left[(9+\sqrt{3})\sin\frac{\pi}{24}(21-\sqrt{3}) + (9-\sqrt{3})\sin\frac{\pi}{24}(21+\sqrt{3})\right] = 0.3820\frac{N_0}{\pi^2}$$

The numerical solution with Gauss quadrature of the line integral is presented in Table 2.3. **GEM** solution with exact evaluation of the line integral essentially reproduced the exact solution, but that with Gauss quadrature has a maximum percentage error of

3.25% which is observed with the solution for h at $x=0.25$ and $x=0.75$.

Table 2.3 Comparison of two **GEM** solutions with exact solution for example 2

Solution		Exact solution (x N_0/π^2)	**GEM** (x N_0/π^2)		
			by exact line integration	Gauss Quadrature	
				Solution	Error (%)
of h at	x=0.25	-0.7071	-0.7071	-0.7301	3.25
	x=0.5	-1.0	-1.0	-1.0230	2.3
	x=0.75	-0.7071	-0.7071	-0.7301	3.25
of φ at	x=0	-3.1416	-3.1416	-3.1185	0.74
	x=0.25	-2.2214	-2.2214	-2.2212	0.01
	x=0.5	0.0	0.0	0.0	0.0
	x=0.75	2.2214	2.2214	2.2212	0.01
	x=1.0	3.1416	3.1416	3.1185	0.74

Example 3

Our third example is again a linear 1-D. Poisson equation with an external recharge comprising a distributed one and a point one. The example is given by

$$\frac{d^2h}{dx^2} = N_0 + Q_0\delta(x-0.75), \quad \frac{dh(x=0)}{dx} \equiv \varphi(x=0) = 1, \; h(x=1) = 2.0 \tag{2.68}$$

where N_0 and Q_0 are constants. The reader may verify that the exact solution to this example is given by

$$h(x) = 1-\frac{1}{4}(2N_0+Q_0)+x+\frac{1}{2}N_0x^2+(x-0.75)H(x-0.75)Q_0 \tag{2.69}$$

The **GEM** calculations are performed on a computational domain discretized into 5 uniform elements so that the length of each element is 0.2. The Green element equations of the differential equation are given by eq. (2.41), and the element matrices of those equations are:

$$R_{ij}^{(e)} = R_{ij} = (-1)^{i+j-1}, \; L_{ij}^{(e)} = \frac{1}{5}\begin{bmatrix} 1 & -2 \\ 2 & -1 \end{bmatrix}$$

$F_1^{(e)}$ and $F_2^{(e)}$ are obtained from eqs. (2.60) and (2.61), and from Table 2.2. It should be noted that the contribution of the point source is accounted for only in the fourth element.)

$$F_1^{(1)} = F_1^{(2)} = F_1^{(3)} = F_1^{(5)} = N_0 l^{(e)}(\tilde{l}+\frac{l^{(e)}}{2}) = \frac{3}{50}N_0$$

$$F_2^{(1)} = F_2^{(2)} = F_2^{(3)} = F_2^{(5)} = N_0 l^{(e)}(\tilde{l}+\frac{l^{(e)}}{2}) = \frac{3}{50}N_0$$

$$F_1^{(4)} = \frac{3}{50}N_0 + (0.75-0.6+l)Q_0 = (\frac{3}{50}+0.35\frac{Q_0}{N_0})N_0$$

$$F_2^{(4)} = \frac{3}{50}N_0 + (0.8-0.75+l)Q_0 = (\frac{3}{50}+0.25\frac{Q_0}{N_0})N_0$$

The global matrix equation for this example now becomes

$$\begin{bmatrix} -1 & 1 & -2/5 & 0 & 0 & 0 & 0 & 0 & 0 & 0 \\ 1 & -1 & -1/5 & 0 & 0 & 0 & 0 & 0 & 0 & 0 \\ 0 & -1 & 1/5 & 1 & -2/5 & 0 & 0 & 0 & 0 & 0 \\ 0 & 1 & 2/5 & -1 & -1/5 & 0 & 0 & 0 & 0 & 0 \\ 0 & 0 & 0 & -1 & 1/5 & 1 & -2/5 & 0 & 0 & 0 \\ 0 & 0 & 0 & 1 & 2/5 & -1 & -1/5 & 0 & 0 & 0 \\ 0 & 0 & 0 & 0 & 0 & -1 & 1/5 & 1 & -2/5 & 0 \\ 0 & 0 & 0 & 0 & 0 & 1 & 2/5 & -1 & -1/5 & 0 \\ 0 & 0 & 0 & 0 & 0 & 0 & 0 & -1 & 1/5 & -2/5 \\ 0 & 0 & 0 & 0 & 0 & 0 & 0 & 1 & 2/5 & -1/5 \end{bmatrix} \begin{Bmatrix} h_1 \\ h_2 \\ \varphi_2 \\ h_3 \\ \varphi_3 \\ h_4 \\ \varphi_4 \\ h_5 \\ \varphi_5 \\ \varphi_6 \end{Bmatrix} = \begin{Bmatrix} -3N_0/50-1/5 \\ -3N_0/50-2/5 \\ -3N_0/50 \\ -3N_0/50 \\ -3N_0/50 \\ -3N_0/50 \\ -(3/50+0.35Q_0/N_0)N_0 \\ -(3/50+0.25Q_0/N_0)N_0 \\ -3N_0/50-2 \\ -3N_0/50+2 \end{Bmatrix}$$

For values of $N_0 = 0.5$ and $Q_0 = 2$, the **GEM** solutions from the global matrix equation are:
$h_1 = h(0) = 0.25$, $h_2 = h(0.2) = 0.46$, $h_3 = h(0.4) = 0.69$, $h_4 = h(0.6) = 0.94$, $h_5 = h(0.8) = 1.31$, $\varphi_2 = \varphi(0.2) = 1.1$, $\varphi_3 = \varphi(0.4) = 1.2$, $\varphi_4 = \varphi(0.6) = 1.3$, $\varphi_5 = \varphi(0.8) = 3.4$, $\varphi_6 = \varphi(1) = 3.5$. It can be readily verified that the **GEM** solutions agree with the exact solutions.

Example 4

Our fourth example is given by

$$\frac{d^2h}{dx^2} = \cos(\frac{\pi x}{5}), \quad 0 \le x \le 10, \; h(x=0) = 2, \; \varphi(x=10) = -1.0 \tag{2.70}$$

It can readily be verified that its exact solution is given by

$$h(x) = \frac{5^2}{\pi^2}[1 - \cos(\frac{\pi x}{5})] - x + 2 \tag{2.71}$$

This example is solved by the executable computer program **GEMLN1D** that has been developed along the theoretical concepts of the Green element method. (**GEMLN1D** is written in such a user-friendly manner that it can either be run from a data file, being guided by the user manual in Appendix A, or interactively by providing the information requested by the program. Data supplied interactively are stored a file for future use, and that file can be edited.)

The computational domain is divided into 20 equal elements so that the spatial element size is uniform, that is $l^{(e)} = l = 0.5$, and the distributed recharge data is read by the program at the nodes because the computer program uses linear interpolation functions to represent the distribution of the recharge data. The data file generated from running **GEMLN1D** interactively is given in Table 2.4a

Table 2.4a Data file ran with **GEMLN1D**

```
Example 4 of Chapter 2
y  Steady problem=y, and unsteady=n
 2  No. of segments into which the domain is  divided
.0000E+00 x-coordinate of the first node
.1000E+02 x-coordinate of the last node
10 Number of elements generated in segment  1
10 Number of elements generated in segment  2
 2  .2000E+01 Node type for node   1;  Boundary data
 1  -.1000E+01 Node type for node  21;  Boundary data
 1  Recharge data distribution = 0 (uniform); = 1 (nonuniform)
 .1000E+01  .9510565E+00  .8090169E+00  .5877852E+00  .309017E+00  .0000E+00
 -.309017E+00 -.5877852E+00 -.8090169E+00 -.9510565E+00 -.1000E+01
 -.9510565E+00 -.8090169E+00 -.5877852E+00 -.309017E+00  .0000E+00
  .309017E+00  .5877852E+00  .8090169E+00  .9510565E+00  .1000E+01
 0 No. of sources/sinks
.1000E+01 Diffusivity or Conductivity
.0000E+00 Characteristic parameter (chap. 4)
.0000E+00 Uniform velocity
.0000E+00 Rate of first-order decay
```

The output from executing **GEMLN1D** with this data can be found in Table 2.4b

Table 2.4b Output from running **GEMLN1D** with data file in Table 2.4a

Example 4 of Chapter 2
...... NUMBER OF NODES = 21
..... STEADY STATE SOLUTIONS .....

Node No.	Location	Primary Variable	Flux
1	.000E+00	.2000E+01	-.1000E+01
2	.500E+00	.1623E+01	-.5122E+00
3	.100E+01	.1480E+01	-.7222E-01
4	.150E+01	.1536E+01	.2770E+00
5	.200E+01	.1736E+01	.5012E+00
6	.250E+01	.2012E+01	.5784E+00
7	.300E+01	.2289E+01	.5012E+00
8	.350E+01	.2489E+01	.2770E+00
9	.400E+01	.2545E+01	-.7222E-01
10	.450E+01	.2402E+01	-.5122E+00
11	.500E+01	.2025E+01	-.1000E+01
12	.550E+01	.1402E+01	-.1488E+01
13	.600E+01	.5448E+00	-.1928E+01
14	.650E+01	-.5110E+00	-.2277E+01
15	.700E+01	-.1711E+01	-.2501E+01
16	.750E+01	-.2988E+01	-.2578E+01
17	.800E+01	-.4264E+01	-.2501E+01
18	.850E+01	-.5464E+01	-.2277E+01
19	.900E+01	-.6520E+01	-.1928E+01
20	.950E+01	-.7377E+01	-.1488E+01
21	.100E+02	-.8000E+01	-.1000E+01

The reader may confirm that the **GEM** and exact solutions are in good agreement, considering the fact that the recharge data is approximated by piecewise linear functions. Graphical presentations of the two solutions are shown in Figures 2.9a and 2.9b.

2.10 Remarks

The linear 1-D. Laplace/Poisson equation has been used in this chapter to present the basic concept and implementation procedure of the Green element method. We also intentionally presented the boundary element formulation for this problem to highlight its overwhelming advantages over **GEM**. For this problem, **BEM** is an explicit scheme; it is not necessary to solve a system of linear equations. The solution at any point is obtained in terms of the complete solution at the end points of the computational domain. The necessity to discretize the computational domain in **BEM** computations arises only when there is an external distributed recharge term to contend with. One would have wished that these pleasant numerical features of **BEM** are preserved as the complexity of the differential equation increases. Unfortunately, that is just not the case. For instance these pleasant numerical features of **BEM** do not extend to the nonlinear 1-D. Laplace/Poisson equation in a heterogeneous medium which is addressed in the next chapter. In fact, to proceed along the lines of **BEM** for the nonlinear problem, there is the need to make some grave simplifications of the original differential equation. Essentially those simplifications result in transforming the nonlinear equation to a linear one similar to eq. (2.6). Such simplifications are valid only when the parameter K is weakly dependent on h, and the variation of h is small over the computational domain. That has been one of the major handicaps of **BEM** — its inability to handle nonlinear heterogeneous problems. However, as we proceed in this text you will come to appreciate that although **GEM** is based on the same singular Fredholm integral theory, it has no limitation in handling nonlinear, heterogeneous problems. The subsequent chapters in which more challenging problems are addressed provide opportunities for exploiting the strengths and versatility of **GEM**.

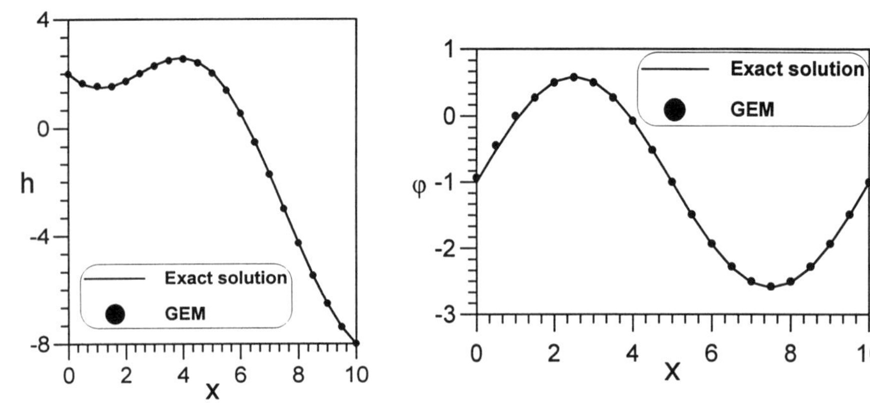

(a) (b)

Figure 2.9 GEM and exact solutions of example 4 — **(a)** solution for h, **(b)** solution for φ.

2.11 Exercises

2.1 Carry out by hand the calculations of the boundary element solutions for the first three numerical examples used to verify the Green element model. Compare the boundary element solutions and those of the Green element solutions.

2.2 Derive the Green element equations for the linear 1-D. Laplace/Poisson equation given by (2.6) by approximating the spatial derivative of h, that is φ, so that at each node there is only one degree of freedom. (*Hint*: This formulation can accept a value of zero for the arbitrary constant k in the free-space Green's function.) How can Dirichlet boundary conditions be handled by this formulation?

2.3 Use the Green element formulation derived in Question 2.2 to solve the first three numerical examples presented in the previous section, and compare the solutions of this formulation to those presented.

2.4 Consider the tight string problem (Figure E2.4) which is governed by

$$T\frac{d^2w}{dx^2} + N(x) = 0, \quad w(0) = w(L) = 0$$

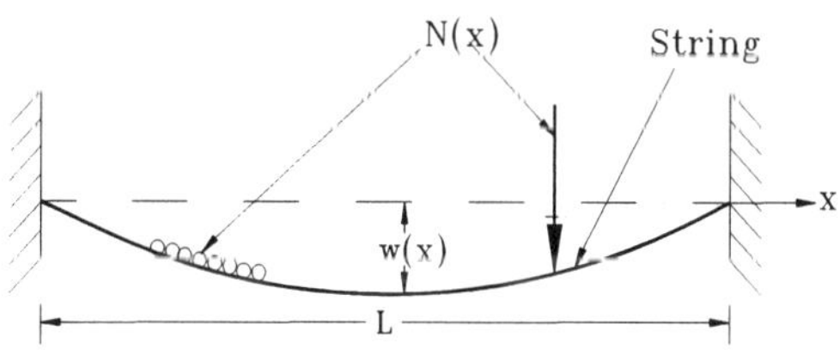

Figure E2.4

where $w(x)$ is the displacement of the string in the vertical direction, and T is the tension of the string. For values of $T = 1$, and $L = 1$, determine the Green element solutions on a 4-element uniform discretization of the computational domain for the following cases of loading on the string:

(a) $N(x) = \delta(x - 0.5)$
(b) $N(x) = 1$
(c) $N(x) = (1 - x^2)$
(d) $N(x) = 1 - x + \delta(x - 1/3) + \delta(x - 2/3)$

Compare the Green element solutions and their exact solutions.

2.12 References

1. Reddy, J.N., *An Introduction to the Finite Element Method*, McGraw-Hill, Inc., New York, 1984.

2. Brebbia, C.A., *The Boundary Element Method for Engineers*, Pentech Press, London, Halsted Press,

New York, 1978.

3. Gipson G.S., *Boundary Element Fundamentals - Basic Concepts and Recent Developments in Poisson Equation*, Computational Mechanics Publications, Southampton, UK and Boston, USA, 1987.

4. Liggett, J.A. and L-F. Liu, *The Boundary Integral Equation Method for Porous Media Flow*, Allen & Unwin, London, UK., 1983.

5. Abramowitz, M. and I.A. Stegun, *Handbook of Mathematical Functions With Formulas, Graphs, and Mathematical Tables*, National Bureau of Standards, 1970.

3 NONLINEAR LAPLACE/POISSON EQUATION

3.1 Introduction

In the previous chapter, we presented the boundary element and Green element formulations for the linear version of the steady 1-D. second-order differential equation given by eq. (2.1) in which the parameter K assumed a constant value. We used the example of flow in a confined aquifer of uniform thickness to derive the linear form of the differential equation. Here we relax that condition and consider cases where K is a function of the spatial variable x (heterogeneous case), and a function of the dependent variable h (nonlinear case). Nonlinear heterogeneous problems are frequently encountered in many engineering applications - heat flux through a material whose thermal properties are nonlinear, infiltration into unsaturated soils, elastic deformation of a bar of varying section subject to axial loads, etc. We elect to use a heat transfer example to derive the nonlinear equation of the form of eq. (2.1).

3.2 Derivation of Differential Equation (Heat Conduction Example)

An example of heat conduction through the material shown in Figure 3.1 is used to derive the 1-D. steady nonlinear Poisson differential equation. It is assumed that the temperature variation across any section is insignificant compared to that along the *bar* or *rod*, so that we can essentially consider the flux of heat as occurring only along the bar. A known distribution of heat input $N(x)$ [J/(m.s)] is also considered; this input may be from an external source of heat or one generated internally like the case of an electric heating wire embodied in the bar. We extract an appropriate control volume of bar of infinitesimal length dx (Figure 3.1b) in which the principle of *conservation of heat* is examined. The principle of conservation of heat within the infinitesimal element under steady condition states that the flux of heat into the control volume equals that out of it. Mathematically the statement of *heat conservation* yields

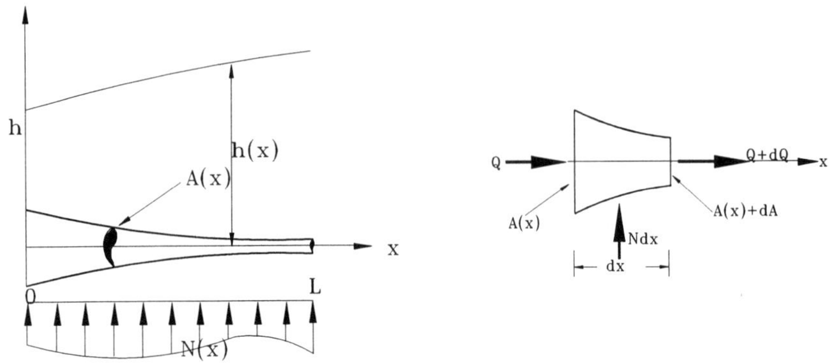

(a) **(b)**

Figure 3.1 Definition sketch of steady heat flux in a bar — **(a)** bar element, **(b)** infinitesimal element.

$$Q(x) + N(x)dx = Q(x+dx) = Q(x) + dQ \tag{3.1}$$

That is

$$\frac{dQ}{dx} = N \tag{3.2}$$

in which Q is the *heat flux* [J/s]. The relationship between the heat flux and the primary variable, the temperature h, is provided by *Fourier law of heat conduction.* Fourier law states that the heat flux is proportional to the gradient of the temperature, and it is given by

$$Q = -k(x,h)A(x)\frac{dh}{dx} \tag{3.3}$$

where $k(x,h)$ is the *heat conduction coefficient* or *thermal conductivity* [J/(m.s.K)]. It is analogous to the *hydraulic conductivity* in groundwater flow or *diffusion coefficient* in contaminant transport. The heat conduction coefficient is a property of the rod which reflects the ability of the rod to conduct heat. Eq. (3.3) is introduced into eq. (3.2) to give

$$\frac{d}{dx}\left(kA\frac{dh}{dx} \right) + N(x) = 0 \tag{3.4}$$

or

$$\frac{d}{dx}\left(K(x,h)\frac{dh}{dx} \right) = f(x) \tag{3.5}$$

where $K = kA$ and $f(x) = -N(x)$.

Most boundary element formulations have been proposed for the 2-D. form of eq. (3.5), that is $\nabla.(K\nabla h) = f(x,y)$, where $\nabla = \boldsymbol{i}\partial/\partial x + \boldsymbol{j}\partial/\partial y$ is the 2-D. gradient vector. The solution of the differential equation has proved to be quite a great challenge to the method. An excellent summary of the various boundary element formulations to the steady 2-D. nonlinear heterogeneous flow was provided by Cheng and Ouaza [1] in Chapter 8 of the text on *Boundary Element Techniques in Geomechanics*, edited by G.D. Manolis and T.G. Davis. These efforts range from assuming that K is uniform throughout the flow region to a formulation that perturbs the solution about that of the linear case (Lafe and Cheng [2]) to a *dual reciprocity formulation* which attempts to avoid domain integrations (Partridge *et al.* [3]), and to a *direct Green's function approach* which uses the free-space Green function for various distributions of the parameter K (Cheng [4]). In all the above applications only the heterogeneous case was studied. The inability of **BEM** to accurately simulate this problem has not been a weakness on the part of the theory, but rather a reluctance on the part of experts in boundary element method to break faith with the boundary-only character of the method (Taigbenu [5]). We must come to terms with the fact that the excellent numerical features of **BEM** for the linear homogeneous case do not translate to the nonlinear heterogeneous case. Our recent research efforts represent the first successful attempt at solving the nonlinear heterogeneous problem with the boundary element theory (Taigbenu and Onyejekwe [6]). Here, we present three formulations of **GEM** for the 1-D. steady nonlinear heterogeneous Poisson problem. The three formulations offer different levels of approximation of the differential equation, and they are presented to demonstrate the flexibility that **GEM** offers its users.

3.3 Integral Representation

We begin the Green element formulation to eq. (3.5) by recasting it in the form

$$\frac{d^2h}{dx^2} = -\frac{1}{K}\frac{dK}{dx}.\frac{dh}{dx} + \frac{f(x)}{K} \tag{3.6}$$

Eq. (3.6) now looks like eq. (2.6) of the previous chapter if the two terms on the right-hand-side are jointly considered as the *recharge term*. We can now follow the procedure which we took in the previous chapter in formulating the Green element formulation for the linear problem. Our complementary differential equation remains

$$\frac{d^2G}{dx^2} = \delta(x - x_i), \quad on \quad -\infty \le x \le \infty \tag{3.7}$$

The solution to eq. (3.7) is still $G(x,x_i) = (|x-x_i|+k)/2$. The arbitrary constant k which, based on the explanation given in section 2.6.1, shall not be set zero to avoid having a singular global coefficient matrix, but to the longest spatial element size $\tilde{l}$. Green's identity given by (2.25) is applied to eqs. (3.6) and (3.7) to obtain the integral equation

$$-\lambda h_i + h\frac{dG}{dx}\bigg|_{x_0}^{x_L} - G\frac{dh}{dx}\bigg|_{x_0}^{x_L} + \int_{x_0}^{x_L} \frac{G}{K}\left(-\frac{dK}{dx}.\frac{dh}{dx} + f(x)\right) dx = 0 \tag{3.8}$$

The flow region $[x_0,x_L]$ is discretized into M line segments and each segment, denoted as e, has two end points whose coordinates are $x_1^{(e)}$ and $x_2^{(e)}$ (see Figure 2.5). Since the original differential equation and the derived integral equation apply to the entire flow domain, they also apply to the subdomain $[x_1^{(e)}, x_2^{(e)}]$. Thus, eq. (3.8) is written for the e-th line segment or element to give

$$-\lambda^{(e)} h_i^{(e)} + h\frac{dG}{dx}\bigg|_{x_1^{(e)}}^{x_2^{(e)}} - G\frac{dh}{dx}\bigg|_{x_1^{(e)}}^{x_2^{(e)}} + \int_{x_1^{(e)}}^{x_2^{(e)}} \frac{G}{K}\left(-\frac{dK}{dx}.\frac{dh}{dx} + f(x)\right) dx = 0 \tag{3.9}$$

3.4 First Green Element Model (GESNH-1)

Introducing the expressions of $G(x,x_i)$ and $dG/dx = [H(x-x_i)-H(x_i-x)]/2$ into eq. (3.9) yields

$$\begin{aligned} &-2\lambda^{(e)} h_i^{(e)} + [H(x_2^{(e)}-x_i^{(e)}) - H(x_i^{(e)}-x_2^{(e)})]h_2^{(e)} - \\ &[H(x_1^{(e)}-x_i^{(e)}) - H(x_i^{(e)}-x_1^{(e)})]h_1^{(e)} - (|x_2^{(e)}-x_i^{(e)}| + \tilde{l})\varphi_2^{(e)} + \\ &(|x_1^{(e)}-x_i^{(e)}| + \tilde{l})\varphi_1^{(e)} + \int_{x_1^{(e)}}^{x_2^{(e)}} \frac{(|x-x_i^{(e)}| + \tilde{l})}{K}\left(-\frac{dK}{dx}.\frac{dh}{dx} + f(x)\right) dx = 0, \;\; i=1,2 \end{aligned} \tag{3.10}$$

We suggest our first approximation of eq. (3.10) by assuming that K is constant or uniform within an element, that is, for the e-th element.

$$K(x,h) = \bar{K}^{(e)} \tag{3.11}$$

where $\bar{K}^{(e)}$ can be taken as an average quantity of K in the element. That average quantity can be computed by the relation

$$\bar{K}^{(e)} = \frac{1}{2}[K(x_1^{(e)},h_1^{(e)}) + K(x_2^{(e)},h_2^{(e)})] = \frac{1}{2}[K_1^{(e)} + K_2^{(e)}] \tag{3.12}$$

By adopting the above approximation $dK/dx = 0$, and eq. (3.10) now becomes

$$\begin{aligned} &-2\lambda^{(e)}h_i^{(e)} + [H(x_2^{(e)}-x_i^{(e)}) - H(x_i^{(e)}-x_2^{(e)})]h_2^{(e)} - \\ &[H(x_1^{(e)}-x_i^{(e)}) - H(x_i^{(e)}-x_1^{(e)})]h_1^{(e)} - (|x_2^{(e)}-x_i^{(e)}| + \tilde{l})\varphi_2^{(e)} + \\ &(|x_1^{(e)}-x_i^{(e)}| + \tilde{l})\varphi_1^{(e)} + \frac{1}{\bar{K}^{(e)}}\int_{x_1^{(e)}}^{x_2^{(e)}} (|x-x_i^{(e)}| + \tilde{l})f(x)\,dx = 0, \quad i=1,2 \end{aligned} \tag{3.13}$$

We shall refer to the formulation arising from the approximation of eq. (3.11) as the first Green element formulation of the steady 1-D. nonlinear Poisson equation for a heterogeneous medium — **GESNH-1**. The numeral postfix 1 is simply used to indicate that this is the first approximation of the differential equation; higher levels of approximation will be proposed later in this chapter. Let us attempt to interpret what we have actually done by the approximation of eq. (3.11). What we have succeeded in doing is to transform the computational domain to a multiply zoned or multiply layered domain with hydraulic conductivities of each layer being $\bar{K}^{(e)}$, $e = 1, 2, ..., M-1, M$. The compatibility relations at a node shared by two elements e and f now become (See Figure 2.6)

$$h_2^{(e)} = h_1^{(f)} \tag{3.14a}$$

and

$$\bar{K}^{(e)}\varphi_2^{(e)} = \bar{K}^{(f)}\varphi_1^{(f)} \tag{3.14b}$$

Proceeding with the **GESNH-1** formulation, eq. (3.13) is split into two: one when $x_i^{(e)}$ is located at $x_1^{(e)}$

$$-h_1^{(e)} + h_2^{(e)} + \tilde{l}\varphi_1^{(e)} - (\tilde{l} + l^{(e)})\varphi_2^{(e)} + \frac{1}{\bar{K}^{(e)}}\int_{x_1^{(e)}}^{x_2^{(e)}} f(x)(x-x_1^{(e)} + \tilde{l})dx = 0 \tag{3.15}$$

and the other when $x_i^{(e)}$ is located at $x_2^{(e)}$

$$h_1^{(e)} - h_2^{(e)} + (\tilde{l} + l^{(e)})\varphi_1^{(e)} - \tilde{l}\varphi_2^{(e)} + \frac{1}{\bar{K}^{(e)}} \int_{x_1^{(e)}}^{x_2^{(e)}} f(x)(x_2^{(e)} - x + \tilde{l})dx = 0 \tag{3.16}$$

Eqs. (3.15) and (3.16) are the element equations for the **GESNH-1** formulation of the differential equation. These element equations provide all the information on the approximate solution. You will come to observe that the core of the Green element method is being able to form the element equations of the differential operator. In matrix form those element equations are of the form

$$\boxed{R_{ij}^{(e)} h_j^{(e)} + L_{ij}^{(e)} \varphi_j^{(e)} + \frac{1}{\bar{K}^{(e)}} F_i^{(e)} = 0, \quad i,j=1,2; \quad e=1,2,\ldots,M} \tag{3.17}$$

where the matrices $R_{ij}^{(e)}$ and $L_{ij}^{(e)}$ are given by eqs. (2.42a) and (2.42b), while the vector $F_i^{(e)}$ is given by eqs. (2.42c) and (2.42d). We can now proceed to assemble the information given by eq. (3.17) for all the M elements as we did in the previous chapter, imposing the compatibility requirements at each inter-element node, and the specified boundary conditions. The global matrix equation is similar to eq. (2.46) and of the form

$$A_{ij} \begin{Bmatrix} h_j \\ \varphi_j \end{Bmatrix} = S_i \tag{3.18}$$

The global coefficient matrix still has a half bandwidth of 2 and a row dimension of $2M$.

GESNH-1 model offers the first level of approximation that the Green element method imposes on the flow problem. We have assumed that $\bar{K}^{(e)}$ in each element is a known quantity, and that is only true when K does not depend on h. We will revisit the **GESNH-1** formulation for the nonlinear case after we have reviewed solution schemes for systems of nonlinear equations. Higher levels of sophistication of the Green element formulation would require knowledge of the theory of interpolations which is herein briefly reviewed.

3.5 Interpolation

The general concept of interpolation of a quantity or parameter or variable is to obtain its functional distribution in terms of a linear combination of *simple* or *elementary functions* or *basis functional vectors* and undetermined coefficients. That can be done by ensuring either that the interpolating function passes through discrete points where the quantity is either known or sought (see Figure 3.2a), or that the error from approximating the quantity by the interpolating function is minimized, say, in the least square sense (Figure 3.2b). In Figures 3.2a and 3.2b we have assumed that the quantity h is known at a number of discrete points and its distribution is approximated by $\hat{h}(x)$. The elementary functions could be polynomials, or trigonometric or exponential or rational functions. General expressions of these interpolating functions for 1-D. spatial dimension are:

$$h(x) \approx \hat{h}(x) = \sum_{j=0}^{n} a_j x^j \tag{3.19a}$$

using *polynomial elementary functions*,

$$h(x) \approx \hat{h}(x) = \sum_{j=0}^{n} [a_j \cos(jx) + b_j \sin(jx)] \tag{3.19b}$$

using *trigonometric elementary functions*, and

$$h(x) \approx \hat{h}(x) = \sum_{j=0}^{n} a_j \exp(b_j x) \tag{3.19c}$$

when we use *exponential functions* as elementary functions. In the above three equations a_j and b_j are constant coefficients that have to be determined on the basis of some conditions imposed on h.

In Green element calculations we are usually faced with the interpolation of unknown quantities like h (temperature, pressure head, or displacement, etc) or its derivative φ and known quantities or medium parameters like K (the hydraulic conductivity, or diffusion coefficient, etc), and these could either be interpolated in a *global manner* or an *element manner*. In the former, the quantity is approximated directly over the entire flow region. There are a number of disadvantages with this approach; one being that the global coefficient matrix will no longer be banded, and the second being that the solutions obtained will be applicable only to problems that have that geometry. For these reasons this approach is less attractive and as such not recommended. The second approach of interpolating quantities has the advantage of being readily amenable to general geometries. In this approach the region is decomposed into several elements or subdomains, and the quantity to be interpolated is then expressed in terms of the chosen elementary functions and the values of the quantity at the nodes. Incidentally this approach is also followed in the finite element method, and the term *finite elements* arose from the *elements* used to discretize the computational region. It still appears open-ended concerning the choice of the elementary functions. One of the criterion that determines the choice of the elementary functions is the ease with which computations can be carried out, and in this respect polynomial functions are easier to deal with compared to trigonometric or exponential functions. The order of polynomial functions employed should be such that when the size of the elements becomes infinitely small the approximation of the quantity should approach its correct distribution. This requirement is referred to as the *convergence requirement* in finite element methodology. Because this subject has been carefully treated by a number of authors it will not be repeated here. We will refer you to the following text: Strang and Fix [7] and Zienkiewicz and Taylor [8].

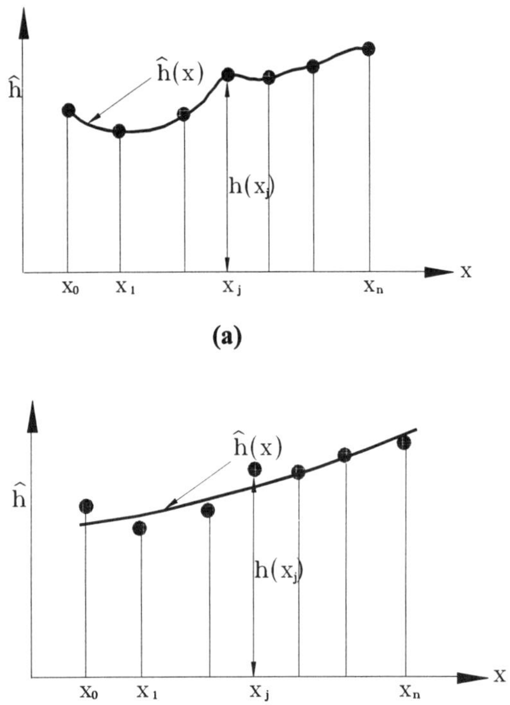

(b)

Figure 3.2 Functional interpolation — **(a)** through data points **(b)** in the least square sense

3.6 One-Dimensional Elements

3.6.1 *Linear Elements*

One of the simplest 1-D. elements is that which approximates a quantity by a linear interpolating function that is given by

$$h \approx \hat{h} = a_0 + a_1 x \tag{3.20}$$

In most Green element calculations we usually employ *conformal elements* which ensures that h is continuous at boundaries between elements so that (see Figure 3.3)

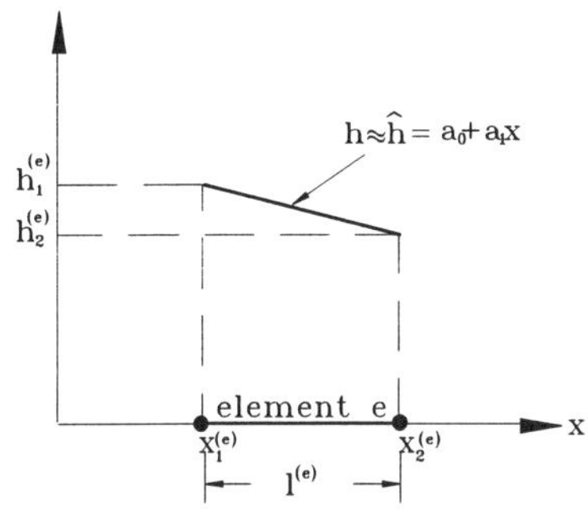

Figure 3.3 Linear conformal element.

$$h(x_1^{(e)}) = h_1^{(e)}, \quad h(x_2^{(e)}) = h_2^{(e)} \tag{3.21}$$

The conditions at the nodes $x_1^{(e)}$ and $x_2^{(e)}$ given by eq. (3.21) are used in eq. (3.20) to determine a_0 and a_1

$$\begin{aligned} h_1^{(e)} &= a_0 + a_1 x_1^{(e)} \\ h_2^{(e)} &= a_0 + a_1(x_1^{(e)} + l^{(e)}) \end{aligned} \tag{3.22}$$

The constant coefficients are obtained from the above two equations

$$a_0 = \left(1 + \frac{x_1^{(e)}}{l^{(e)}}\right) h_1^{(e)} - \frac{x_1^{(e)}}{l^{(e)}} h_2^{(e)}, \quad a_1 = \frac{h_2^{(e)} - h_1^{(e)}}{l^{(e)}} \tag{3.23}$$

Eq. (3.20) can now be expressed as

$$h \approx \hat{h} = \Omega_1^{(e)} h_1^{(e)} + \Omega_2^{(e)} h_2^{(e)} \tag{3.24}$$

where

$$\Omega_1^{(e)} = \frac{x_2^{(e)} - x}{l^{(e)}}$$
$$\Omega_2^{(e)} = \frac{x - x_1^{(e)}}{l^{(e)}} \tag{3.25}$$

The functions $\Omega_1^{(e)}$ and $\Omega_2^{(e)}$ are referred to as *element shape functions* or *element interpolating functions* or *element basis functions*. Because of the role that the source and field nodes play in Green element calculations, it is more convenient to introduce a local coordinate system ζ whose origin is at $x_1^{(e)}$. It is expressed as

$$\zeta = \frac{x - x_1^{(e)}}{l^{(e)}} \tag{3.26}$$

With this local coordinate system, the element interpolation functions for linear 1-D. element are given by

$$\boxed{\Omega_1^{(e)} = 1 - \zeta, \quad \Omega_2^{(e)} = \zeta} \tag{3.27}$$

These element basis functions are graphically presented in Figure 3.4, and they have the following properties

$$\Omega_i^{(e)}(\zeta_j) = \begin{cases} 1, & i=j \\ 0, & i \neq j \end{cases} \tag{3.28a}$$

and

$$\Omega_1^{(e)}(\zeta) + \Omega_2^{(e)}(\zeta) = 1 \tag{3.28b}$$

Element interpolation functions that satisfy the properties of eqs. (3.28a) and (3.28b) are said to be of the *Lagrange family*. In general, element interpolation functions of the Lagrange family can be constructed for an element with n number of nodes. Such an interpolating function should be of $n - 1$ degree and has the expression

$$\Omega_i(\zeta) = \frac{(\zeta - \zeta_1)(\zeta - \zeta_2)\dots(\zeta - \zeta_{i-1})(\zeta - \zeta_{i+1})\dots(\zeta - \zeta_n)}{(\zeta_i - \zeta)(\zeta_i - \zeta_2)\dots(\zeta_i - \zeta_{i-1})(\zeta_i - \zeta_{i+1})\dots(\zeta_i - \zeta_n)} \tag{3.29}$$

3.6.2 *Quadratic Elements*

We shall use eq. (3.29) to construct the element interpolating functions of the Lagrange family for a quadratic 1-D. element ($n = 3$). Since we will not be using more than quadratic interpolating functions throughout this text we shall leave the reader who may be interested in higher interpolating functions of the Lagrange family with the following references: Zienkiewicz [9], Reddy [10], Rao [11], and Ottosen and Peterson [12]. Using the 1-D. quadratic element shown in Figure 3.5, and the fact that the local coordinate ζ has values of 0, 0.5, and 1 at nodes 1, 2, and 3 of the element, respectively, then

$$\Omega_1(\zeta) = \frac{(\zeta - \frac{1}{2})(\zeta - 1)}{(-\frac{1}{2})(-1)} = (2\zeta - 1)(\zeta - 1) \tag{3.30a}$$

$$\Omega_2(\zeta) = \frac{(\zeta - 0)(\zeta - 1)}{(\frac{1}{2})(\frac{1}{2} - 1)} = -4\zeta(\zeta - 1) \tag{3.30b}$$

$$\Omega_3(\zeta) = \frac{(\zeta - 0)(\zeta - \frac{1}{2})}{(1 - 0)(1 - \frac{1}{2})} = \zeta(2\zeta - 1) \tag{3.30c}$$

Their representations are also shown in Figure 3.5.

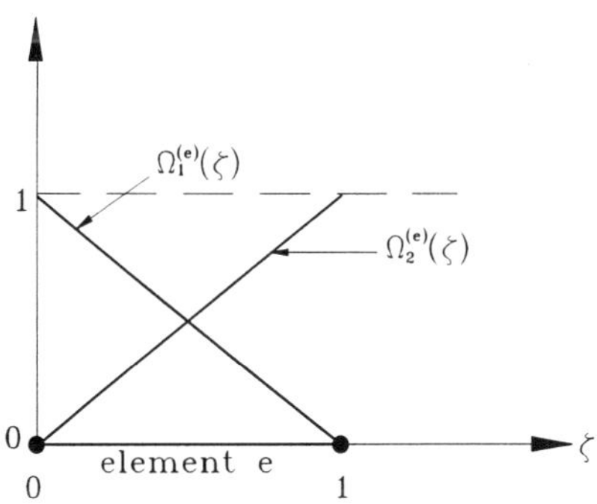

Figure 3.4 1-D. Linear element shape functions

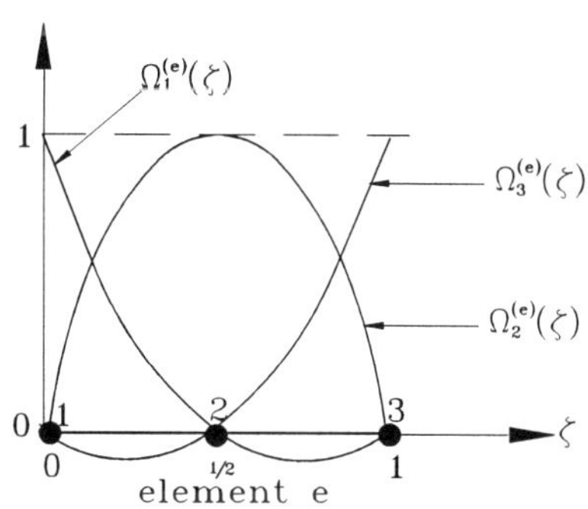

Figure 3.5 1-D. Quadratic element shape functions.

Having introduced the concept of interpolation within an element, we are in a position to give a different treatment to the line integrals of the distributed recharge $f(x)$ which we had denoted as $F_i^{(e)}$, $i = 1,2$ in eq. (3.17). In Table 2.2 of chapter 2 we supplied the analytical expressions for those line integrals when $f(x)$ has a simple functional expression. (It may be pointed out that such a simple functional expression of $f(x)$ could have arisen from a global functional approximation of the recharge data using simple elementary interpolating functions.) As stated earlier it is easier to deal with elemental approximation of $f(x)$ than global approximation, and for that reason we approximate $f(x)$ in an element sense. Using linear interpolating functions, $f(x)$ is approximated within a typical element as

$$f(x) = f(\zeta) \approx \Omega_1^{(e)} f_1^{(e)} + \Omega_2^{(e)} f_2^{(e)} \tag{3.31}$$

We can now introduce eq. (3.31) for the recharge term into the expressions in eqs. (2.42c) and (2.42d) for the vectors $F_i^{(e)}$ to yield

$$F_1^{(e)} = \int_{x_1^{(e)}}^{x_2^{(e)}} \Omega_j^{(e)} f_j^{(e)} (x - x_1^{(e)} + \tilde{l}) dx = T_{1j}^{(e)} f_j^{(e)}, \quad j=1,2 \tag{3.32a}$$

and

$$F_2^{(e)} = \int_{x_1^{(e)}}^{x_2^{(e)}} \Omega_j^{(e)} f_j^{(e)} (x_2^{(e)} - x + \tilde{l}) dx = T_{2j}^{(e)} f_j^{(e)}, \quad j=1,2 \tag{3.32b}$$

where

$$\begin{aligned}
T_{11}^{(e)} &= l^{(e)} \int_0^1 (1-\zeta)(\tilde{l} + l^{(e)}\zeta) d\zeta = \\
T_{22}^{(e)} &= l^{(e)} \int_0^1 \zeta[\tilde{l} + l^{(e)}(1-\zeta)] d\zeta = \frac{l^{(e)}(3\tilde{l} + l^{(e)})}{6} \\
T_{12}^{(e)} &= l^{(e)} \int_0^1 \zeta(\tilde{l} + l^{(e)}\zeta) d\zeta = \\
T_{21} &= l^{(e)} \int_0^1 (1-\zeta)[\tilde{l} + l^{(e)}(1-\zeta)] d\zeta = \frac{l^{(e)}(3\tilde{l} + 2l^{(e)})}{6}
\end{aligned} \tag{3.33}$$

3.7 Second Green Element Model (GESNH-2)

Having introduced the subject of interpolating functions, we are now in a position to derive a second Green element formulation of the 1-D. nonlinear Poisson equation for a heterogeneous medium. We recognize that

$$\frac{1}{K}\frac{dK}{dx} = \frac{d\ln K}{dx} \tag{3.34}$$

so that the integral equation (3.10) now becomes

$$\begin{aligned}
&-2\lambda^{(e)}h_i^{(e)} + [H(x_2^{(e)}-x_i^{(e)}) - H(x_i^{(e)}-x_2^{(e)})]h_2^{(e)} - \\
&[H(x_1^{(e)}-x_i^{(e)}) - H(x_i^{(e)}-x_1^{(e)})]h_1^{(e)} - (|x_2^{(e)}-x_i^{(e)}| + \tilde{l})\varphi_2^{(e)} + \\
&(|x_1^{(e)}-x_i^{(e)}| + \tilde{l})\varphi_1^{(e)} + \int_{x_1^{(e)}}^{x_2^{(e)}} (|x-x_i^{(e)}| + \tilde{l})\left(\frac{f(x)}{K} - \frac{d\ln K}{dx}\frac{dh}{dx}\right) dx = 0, \ \ i=1,2
\end{aligned} \tag{3.35}$$

In practice, the parameter K is usually determined from experimental or field surveys at a number of spatial positions, and it is from those data points a functional relationship, K=$K(x)$ is constructed either by the Lagrange or least square or other appropriate interpolation method. For the nonlinear case the parameter K is experimentally evaluated for various values of the primary variable h, and then some interpolation method is used to determine a functional relation K=$K[h(x)]$. Also, as we stated in chapter 2, the functional expression for $f(x)$ is not usually given but its values at a number of spatial positions so its functional expression is also obtained by interpolation. Within the element e we have liberty to interpolate the reciprocal of K, the logarithm of K, φ, and $f(x)$ with respect to x using appropriate element interpolation functions. We elect here to use linear interpolation functions to approximate all those quantities, that is

$$\ln K \equiv \Theta \approx \Omega_1^{(e)}\Theta_1^{(e)} + \Omega_2^{(e)}\Theta_2^{(e)} \tag{3.36a}$$

$$\frac{1}{K} \equiv \Psi \approx \Omega_1^{(e)}\Psi_1^{(e)} + \Omega_2^{(e)}\Psi_2^{(e)} \tag{3.36b}$$

$$\frac{dh}{dx} \equiv \varphi \approx \Omega_1^{(e)}\varphi_1^{(e)} + \Omega_2^{(e)}\varphi_2^{(e)} \tag{3.36c}$$

$$f(x) \approx \Omega_1^{(e)}f_1^{(e)} + \Omega_2^{(e)}f_2^{(e)} \tag{3.36d}$$

Essentially, eqs. (3.36a) to (3.36d) ensure that $\ln K$ or Θ, $1/K$ or Ψ, φ, and $f(x)$ are piecewise continuous at each node or they possess C^0 continuity. The derivative of the logarithm of K with respect to x is given by

$$\frac{d\ln K}{dx} = \frac{d\Theta}{dx} = \frac{d\Theta}{d\zeta}\frac{d\zeta}{dx} \approx \frac{\Theta_2^{(e)} - \Theta_1^{(e)}}{l^{(e)}} \tag{3.37}$$

The integral equation (3.35) is simplified further by employing the relations in eqs. (3.36) and (3.37) to give

$$\begin{aligned}
&-2\lambda^{(e)}h_i^{(e)} + [H(x_2^{(e)}-x_i^{(e)}) - H(x_i^{(e)}-x_2^{(e)})]h_2^{(e)} - \\
&[H(x_1^{(e)}-x_i^{(e)}) - H(x_i^{(e)}-x_1^{(e)})]h_1^{(e)} - \\
&(|x_2^{(e)}-x_i^{(e)}| + \tilde{l})\varphi_2^{(e)} + (|x_1^{(e)}-x_i^{(e)}| + \tilde{l})\varphi_1^{(e)} - \\
&\frac{1}{l^{(e)}}\int_{x_1^{(e)}}^{x_2^{(e)}} (|x-x_i^{(e)}| + \tilde{l})(\Theta_2^{(e)} - \Theta_1^{(e)})(\Omega_1^{(e)}\varphi_1^{(e)} + \Omega_2^{(e)}\varphi_2^{(e)})dx + \\
&\int_{x_1^{(e)}}^{x_2^{(e)}} (|x-x_i^{(e)}| + \tilde{l})[\Omega_1^{(e)}\Psi_1^{(e)} + \Omega_2^{(e)}\Psi_2^{(e)}][\Omega_1^{(e)}f_1^{(e)} + \Omega_2^{(e)}f_2^{(e)}]dx = 0, \quad i=1,2
\end{aligned} \tag{3.38}$$

The matrix equation for the element is now constructed by separately considering $i=1$ and $i=2$. The outcome is

$$\boxed{R_{ij}^{(e)}h_j^{(e)} + (L_{ij}^{(e)} - V_{inj}^{(e)}\Theta_n^{(e)})\varphi_j^{(e)} + U_{ijn}^{(e)}\Psi_j^{(e)}f_n^{(e)} = 0; \quad i,j,n = 1,2} \tag{3.39}$$

We are already quite familiar with the two element matrices $R_{ij}^{(e)}$ and $L_{ij}^{(e)}$ from chapter 2; the two new matrices now being encountered are $V_{inj}^{(e)}$ and $U_{ijn}^{(e)}$,

$$V_{inj}^{(e)} = \frac{1}{l^{(e)}}\int_{x_1^{(e)}}^{x_2^{(e)}} \frac{d\Omega_n^{(e)}}{d\zeta}\Omega_j^{(e)}(|x - x_i^{(e)}| + \tilde{l})dx; \quad i,n,j = 1,2 \tag{3.40}$$

We can introduce the local coordinate $\zeta = (x - x_1^{(e)})/l^{(e)}$ into eq. (3.40) to have

$$V_{1nj}^{(e)} = \int_0^1 \frac{d\Omega_n^{(e)}}{d\zeta}\Omega_j^{(e)}(\tilde{l} + l^{(e)}\zeta)d\zeta; \quad n,j = 1,2 \tag{3.41a}$$

$$V_{2nj}^{(e)} = \int_0^1 \frac{d\Omega_n^{(e)}}{d\zeta}\Omega_j^{(e)}[\tilde{l} + l^{(e)}(1-\zeta)]d\zeta; \quad n,j = 1,2 \tag{3.41b}$$

We can introduce the expressions for $\Omega_j^{(e)}$ from eq. (3.27) into eqs. (3.41a) and (3.41b), and observe that those integrals can be evaluated in an exact manner

$$V_{111}^{(e)} = \int_0^1 (\zeta-1)(\tilde{l}+l^{(e)}\zeta)d\zeta = -V_{121}^{(e)} = \frac{-(3\tilde{l}+l^{(e)})}{6}$$

$$V_{112}^{(e)} = \int_0^1 -\zeta(\tilde{l}+l^{(e)}\zeta)d\zeta = -V_{122} = \frac{-(3\tilde{l}+2l^{(e)})}{6} \tag{3.42a}$$

$$V_{211}^{(e)} = \int_0^1 (\zeta-1)[\tilde{l}+l^{(e)}(1-\zeta)]d\zeta = -V_{221}^{(e)} = \frac{-(3\tilde{l}+2l^{(e)})}{6}$$

$$V_{212}^{(e)} - \int_0^1 -\zeta[\tilde{l}+l^{(e)}(1-\zeta)]d\zeta = -V_{222} = \frac{-(3\tilde{l}+l^{(e)})}{6} \tag{3.42b}$$

A summary of the expressions for $V_{ijn}^{(e)}$ is given as

$$V_{1jn}^{(e)} = \frac{1}{6}\begin{bmatrix} -(3\tilde{l}+l^{(e)}) & -(3\tilde{l}+2l^{(e)}) \\ 3\tilde{l}+l^{(e)} & 3\tilde{l}+2l^{(e)} \end{bmatrix}, \text{ and}$$

$$V_{2jn}^{(e)} = \frac{1}{6}\begin{bmatrix} -(3\tilde{l}+2l^{(e)}) & -(3\tilde{l}+l^{(e)}) \\ 3\tilde{l}+2l^{(e)} & 3\tilde{l}+l^{(e)} \end{bmatrix} \tag{3.42c}$$

And the matrix $U_{ijn}^{(e)}$ is given by

$$U_{ijn}^{(e)} = \int_{x_1^{(e)}}^{x_2^{(e)}} \Omega_j^{(e)}\Omega_l^{(e)}(|x - x_i^{(e)}| + \tilde{l})dx; \quad i,j,n = 1,2 \tag{3.43}$$

We observe the symmetry of $U_{ijn}^{(e)}$, that is $U_{ijn}^{(e)} = U_{inj}^{(e)}$. The expression for the local coordinate $\zeta = (x - x_1^{(e)})/l^{(e)}$ is introduced into eq. (3.43) to give

$$U_{1jn}^{(e)} = l^{(e)}\int_0^1 \Omega_j^{(e)}\Omega_n^{(e)}(\tilde{l}+l^{(e)}\zeta)d\zeta; \quad j,n=1,2 \tag{3.44a}$$

$$U_{2jn}^{(e)} = l^{(e)}\int_0^1 \Omega_j^{(e)}\Omega_n^{(e)}[\tilde{l}+l^{(e)}(1-\zeta)]d\zeta; \quad j,n=1,2 \tag{3.44b}$$

Introducing the expressions for $\Omega_j^{(e)}$ from eq. (3.27) into eqs. (3.44a) and (3.44b), and performing the integrations yield

$$U_{111}^{(e)} = l^{(e)}\int_0^1 (1-\zeta)^2(\tilde{l}+l^{(e)}\zeta)d\zeta = \frac{l^{(e)}(4\tilde{l}+l^{(e)})}{12}$$

$$U_{112}^{(e)} = l^{(e)}\int_0^1 \zeta(1-\zeta)(\tilde{l}+l^{(e)}\zeta)d\zeta = U_{121} = \frac{l^{(e)}(2\tilde{l}+l^{(e)})}{12} \tag{3.45a}$$

$$U_{122}^{(e)} = l^{(e)}\int_0^1 \zeta^2(\tilde{l}+l^{(e)}\zeta)d\zeta = \frac{l^{(e)}(4\tilde{l}+3l^{(e)})}{12}$$

$$U_{211}^{(e)} = l^{(e)}\int_0^1 (1-\zeta)^2[\tilde{l}+l^{(e)}(1-\zeta)]\zeta)d\zeta = \frac{l^{(e)}(4\tilde{l}+3l^{(e)})}{12}$$

$$U_{212}^{(e)} = l^{(e)}\int_0^1 \zeta(1-\zeta)[\tilde{l}+l^{(e)}(\tilde{l}-\zeta)]\zeta)d\zeta = U_{221} = \frac{l^{(e)}(2\tilde{l}+l^{(e)})}{12} \tag{3.45b}$$

$$U_{222}^{(e)} = l^{(e)}\int_0^1 \zeta^2[\tilde{l}+l^{(e)}(1-\zeta)]d\zeta = \frac{l^{(e)}(4\tilde{l}+l^{(e)})}{12}$$

Summarizing the expressions for the element matrix $U_{ijn}^{(e)}$ yields

$$U_{1jn}^{(e)} = \frac{l^{(e)}}{12}\begin{bmatrix} 4\tilde{l}+l^{(e)} & 2\tilde{l}+l^{(e)} \\ 2\tilde{l}+l^{(e)} & 4\tilde{l}+3l^{(e)} \end{bmatrix}, \quad U_{2jn}^{(e)} = \frac{l^{(e)}}{12}\begin{bmatrix} 4\tilde{l}+3l^{(e)} & 2\tilde{l}+l^{(e)} \\ 2\tilde{l}+l^{(e)} & 4\tilde{l}+l^{(e)} \end{bmatrix} \tag{3.45c}$$

Eq. (3.39) is the system of element equations from a second Green element formulation for the 1-D. nonlinear Poisson equation in a heterogeneous medium. For ease of reference, this formulation is considered as a second level approximation of the differential equation, and herein denoted with the code name **GESNH-2**. It will be of interest to evaluate its performance against that of **GESNH-1** earlier presented.

The steps that are needed to take us from eq. (3.39) to the global matrix equation are as outlined in chapter 2 for the linear homogeneous problem. The steps simply involve assembling eq. (3.39) for all the M elements, ensuring that the specified boundary conditions are properly accounted for. The nonlinear case shall be addressed after strategies for solving systems of nonlinear equations have been reviewed.

You may have observed the considerable flexibility offered by interpolating a quantity or function within the element as indicated by eqs. (3.36a) to (3.36d). By expressing $\ln K$ or Θ and $1/K$ or Ψ the way we have done in eqs. (3.36a) and (3.36b), it appears that $\ln K$ and $1/K$ are independent of each other. We simply treated them as any other functional quantity, ensuring, by the linear interpolation employed, that they be piecewise continuous at the nodal points. But $\ln K$ and $1/K$ are not independent functional quantities; they are related by

$$\frac{1}{K} = K^{-1} = \exp[-\ln K] \tag{3.46}$$

The interpolation of eqs. (3.36a) and (3.36b) only guarantees that the relationship between $\ln K$ and $1/K$ given by eq. (3.46) holds at the nodes and no where else. How much we violate eq. (3.46) between the nodes will depend on how widely spaced the nodes are from themselves, and on the variability of K.

3.8 Third Green Element Model (GESNH-3)

In achieving our third level of approximation to the original differential equation, we take the liberty to approximate $1/K$ and K by linear piecewise interpolation functions within the element, that is

$$K \approx \Omega_1^{(e)} K_1^{(e)} + \Omega_2^{(e)} K_2^{(e)} \tag{3.47}$$

Eq. (3.47) is combined with eq. (3.36b) so that

$$\frac{1}{K}\frac{dK}{dx} \approx \frac{1}{l^{(e)}} \Omega_j^{(e)} \Psi_j^{(e)} \frac{d\Omega_n^{(e)}}{d\zeta} K_n^{(e)} \tag{3.48}$$

Using eq. (3.48) to simplify the heterogeneous term in the line integral, and introducing it into eq. (3.35), yields the element matrix equation

$$\boxed{R_{ij}^{(e)} h_j^{(e)} + (L_{ij}^{(e)} - S_{imnj}^{(e)} K_m^{(e)} \Psi_n^{(e)}) \varphi_j^{(e)} + U_{ijm}^{(e)} \Psi_j^{(e)} f_m^{(e)} = 0; \quad i,j,m,n = 1,2} \tag{3.49}$$

The only matrix that is yet familiar is $S_{imnj}^{(e)}$. It is given by

$$S_{imnj}^{(e)} = \frac{1}{l^{(e)}}\int_{x_1^{(e)}}^{x_2^{(e)}} \frac{d\Omega_m^{(e)}}{d\zeta}\Omega_n^{(e)}\Omega_j^{(e)}(|x - x_i^{(e}| + \tilde{l})dx; \quad i,j,m,n = 1,2 \tag{3.50}$$

Its symmetric property can be observed, that is $S_{imnj}^{(e)} = S_{imjn}^{(e)}$. The expression for the local coordinate $\zeta = (x - x_1^{(e)})/l^{(e)}$ is introduced into eq. (3.50) to give

$$S_{1mnj}^{(e)} = \int_0^1 \frac{d\Omega_m^{(e)}}{d\zeta}\Omega_n^{(e)}\Omega_j^{(e)}(\tilde{l} + l^{(e)}\zeta)d\zeta; \quad j,m,n = 1,2 \tag{3.51a}$$

$$S_{2mnj}^{(e)} = \int_0^1 \frac{d\Omega_m^{(e)}}{d\zeta}\Omega_n^{(e)}\Omega_j^{(e)}[\tilde{l} + l^{(e)}(1 - \zeta)]d\zeta; \quad j,m,n = 1,2 \tag{3.51b}$$

Introducing the expressions for $\Omega_j^{(e)}$ from eq. (3.27) into eqs. (3.51a) and (3.51b), and carrying out the integrations yield

$$S_{1111}^{(e)} = \int_0^1 -(1 - \zeta)^2(\tilde{l} + l^{(e)}\zeta)d\zeta = -S_{1211}^{(e)} = \frac{-(4\tilde{l} + l^{(e)})}{12}$$

$$S_{1112}^{(e)} = \int_0^1 -\zeta(1 - \zeta)(\tilde{l} + l^{(e)}\zeta)d\zeta = -S_{1212}^{(e)} = S_{1121}^{(e)} = -S_{1221}^{(e)} = \frac{-(2\tilde{l} + l^{(e)})}{12} \tag{3.51a}$$

$$S_{1122}^{(e)} = \int_0^1 -\zeta^2(\tilde{l} + l^{(e)}\zeta)d\zeta = -S_{1222}^{(e)} = \frac{-(4\tilde{l} + 3l^{(e)})}{12}$$

$$S_{2111}^{(e)} = \int_0^1 -(1 - \zeta)^2[\tilde{l} + l^{(e)}(1 - \zeta)]d\zeta = -S_{2211}^{(e)} = \frac{-(4\tilde{l} + 3l^{(e)})}{12}$$

$$S_{2112}^{(e)} = -\int_0^1 \zeta(1 - \zeta)[\tilde{l} + l^{(e)}(1 - \zeta)]d\zeta = -S_{2212}^{(e)} = S_{2121}^{(e)} = -S_{2221}^{(e)} = \frac{-(2\tilde{l} + l^{(e)})}{12} \tag{3.51b}$$

$$S_{2122}^{(e)} = \int_0^1 -\zeta^2[\tilde{l} + l^{(e)}(1 - \zeta)]d\zeta = -S_{2222}^{(e)} = \frac{-(4\tilde{l} + l^{(e)})}{12}$$

We summarize the expressions for $S_{imnj}^{(e)}$ as

$$S_{11nj}^{(e)} = -S_{12nj}^{(e)} = \frac{-1}{12}\begin{bmatrix} 4\tilde{l}+l^{(e)} & 2\tilde{l}+l^{(e)} \\ 2\tilde{l}+l^{(e)} & 4\tilde{l}+3l^{(e)} \end{bmatrix} = -l^{(e)}U_{1nj}^{(e)},$$

$$S_{21nj}^{(e)} = -S_{22nj}^{(e)} = \frac{-1}{12}\begin{bmatrix} 4\tilde{l}+3l^{(e)} & 2\tilde{l}+l^{(e)} \\ 2\tilde{l}+l^{(e)} & 4\tilde{l}+l^{(e)} \end{bmatrix} = -l^{(e)}U_{2nj}^{(e)} \tag{3.51c}$$

Eq. (3.49) is the third system of element equations offered by the Green element method to solve the 1-D. nonlinear heterogeneous Poisson equation. It is herein denoted **GESNH-3**. The three **GE** formulations shall be applied first to some linear heterogeneous problems to assess their accuracies.

3.9 Examples on Linear Heterogeneous Laplace/Poisson Problems

Example 1

Our first example of steady 1-D. linear heterogeneous problem is one which is governed by

$$\frac{d}{dx}\left(K\frac{dh}{dx}\right) = 0 \tag{3.52a}$$

with the boundary conditions

$$h(x=0) = 0 \quad \text{and} \quad h(x=1) = 1 \tag{3.52b}$$

We examine cases of the parameter K having the expressions: (a) $K = 1+x$, (b) $K = 1+x^2$, (c) $K = e^x$, and (d) $K = 1+(x-0.5)^3$. The general solution to the differential equation is given as

$$h(x) = \int_0^x \frac{\alpha_2}{K(s)}ds + \alpha_1 \tag{3.53}$$

where α_1 and α_2 are constants that are determined from the boundary conditions at $x = 0$ and at $x = 1$.

Case a: When $K = 1+x$, then

$$h(x) = \int_0^x \frac{\alpha_2}{(1+s)}ds + \alpha_1 = \alpha_2 \ln(1+x) + \alpha_1$$

$h(x=0) = 0$ implies that $\alpha_1 = 0$, and $h(x=1) = 1$ implies that $\alpha_2 = 1/\ln(2)$. Therefore the particular solution is

$$h(x) = \frac{\ln(1+x)}{\ln 2} \text{ and } \frac{dh}{dx} = \frac{1}{(1+x)\ln 2} \tag{3.54}$$

Case b: When $K = 1+x^2$, then

$$h(x) = \int_0^x \frac{\alpha_2}{(1+s^2)} ds + \alpha_1 = \alpha_2 \tan^{-1}x + \alpha_1$$

$h(x=0) = 0$ implies that $\alpha_1 = 0$, and $h(x=1) = 1$ implies that $\alpha_2 = 1/\tan^{-1}1$. Therefore the particular solution is

$$h(x) = \frac{\tan^{-1}x}{\tan^{-1}1} \text{ and } \frac{dh}{dx} = \frac{1}{(1+x^2)\tan^{-1}1} \tag{3.55}$$

Case c: $K = e^x$, then

$$h(x) = \int_0^x \frac{\alpha_2}{e^s} ds + \alpha_1 = -\alpha_2 e^{-x} + \alpha_1$$

$h(x=0) = 0$ implies that $\alpha_1 = \alpha_2$, and $h(x=1) = 1$ implies that $\alpha_2 = 1/(1-e^{-1})$. Therefore the particular solution is

$$h(x) = \frac{(e^{-x}-1)}{(e^{-1}-1)} \text{ and } \frac{dh}{dx} = \frac{e^{-x}}{(1-e^{-1})} \tag{3.56}$$

Case d: When $K = 1+(x-1/2)^3$, then

$$\begin{aligned} h(x) &= \int^x \frac{\alpha_2}{[1+(s-1/2)^3]} ds + \alpha_1 \\ &= \alpha_2 \left[\frac{1}{6} \ln\left(\frac{(x+0.5)^2}{x^2-2x+1.75} \right) + \frac{1}{\sqrt{3}} \tan^{-1}\left(\frac{2(x-1)}{\sqrt{3}} \right) \right] + \alpha_1 \end{aligned} \tag{3.57a}$$

$h(x=0) = 0$ implies that $\alpha_1 = [\ln(1/7)/6 + \tan^{-1}(-2/\sqrt{3})/\sqrt{3}]\alpha_2 + \alpha_1 = 0$, and $h(x=1) = 1$ implies that $\alpha_2 = [\ln(3)/6]\alpha_2 + \alpha_1 = 1$. We now solve for α_1 and α_2 to give $\alpha_1 = 0.817309$, and $\alpha_2 = 0.997754$. In addition

$$\frac{dh(x)}{dx} = \varphi(x) = \left[\frac{1}{3}\left(\frac{1}{x+0.5} + \frac{1-x}{x^2-2x+1.75}\right) + \frac{2}{3+4(x-1)^2}\right]\alpha_2 \tag{3.57b}$$

The three different Green element formulations (**GESNH-1, GESNH-2**, and **GESNH-3**) are first implemented using hand calculations of *case b* ($K=1+x^2$). A 3-element discretization is used, and is shown in Figure 3.6.

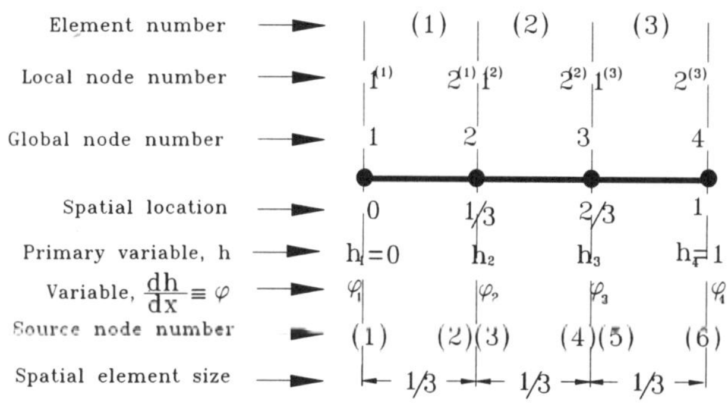

Figure 3.6 3-element discretization of domain for hand computations.

3.9.1 *Implementation of GESNH-1 on a 3-Element Computational Region*

The relevant system of discrete element equations of the **GESNH-1** formulation is eq. (3.17), and restated below

$$R_{ij}^{(e)}h_j^{(e)} + L_{ij}^{(e)}\varphi_j^{(e)} + \frac{1}{\bar{K}^{(e)}}F_i^{(e)} = 0, \quad i,j=1,2; \quad e=1,2,,...,M \tag{3.17}$$

Since there is no recharge in our example, $F_i^{(e)} = 0$. The three elements used to discretize the flow region are of equal size so that $l^{(e)} = l = \tilde{l} = 1/3$. The element matrices are

$$R_{ij}^{(e)} = R_{ij} = (-1)^{i+j-1}, \quad L_{ij}^{(e)} = L_{ij} = \frac{1}{3}\begin{bmatrix} 1 & -2 \\ 2 & -1 \end{bmatrix}$$

for all 3 elements. To account for the heterogeneous nature of the flow, we compute $\bar{K}^{(e)}$ for each element. Since $K=1+x^2$, then $K_1 = 1$, $K_2=10/9$, $K_3 = 13/9$, and $K_4 = 2$, and the mean of these K values for each element are: $\bar{K}^{(1)} = (K_1+K_2)/2 = 19/18$, $\bar{K}^{(2)} = (K_2+K_3)/2 = 23/18$, and $\bar{K}^{(3)} = (K_3+K_4)/2 = 31/18$.

In implementing this formulation, we must take into account the compatibility equations given by eqs. (3.14a) and (3.14b). In our example these equations yield

$$h_2^{(1)} = h_1^{(2)} = h_2; \quad \bar{K}^{(1)}\varphi_2^{(1)} = \bar{K}^{(2)}\varphi_1^{(2)}$$

$$h_2^{(2)} = h_1^{(3)} = h_3; \quad \bar{K}^{(2)}\varphi_2^{(2)} = \bar{K}^{(3)}\varphi_1^{(3)}$$

We are now in a position to assemble the global matrix equation for the **GESNH-1** formulation, and it is given by

$$\begin{bmatrix} 1/3 & 1 & -46/57 & 0 & 0 & 0 \\ 2/3 & -1 & -23/57 & 0 & 0 & 0 \\ 0 & -1 & 1/3 & 1 & -62/69 & 0 \\ 0 & 1 & 2/3 & -1 & -31/69 & 0 \\ 0 & 0 & 0 & -1 & 1/3 & -2/3 \\ 0 & 0 & 0 & 1 & 2/3 & -1/3 \end{bmatrix} \begin{Bmatrix} \varphi_1 \\ h_2 \\ \varphi_2 \\ h_3 \\ \varphi_3 \\ \varphi_4 \end{Bmatrix} = \begin{Bmatrix} 0 \\ 0 \\ 0 \\ 0 \\ -1 \\ 1 \end{Bmatrix}$$

The solution to the above matrix equation was obtained for the unknown quantities by a matrix solver. It is presented in Table 3.1.

3.9.2 *Implementation of GESNH-2 on a 3-Element Computational Region*

The system of discrete element equations of the **GESNH-2** formulation is given by (3.39), and restated below

$$R_{ij}^{(e)}h_j^{(e)} + (L_{ij}^{(e)} - V_{inj}^{(e)}\Theta_n^{(e)})\varphi_j^{(e)} + U_{ijn}^{(e)}\Psi_j^{(e)}f_n^{(e)} = 0; \quad i,j,n = 1,2 \tag{3.39}$$

$R_{ij}^{(e)}$ and $L_{ij}^{(e)}$ are the same as in the **GESNH-1** formulation, while the last term on the left hand side of eq. (3.39) is zero. The only term that needs to be evaluated in order to implement the **GESNH-2** formulation is $V_{inj}^{(e)}\Theta_n^{(e)}$ or $V_{inj}^{(e)}(\ln K)_n^{(e)}$. The values of $\Theta_j^{(e)}$ in each element are: $\Theta_1^{(1)} = 0$, $\Theta_2^{(1)} = \Theta_1^{(2)} = \ln(10/9)$, $\Theta_2^{(2)} = \Theta_1^{(3)} = \ln(13/9)$, $\Theta_2^{(3)} = \ln 2$. With these

$$V_{1nj}^{(e)} = \frac{1}{18}\begin{bmatrix} -7 & -11 \\ 7 & 11 \end{bmatrix} \text{ and } V_{2nj}^{(e)} = \frac{1}{18}\begin{bmatrix} -11 & -7 \\ 11 & 7 \end{bmatrix}$$

We define $P_{ij}^{(e)} = V_{inj}^{(e)}\Theta_n^{(e)}$, then

$$P_{ij}^{(1)} = \frac{\ln(10/9)}{18}\begin{bmatrix} 7 & 11 \\ 11 & 7 \end{bmatrix},\ P_{ij}^{(2)} = \frac{\ln(13/10)}{18}\begin{bmatrix} 7 & 11 \\ 11 & 7 \end{bmatrix},\ \text{and } P_{ij}^{(3)} = \frac{\ln(18/13)}{18}\begin{bmatrix} 7 & 11 \\ 11 & 7 \end{bmatrix}.$$

The global coefficient matrix of the **GESNH-2** formulation for this 3-element discretization of the flow problem now becomes

$$\begin{bmatrix} 1-\frac{7}{18}\ln\frac{10}{9} & 1 & -\frac{4}{3}-\frac{11}{18}\ln\frac{10}{9} & 0 & 0 & 0 \\ \frac{4}{3}-\frac{11}{18}\ln\frac{10}{9} & -1 & -1-\frac{7}{18}\ln\frac{10}{9} & 0 & 0 & 0 \\ 0 & -1 & 1-\frac{7}{18}\ln\frac{13}{10} & 1 & -\frac{4}{3}-\frac{11}{18}\ln\frac{13}{10} & 0 \\ 0 & 1 & \frac{4}{3}-\frac{11}{18}\ln\frac{13}{10} & -1 & -1-\frac{7}{18}\ln\frac{13}{10} & 0 \\ 0 & 0 & 0 & -1 & 1-\frac{7}{18}\ln\frac{18}{13} & -\frac{4}{3}-\frac{11}{18}\ln\frac{18}{13} \\ 0 & 0 & 0 & 1 & \frac{4}{3}-\frac{11}{18}\ln\frac{18}{13} & -1-\frac{7}{18}\ln\frac{18}{13} \end{bmatrix}\begin{Bmatrix} \varphi_1 \\ h_2 \\ \varphi_2 \\ h_3 \\ \varphi_3 \\ \varphi_4 \end{Bmatrix} = \begin{Bmatrix} 0 \\ 0 \\ 0 \\ 0 \\ -1 \\ 1 \end{Bmatrix}$$

or

$$\begin{bmatrix} 0.9590 & 1 & -1.3977 & 0 & 0 & 0 \\ 1.2690 & -1 & -1.0410 & 0 & 0 & 0 \\ 0 & -1 & 0.8980 & 1 & -1.4937 & 0 \\ 0 & 1 & 1.1730 & -1 & -1.1020 & 0 \\ 0 & 0 & 0 & -1 & 0.8734 & -1.5322 \\ 0 & 0 & 0 & 1 & 1.1345 & -1.1266 \end{bmatrix}\begin{pmatrix} \varphi_1 \\ h_2 \\ \varphi_2 \\ h_3 \\ \varphi_3 \\ \varphi_4 \end{pmatrix} = \begin{pmatrix} 0 \\ 0 \\ 0 \\ 0 \\ -1 \\ 1 \end{pmatrix}$$

The solution of the above matrix equation is presented in Table 3.1.

3.9.3 *Implementation of GESNH-3 on the 3-Element Computational Region*

The system of discrete element equations of the **GESNH-3** model of the heterogeneous problem with no external recharge is obtained from eq. (3.49) and stated below

$$R_{ij}^{(e)}h_j^{(e)} + (L_{ij}^{(e)} - S_{imnj}^{(e)}K_m^{(e)}\Psi_n^{(e)})\varphi_j^{(e)} = 0; \quad i,j,m,n = 1,2$$

The element matrices $R_{ij}^{(e)}$ and $L_{ij}^{(e)}$ remain the same as in the **GESNH-1** and **GESNH-2** models. The only term that is different from the earlier two models is $S_{imnj}^{(e)}K_m^{(e)}\Psi_n^{(e)}\varphi_j^{(e)}$. To evaluate that term we make use of the following information: $l^{(e)} = l = \tilde{l} = 1/3$; $K_1^{(1)} = 1$, $K_2^{(1)} = K_1^{(2)} = 10/9$, $K_2^{(2)} = K_1^{(3)} = 13/9$, and $K_2^{(3)} = 2$; $\Psi_1^{(1)} = 1$, $\Psi_2^{(1)} = \Psi_1^{(2)} = 9/10$, $\Psi_2^{(2)} = \Psi_1^{(3)} = 9/13$, and $\Psi_2^{(3)} = 1/2$. The element matrix $S_{imnj}^{(e)}$ for this 3-element discretization is obtained from eq. (3.51c), and given below:

$$S_{1mn1}^{(e)} = \frac{1}{36}\begin{bmatrix} -5 & -3 \\ 5 & 3 \end{bmatrix}, S_{1mn2}^{(e)} = \frac{1}{36}\begin{bmatrix} -3 & -7 \\ 3 & 7 \end{bmatrix}, S_{2mn1}^{(e)} = \frac{1}{36}\begin{bmatrix} -7 & -3 \\ 7 & 3 \end{bmatrix}, S_{2mn2}^{(e)} = \frac{1}{36}\begin{bmatrix} -3 & -5 \\ 3 & 5 \end{bmatrix}.$$

Because of the uniform discretization of the computational domain, the element matrix $S_{imnj}^{(e)}$ is the same for all the elements. We define $P_{ij}^{(e)} = S_{imnj}^{(e)}K_m^{(e)}\Psi_n^{(e)}$, then

$$P_{ij}^{(1)} = \frac{1}{3240}\begin{bmatrix} 77 & 93 \\ 97 & 75 \end{bmatrix}, P_{ij}^{(2)} = \frac{1}{4680}\begin{bmatrix} 285 & 327 \\ 363 & 267 \end{bmatrix}, P_{ij}^{(e)} = \frac{1}{8424}\begin{bmatrix} 645 & 725 \\ 825 & 595 \end{bmatrix}.$$

With $P_{ij}^{(e)}$ computed for these three elements we are now in a position to form the global matrix equation which is given by

$$\begin{bmatrix} \frac{1}{3}-\frac{77}{3240} & 1 & -\frac{2}{3}-\frac{93}{3240} & 0 & 0 & 0 \\ \frac{2}{3}-\frac{97}{3240} & -1 & -\frac{1}{3}-\frac{75}{3240} & 0 & 0 & 0 \\ 0 & -1 & \frac{1}{3}-\frac{285}{4680} & 1 & -\frac{2}{3}-\frac{327}{4680} & 0 \\ 0 & 1 & \frac{2}{3}-\frac{363}{4680} & -1 & -\frac{1}{3}-\frac{267}{4680} & 0 \\ 0 & 0 & 0 & -1 & \frac{1}{3}-\frac{645}{8424} & -\frac{2}{3}-\frac{725}{8424} \\ 0 & 0 & 0 & 1 & \frac{2}{3}-\frac{825}{8424} & -\frac{1}{3}-\frac{595}{8424} \end{bmatrix} \begin{Bmatrix} \varphi_1 \\ h_2 \\ \varphi_2 \\ h_3 \\ \varphi_3 \\ \varphi_4 \end{Bmatrix} = \begin{Bmatrix} 0 \\ 0 \\ 0 \\ 0 \\ -1 \\ 1 \end{Bmatrix}$$

or

$$\begin{bmatrix} 0.3096 & 1 & -0.6954 & 0 & 0 & 0 \\ 0.6367 & -1 & -0.3565 & 0 & 0 & 0 \\ 0 & -1 & 0.2724 & 1 & -0.7365 & 0 \\ 0 & 1 & 0.5891 & -1 & -0.3904 & 0 \\ 0 & 0 & 0 & -1 & 0.2568 & -0.7527 \\ 0 & 0 & 0 & 1 & 0.5687 & -0.4040 \end{bmatrix} \begin{Bmatrix} \varphi_1 \\ h_2 \\ \varphi_2 \\ h_3 \\ \varphi_3 \\ \varphi_4 \end{Bmatrix} = \begin{Bmatrix} 0 \\ 0 \\ 0 \\ 0 \\ -1 \\ 1 \end{Bmatrix}$$

The solution to the global matrix equation is presented in Table 3.1. The results from the three models show that they all closely reproduce the exact solution, except the **GESNH-1** model solution for φ which has discontinuities at the nodes because the model approximates the domain as multiply layered with each layer having uniform value of the parameter K.

All four cases of different functional expressions of the parameter K for example 1 were simulated with the three **GEM** models, and all the simulations were carried using five uniform linear elements. Those solutions and their exact solutions are presented in Tables 3.2a through 3.2d.

Table 3.1: Solutions for example 1b from three **GE** models (**GESNH-1, GESNH-2,** and **GESNH-3**) obtained by hand computations using a 3-element discretization.

Solution		Exact	**GESNH-1**	**GESNH-2**	**GESNH-3**
of h at	x = 1/3	0.4097	0.4100	0.4075	0.4102
	x = 2/3	0.7487	0.7489	0.7470	0.7496
of φ at	x = 0	1.2732	1.2300*	1.2880	1.2981
	x = 1/3	1.1459	1.0161*	1.1590	1.1678
	x = 2/3	0.8815	0.7539*	0.8902	0.8928
	x = 1.0	0.6366	0.7539	0.6411	0.6372

** Solution for φ computed just to the right of the node because of the discontinuity in φ inherent in **GESNH-1**'s approximation of the original equation.*

Table 3.2a: Exact and **GEM** solutions of example 1a ($K = 1 + x$)

Solution		Exact	**GESNH-1**	**GESNH-2**	**GESNH-3**
of h at	x=0.2	0.2630	0.2628	0.2634	0.2636
	x=0.4	0.4854	0.4851	0.4859	0.4860
	x=0.6	0.6781	0.6778	0.6784	0.6786
	x=0.8	0.8480	0.8479	0.8482	0.8483
of φ at	x=0	1.4427	1.3139	1.4412	1.4471
	x=0.2	1.2022	1.1118	1.2004	1.2034
	x=0.4	1.0305	0.9635	1.0286	1.0302
	x=0.6	0.9017	0.8502	0.8998	0.9007
	x=0.8	0.8015	0.7607	0.7997	0.8002
	x=1.0	0.7213	0.7607	0.7197	0.7199

Table 3.2b: Exact and **GEM** solutions of example 1b ($K = 1 + x^2$)

Solution		Exact	**GESNH-1**	**GESNH-2**	**GESNH-3**
of h at	x=0.2	0.2513	0.2514	0.2508	0.2514
	x=0.4	0.4845	0.4846	0.4837	0.4847
	x=0.6	0.6881	0.6881	0.6874	0.6884
	x=0.8	0.8591	0.8591	0.8588	0.8594
of φ at	x=0	1.2732	1.2572	1.2785	1.2821
	x=0.2	1.0610	1.1657	1.2294	1.2328
	x=0.4	0.9095	1.0177	1.1021	1.1048
	x=0.6	0.7958	0.8549	0.9397	0.9410
	x=0.8	0.7074	0.7046	0.7788	0.7787
	x=1.0	0.6366	0.7046	0.6382	0.6368

Table 3.2c: Exact and **GEM** solutions of example 1c ($K = e^x$)

Solution		Exact	**GESNH-1**	**GESNH-2**	**GESNH-3**
of h at	x=0.2	0.2868	0.2868	0.2871	0.2880
	x=0.4	0.5215	0.5215	0.5215	0.5232
	x=0.6	0.7138	0.7138	0.7138	0.7152
	x=0.8	0.8711	0.8711	0.8714	0.8720
of φ at	x=0	1.5820	1.4338	1.5842	1.5963
	x=0.2	1.2952	1.1739	1.2961	1.3034
	x=0.4	1.0604	0.9611	1.0605	1.0642
	x=0.6	0.8682	0.7869	0.8677	0.8690
	x=0.8	0.7108	0.6443	0.7099	0.7095
	x=1.0	0.5820	0.6443	0.5808	0.5793

Table 3.2d: Exact and **GEM** solutions of example 1d ($K=1+(x-0.5)^3$)

Solution		Exact	**GESNH-1**	**GESNH-2**	**GESNH-3**
of *h* at	x=0.2	0.2143	0.2159	0.2162	0.2161
	x=0.4	0.4159	0.4183	0.4184	0.4184
	x=0.6	0.6154	0.6178	0.6178	0.6178
	x=0.8	0.8130	0.8146	0.8145	0.8145
of φ at	x=0	1.1403	1.0797	1.1394	1.1402
	x=0.2	1.0254	1.0118	1.0246	1.0249
	x=0.4	0.9988	0.9976	0.9979	0.9982
	x=0.6	0.9968	0.9838	0.9959	0.9962
	x=0.8	0.9715	0.9271	0.9707	0.9710
	x=1.0	0.8869	0.9271	0.8861	0.8862

3.10 Nonlinear Solution Strategies

There are numerical solution procedures which have been developed for solving single nonlinear equations or systems of nonlinear simultaneous equations. The two most popular ones are the *linear iteration* or *Picard scheme* and the *Newton-Raphson* scheme. Although variations of these schemes exist, we focus only on the classical procedures of implementing them in this text.

3.10.1 *Linear Iteration or Picard Scheme*

We develop the Picard scheme first for a single nonlinear equation given by

$$g(x) = 0 \tag{3.58}$$

Our goal is to find, in a number of well-defined steps, a value of x that satisfies eq. (3.58). An equivalent equation is derived from eq. (3.58) of the form

$$g(x) = p(x) - x = 0 \tag{3.59}$$

so that the solution to eq. (3.59) is also a solution to eq (3.58). The linear iteration or Picard procedure of finding the solution is based on the iteration scheme

$$x^{(k+1)} = p(x^{(k)}) \tag{3.60}$$

in which k is the iteration number. $x^{(0)}$ is the first solution that is proposed, and $x^{(1)}$, $x^{(2)}$, $x^{(3)}$, ... are computed by eq. (3.60) which is graphically illustrated in Figure 3.7. The choice of the first guess of x is very important in the process of achieving convergence of the algorithm to the correct solution. A prudent choice of $x^{(0)}$ is the solution of the linearized form of $g(x) = 0$. The iteration process has to be terminated at some point. Convergence of the *Picard algorithm* could be based on the absolute error

$$|x^{(k+1)} - x^{(k)}| \leq \varepsilon \tag{3.61}$$

where ε is a predetermined small value or *convergence tolerance*. Its magnitude depends on the level of accuracy of the solution desired. Convergence could be based on the relative absolute error

$$\left|\frac{x^{(k+1)} - x^{(k)}}{x^{(k+1)}}\right| \leq \varepsilon \tag{3.62}$$

or on the error estimate

$$|g(x^{(k+1)})| \leq \varepsilon \tag{3.63}$$

The Picard scheme is bound to converge under some simple restrictive conditions on $p(x)$, and its convergence rate is linear.

For a system of nonlinear equations given by

$$G(X) = g_i(x_j) = 0, \quad i,j = 1,2,\ldots,n \tag{3.64}$$

the Picard scheme can be derived in a manner similar to that of the single nonlinear equation. Eq. (3.64) is first expressed as

$$G(X) = g_i(x_j) = \delta_{ij}x_j - p_i(x_j) = 0 \tag{3.65}$$

Here δ_{ij} denotes the identity matrix. The Picard iteration algorithm is then

$$\delta_{ij}x_j^{(k+1)} = p_i(x_j^{(k)}) \tag{3.66}$$

In Green element calculations of nonlinear differential operators, we deal with systems of discrete nonlinear equations which usually cannot be expressed in the form of eq. (3.66). In that case, the Picard scheme is modified so that instead of the identity matrix δ_{ij} being the pre-multiplying matrix another matrix which is obtained from the element equations is prudently chosen as the pre-multiplying matrix of $x_j^{(k+1)}$. Thus, the modified Picard scheme of (3.64) is

$$A_{ij}^{(k)}x_j^{(k+1)} = p_i(x_j^{(k)}) \tag{3.67}$$

The matrix $A_{ij}^{(k)}$ comprises elements that depend exclusively on the solution at the previous iteration $x_j^{(k)}$. By matrix inversion the Picard algorithm for a system of nonlinear discrete equations is

$$x_j^{(k+1)} = [A_{ij}^{(k)}]^{-1} p_i(x_j^{(k)}) \tag{3.68}$$

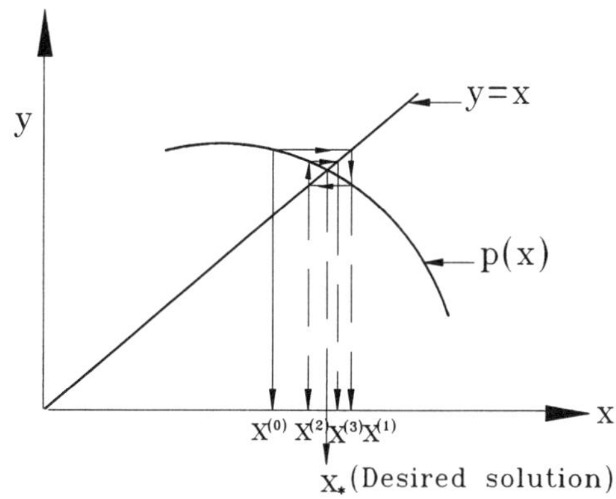

Figure 3.7 Solution procedure of the Picard scheme.

3.10.2 *The Newton-Raphson Solution Algorithm*

The Newton-Raphson solution scheme for a single nonlinear equation

$$g(x) = 0 \tag{3.69}$$

is derived by first proposing a solution $x^{(k)}$ which, in most instances, would not satisfy the equation, and then an update to that initial guess $x^{(k+1)}$ which is presumed to be the solution to the nonlinear equation. That update or improvement on the initial guess is expressed as

$$x^{(k+1)} = x^{(k)} + \Delta x^{(k+1)} \tag{3.70}$$

in which $\Delta x^{(k+1)}$ is referred to as the *increment* on the initial solution, and assumed small in magnitude relative to $x^{(k)}$. By *Taylor's series expansion*

$$g(x^{(k+1)}) = g(x^{(k)}) + \Delta x^{(k+1)} \left.\frac{dg}{dx}\right|_{x=x^{(k)}} + O(\Delta x^{(k+1)})^2 = 0 \tag{3.71}$$

Neglecting powers of $\Delta x^{(k+1)}$ of order 2 and higher values, an expression for the increment $\Delta x^{(k+1)}$ is obtained as

$$\Delta x^{(k+1)} = -\frac{g(x^{(k)})}{g'(x^{(k)})} \tag{3.72}$$

where $g'(x^{(k)})$ is the first derivative of $g(x)$ evaluated at $x^{(k)}$. The convergence conditions given by eqs. (3.61 - 3.63) could be used to terminate the iteration process of the Newton-Raphson scheme. For well-behaved functions, the Newton-Raphson scheme exhibits quadratic convergence characteristics. The underlying theory of the Newton-Raphson algorithm given by eq. (3.72) is illustrated graphically in Figure 3.8. The slope of the curve $g(x)$ at $x^{(0)}$ is given by

$$g'(x^{(0)}) = \frac{g(x^{(0)})}{(x^{(0)} - x^{(1)})} \tag{3.73}$$

Hence

$$x^{(1)} = x^{(0)} - \frac{g(x^{(0)})}{g'(x^{(0)})} \tag{3.74}$$

Eq. (3.74) gives the first iteration of the Newton-Raphson scheme earlier derived in eq. (3.72); higher iteration values of x are seen to approach the true solution x_* in Figure 3.8.

The convergence process of the Newton-Raphson scheme runs into difficulties for functions with roots of high order of multiplicity, those with multiple roots where the process may converge to a solution not desired, and for those functions that have an *S-shape* as shown in Figure 3.9. As shown in Figure 3.9 the iteration process certainly does not show any signs of converging to the right solution. This situation should be expected for problems of unsaturated flow where the hydraulic conductivity parameter variation with respect to either the pressure head or moisture content sometimes exhibits an *S-profile*. Without getting involved in ways of correcting some of these convergence problems of the Newton-Raphson scheme (refer to Hornbeck [12]), we present the Newton-Raphson scheme for a system of nonlinear equations.

$$G(X) = g_i(x_j) = 0, \quad i,j = 1,2,\dots,n \tag{3.75}$$

We propose an initial solution $x_j^{(k)}$ to eq. (3.75), and an improvement that is given by

$$x_j^{(k+1)} = x_j^{(k)} + \Delta x_j^{(k+1)} \tag{3.76}$$

The improved solution given by eq. (3.76) is assumed to satisfy the system of nonlinear equations, and employing Taylor's series expansion, eq. (3.75) becomes

$$g_i(x_j^{(k+1)}) = g_i(x_j^{(k)}+\Delta x_j^{(k+1)}) = g_i(x_j^{(k)}) + \left.\frac{\partial g_i}{\partial x_j}\right|_{x_j=x_j^{(k)}} \Delta x_j^{(k+1)} + O[\Delta x_j^{(k+1)}]^2 = 0 \tag{3.77}$$

Powers of $\Delta x_j^{(k+1)}$ of order 2 and higher are neglected in eq. (3.77) so that the increment $\Delta x_j^{(k+1)}$ is obtained from the matrix equation

$$\left.\frac{\partial g_i}{\partial x_j}\right|_{x_j=x_j^{(k)}} \Delta x_j^{(k+1)} = -g_i(x_j^{(k)})$$

or

$$J_{ij}^{(k)} \Delta x_j^{(k+1)} = -g_i(x_j^{(k)}) \tag{3.78}$$

The matrix $J_{ij}^{(k)}$ is commonly referred to as the *Jacobian matrix* and it is obtained from the partial derivatives of the original system of nonlinear equations with respect to the solution variables, and evaluated at the previous iteration level. When the Newton-Raphson scheme is applied in Green element calculations, efforts are usually directed at accurate evaluation of the Jacobian matrix so that convergence of the scheme can be guaranteed. Convergence is said to be achieved when the L_2 norm of $\Delta x_j^{(k+1)}$ becomes less than a specified tolerance value ε. That is

$$\left\{\Delta x_j^{(k+1)}\right\}^T . \Delta x_j^{(k+1)} \leq \varepsilon \tag{3.79}$$

where $\{\Delta x_j^{(k+1)}\}^T$ is the transpose of the *increment vector* $\Delta x_j^{(k+1)}$.

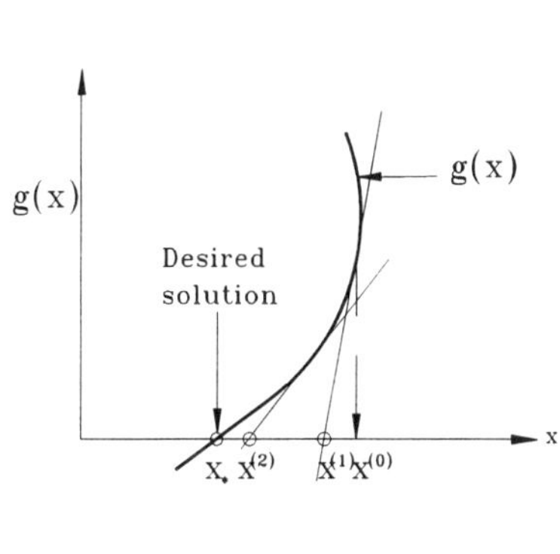

Figure 3.8 Solution procedure of the Newton-Raphson scheme.

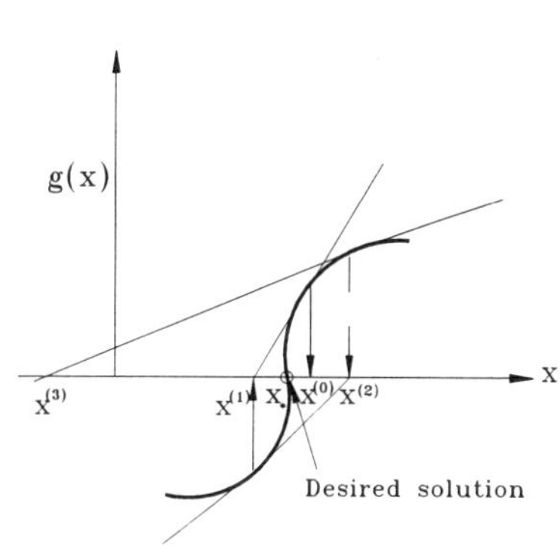

Figure 3.9 Case of a diverging Newton-Raphson iteration process.

3.11 Applications of Picard and Newton-Raphson Algorithms in Green Element Calculations

We are now in a position to incorporate the linear iteration or Picard and the Newton-Raphson schemes into the **GESNH-1**, **GESNH-2**, and **GESNH-3** models which we derived earlier for the steady 1-D. nonlinear heterogeneous problem. The starting point in implementing these nonlinear solution schemes for the **GESNH-1** model is its element equations which were earlier derived, and re-stated here.

$$R_{ij}^{(e)}\varphi_j^{(e)} + L_{ij}^{(e)}\varphi_j^{(e)} + \frac{1}{\bar{K}^{(e)}}T_{ij}^{(e)}f_j^{(e)} = 0, \quad i,j=1,2; \quad e=1,2,\ldots,M$$

(Note that the recharge term has been approximated by linear interpolating functions.) Applying the Picard algorithm to the above equation entails evaluating the first two terms at the current iteration $k+1$, while the third term is evaluated at the previous iteration k. The discrete system of equations with the Picard algorithm now becomes

$$R_{ij}^{(e)}h_j^{(e,k+1)} + L_{ij}^{(e)}\varphi_j^{(e,k+1)} = -\frac{1}{\bar{K}^{(e,k)}}T_{ij}^{(e)}f_j^{(e)}, \; i,j=1,2; \quad e=1,2,\ldots,M \tag{3.80}$$

where k is the iteration number and

$$\bar{K}^{(e,k)} = K\left(\bar{h}^{(e,k)}\right) = K\left(\frac{h_1^{(e,k)} + h_2^{(e,k)}}{2}\right) \tag{3.81}$$

Eq. (3.80) can now be assembled for all the M number of linear elements used in discretizing the flow region to form the global matrix equation from which $h_j^{(e,k+1)}$ and $\varphi_j^{(e,k+1)}$ are computed at all positions where they are not specified. The iteration process is repeated until convergence is achieved. Incorporating the Newton-Raphson algorithm into the **GESNH-1** involves the formation of the Jacobian element matrix equation of the form

$$J_{ij}^{(e,k)}\Delta u_j^{(e,k+1)} = -g_i^{(e,k)} \tag{3.82}$$

in which the right hand side vector of eq. (3.82) is the system of element equations of the **GESNH-1** formulation evaluated at the previous iteration. That is

$$g_i^{(e,k)} = R_{ij}^{(e)}h_j^{(e,k)} + L_{ij}^{(e)}\varphi_j^{(e,k)} + \frac{1}{\bar{K}^{(e,k)}}T_{ij}^{(e)}f_j^{(e)}, \quad i,j=1,2; \; e=1,2,\ldots,M \tag{3.83}$$

and $\Delta u_j^{(e,k+1)}$ is a mixed vector representing the increments of the unknown quantities, that is

$$\Delta u_j^{(e,k+1)} = \begin{Bmatrix} \Delta h_j^{(e,k+1)} \\ \Delta \varphi_j^{(e,k+1)} \end{Bmatrix} \tag{3.84}$$

and the Jacobian matrix is given by

$$J_{ij}^{(e,k)} = \begin{cases} \left.\dfrac{\partial g_i}{\partial h_j}\right|_{h_j = h_j^{(e,k)}} = R_{ij}^{(e)} - \dfrac{1}{(\bar{K}^{(e,k)})^2} T_{in}^{(e)} f_n^{(e)} \dfrac{\partial \bar{K}^{(e,k)}}{\partial h_j} \\ \left.\dfrac{\partial g_i}{\partial \varphi_j}\right|_{\varphi_j = \varphi_j^{(e,k)}} = L_{ij}^{(e)} \end{cases} \tag{3.85}$$

When the Jacobian matrix is properly formed for each element, the assembling of the global matrix follows in a straight forward fashion as described in chapter 2. The global coefficient matrix is solved for the increments of the solution vector. The iteration process is repeated until convergence as stated by eq. (3.79) is satisfied.

For the remaining two Green element models of the 1-D. steady nonlinear heterogeneous problem (**GESNH-2** and **GESNH-3**), we shall simply state their systems of element equations which incorporate the Picard and Newton-Raphson algorithms. Very little explanation shall be offered, but the procedure involved, which can be verified by the reader, is essentially along the same line of approach used for the **GESNH-1** model. The formulation of the **GESNH-2** model is eq. (3.39) and restated below

$$R_{ij}^{(e)} h_j^{(e)} + (L_{ij}^{(e)} - V_{inj}^{(e)} \Theta_n^{(e)}) \varphi_j^{(e)} + U_{ijn}^{(e)} \Psi_j^{(e)} f_n^{(e)} \equiv g_i^{(e)} = 0; \quad i,j,n=1,2 \tag{3.39}$$

The system of discrete element equations which incorporates the Picard algorithm is

$$R_{ij}^{(e)} h_j^{(e,k+1)} + (L_{ij}^{(e)} - V_{inj}^{(e)} \Theta_n^{(e,k)}) \varphi_j^{(e,k+1)} = -U_{ijn}^{(e)} \Psi_j^{(e,k)} f_n^{(e)}; \quad i,j,n=1,2 \tag{3.86}$$

The most crucial step in the Newton-Raphson algorithm is the correct derivation of the Jacobian matrix. For the **GESNH-2** model, the Jacobian matrix is given by

$$J_{ij}^{(e,k)} = \begin{cases} \left.\dfrac{\partial g_i}{\partial h_j}\right|_{h_j=h_j^{(e,k)}} = R_{ij}^{(e)} - V_{inm}^{(e)}\varphi_m^{(e,k)}\dfrac{d\Theta_n}{dh_j} + U_{imn}^{(e)}f_n^{(e)}\dfrac{d\Psi_m}{dh_j} \\ \left.\dfrac{\partial g_i}{\partial \varphi_j}\right|_{\varphi_j=\varphi_j^{(e,k)}} = L_{ij}^{(e)} - V_{inj}^{(e)}\Theta_n^{(e,k)} \end{cases} \tag{3.87}$$

The starting point for incorporating the Picard and Newton-Raphson algorithms into the **GESNH-3** model is the element equations of the model given by eq. (3.49) and restated below

$$R_{ij}^{(e)}h_j^{(e)} + [L_{ij}^{(e)} - S_{imnj}^{(e)}K_m^{(e)}\Psi_n^{(e)}]\varphi_j^{(e)} + U_{ijm}^{(e)}\Psi_j^{(e)}f_m^{(e)} \equiv g_i = 0; \;\; i,j,m,n=1,2 \tag{3.49}$$

The Picard algorithm of eq. (3.49) is

$$R_{ij}^{(e)}h_j^{(e,k+1)} + [L_{ij}^{(e)} - S_{ilnj}^{(e)}K_l^{(e,k)}\Psi_n^{(e,k)}]\varphi_j^{(e,k+1)} = -U_{ijl}^{(e)}\Psi_j^{(e,k)}f_l^{(e)} \tag{3.88}$$

while the Jacobian matrix of the Newton-Raphson algorithm for this model is

$$J_{ij}^{(e,k)} = \begin{cases} \left.\dfrac{\partial g_i}{\partial h_j}\right|_{h_j=h_j^{(e,k)}} = R_{ij}^{(e)} - S_{inml}^{(e)}\Psi_m^{(e,k)}\varphi_l^{(e,k)}\dfrac{dK_n}{dh_j} \\ \qquad - S_{ilnm}K_l^{(e,k)}\varphi_m^{(e,k)}\dfrac{d\Psi_n}{dh_j} + U_{iml}^{(e)}f_l^{(e)}\dfrac{d\Psi_m}{dh_j} \\ \left.\dfrac{\partial g_i}{\partial \varphi_j}\right|_{\varphi_j=\varphi_j^{(e,k)}} = L_{ij}^{(e)} - S_{ilnj}^{(e)}K_l^{(e,k)}\Psi_n^{(e,k)} \end{cases} \tag{3.89}$$

3.12 Example 2 — Nonlinear Example

The one dimensional flow between two reservoirs in an unconfined aquifer receiving recharge for which the *Dupuit-Forcheimmer* assumptions are valid is given by

$$\frac{d}{dx}\left(h\frac{dh}{dx}\right) = N \tag{3.90}$$

Comparing eq. (3.90) with (3.5) it means that $K \equiv h$, and $f(x) = N$. The boundary conditions are:

$$h(x{=}0) = h_0 \text{ and } h(x{=}L) = h_L \tag{3.91}$$

The analytical solution to the problem is (reader should verify this)

$$h(x)^2 = N(x-L)x + h_L^2\frac{x}{L} + h_0^2(1-\frac{x}{L}) \tag{3.92a}$$

and

$$2h\frac{dh}{dx} = N(2x-L) + \frac{1}{L}(h_L^2 - h_0^2) \tag{3.92b}$$

The following parameter values are used in the numerical calculations: $L=1$, $h_0=2$, $h_L=0.5$, and $N=1$. The Green element calculations are carried out with the **GESNH-1**, **GESNH-2**, and **GESNH-3** models, all of which incorporate the Newton-Raphson algorithm. (The reader may wish to carry out similar calculations, incorporating the Picard algorithm into those models and comparing with our results). The flow domain is discretized into 10 uniform linear elements so that the length of each element is 0.1. As earlier pointed out, the major task in correctly implementing the Newton-Raphson algorithm is calculating the Jacobian matrix. When that is achieved, the solution process becomes straightforward. In all the models, that step involves incorporating the expression for the parameter K into eqs. (3.85), (3.87) and (3.89). The initial estimate of the solution $h_j^{(0)}$ and $\varphi_j^{(0)}$ could be obtained from the solution of the linearized differential equation.

$$\frac{d}{dx}\left(\bar{h}\frac{dh}{dx}\right) = 1, \quad h(0)=2, \quad h(1)=0.5 \tag{3.93}$$

where $\bar{h} = [h(0)+h(1)]/2 = 5/4$. The solution to eq. (3.93) is

$$h(x) = \frac{2}{7}x^2 - \frac{25}{14}x + 2 \text{ and } \frac{dh}{dx} \equiv \varphi(x) = \frac{4}{7}x - \frac{25}{14} \tag{3.94}$$

The results from the computer programs based on the three models **GESNH-1**, **GESNH-2**, and **GESNH-3** are presented along side the exact solution in Table 3.3 and Figures 3.10a and 3.10b

Table 3.3: Exact and **GEM** solutions of example 2

Position x	Exact h	Exact φ	**GESNH-1** h	**GESNH-1** φ	**GESNH-2** h	**GESNH-2** φ	**GESNH-2** h	**GESNH-2** φ
0	2.000	-1.188	2.000	-1.224	2.000	-1.185	2.000	-1.185
0.1	1.880	-1.210	1.880	-1.251	1.880	-1.207	1.880	-1.207
0.2	1.758	-1.237	1.758	-1.283	1.758	-1.233	1.758	-1.234
0.3	1.632	-1.271	1.633	-1.323	1.633	-1.267	1.633	-1.267
0.4	1.503	-1.314	1.503	-1.375	1.505	-1.308	1.505	-1.309
0.5	1.369	-1.369	1.369	-1.443	1.371	-1.363	1.371	-1.365
0.6	1.229	-1.444	1.229	-1.538	1.231	-1.437	1.231	-1.438
0.7	1.079	-1.552	1.079	-1.679	1.082	-1.542	1.082	-1.544
0.8	0.917	-1.718	0.917	-1.912	0.920	-1.705	0.920	-1.709
0.9	0.731	-2.017	0.732	-2.396	0.735	-1.998	0.735	-2.008
1	0.500	-2.750	0.500	-2.233	0.500	-2.743	0.500	-2.801

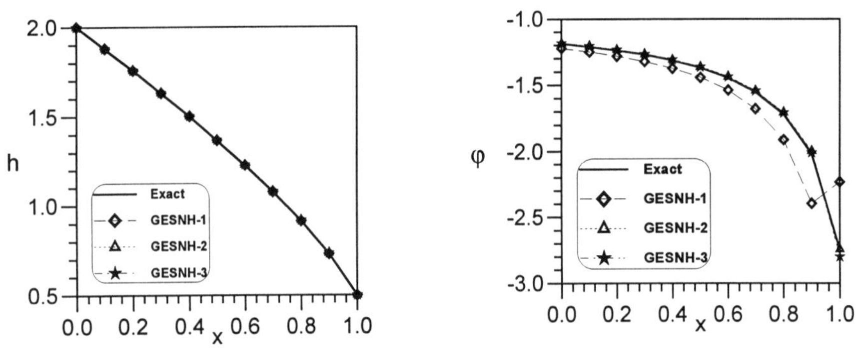

(a) **(b)**

Figure 3.10 GEM and exact solutions of example 2 — **(a)** solution for h, **(b)** solution for φ

3.13 Application to Structural Analysis of Beams

The elastic deflection of a beam in a direction transverse to its longitudinal direction is governed by the fourth-order equation given by

$$\frac{d^2}{dx^2}\left(EI\frac{d^2w}{dx^2}\right) + f = 0, \quad 0 \le x \le L \tag{3.95}$$

in which, as shown in Figure 3.11, w is the dependent variable which is the deflection, E is the *Young's modulus of elasticity* [kN/m^2], I is the moment of inertia of a section of the beam [m^4], f is the loading intensity on the beam (considered positive when the load is in the upward direction or positive z-direction) [kN/m]. In the derivation of eq. (3.95) it has been assumed that plane sections normal to the beam axis or neutral axis remain plane and normal to the beam axis during deflection (*Bernoulli's assumption*), and that the material of which the beam is made deforms according to linear elasticity theory which is expressed by *Hooke's law*. The product of E and I is denoted as K and, in general, is a function of x. Usually these assumptions are valid when the ratio of the longitudinal dimension of the beam to its lateral dimension is greater than 5. The fourth-order differential equation can be split into two second-order differential equations

$$\frac{d^2M}{dx^2} + f = 0 \tag{3.96a}$$

and

$$K(x)\frac{d^2w}{dx^2} = M \tag{3.96b}$$

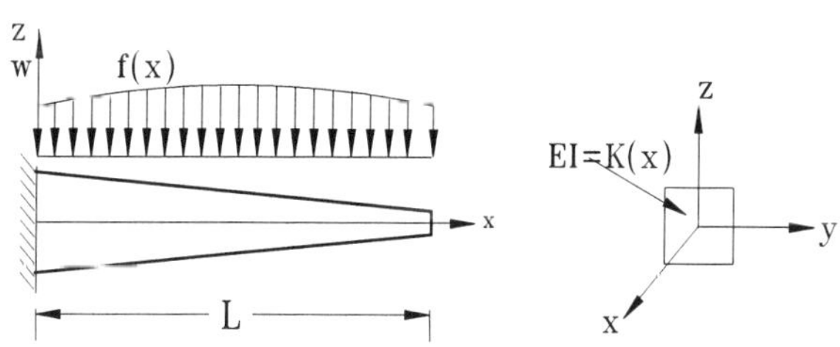

Figure 3.11 Definition sketch of a beam under flexural loading.

Essentially we have reduced the fourth-order to a system of 1-D. Poisson equations for which their Green element approximations had earlier been derived. In structural analysis, the variable M refers to the *moment* about an axis on the plane in which lies the beam. Depending on the ways the ends of the beam are fixed, eq. (3.96a) can be solved first for M, and then used as input for the solution of eq. (3.96b). That can only be done if two appropriate boundary conditions for M and/or its spatial derivative denoted as V (the shear force) are specified. When that is not the case, then the solutions of the two equations have to be achieved in one swoop. Examples of typical boundary conditions or end support conditions are illustrated in Figure 3.12. The arrangement in Figure 3.12a can be handled with eq. (3.96a) being solved first, and then

(3.96b) later. But the support arrangement in Figure 3.12b demands that both equations be solved at one go. Both solution procedures are demonstrated here. In structural analysis, the spatial derivative of the deflection is related to the *rotation* of the member, and it is denoted as θ.

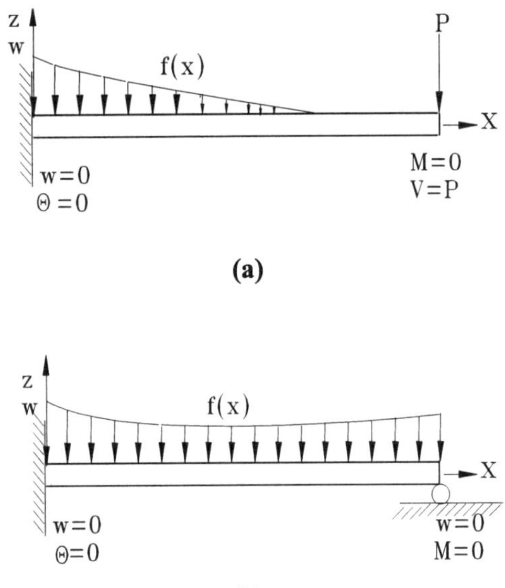

Figure 3.12: Typical end support arrangements.

3.13.1 *Example 3*

This example is a cantilevered beam which is subjected to a triangular distributed load, and a point load $P = 35$kN at the free end (see Figure 3.13). The boundary conditions are thus

$$w(0) = 0, \quad \frac{dw(0)}{dx} = \theta(0) = 0, \quad M(L) = 0, \text{ and } \frac{dM(L)}{dx} = V(L) = P \tag{3.97}$$

The reader can readily verify that the exact solution to this example is given by

$$EI\,w(x) = -f_0(1 - \frac{x}{5L})\frac{x^4}{24} + (2P + f_0L)\frac{x^3}{12} - (f_0L + 6P)\frac{Lx^2}{12} \tag{3.98a}$$

$$EI\theta(x) = -f_0\left(1-\frac{x}{4L}\right)\frac{x^3}{6}+(2P+f_0L)\frac{x^2}{4}-(f_0L+6P)\frac{Lx}{6} \tag{3.98b}$$

and

$$M(x) = -f_0\left(1-\frac{x}{3L}\right)\frac{x^2}{2}+(2P+f_0L)\frac{x}{2}-(f_0L+6P)\frac{L}{6} \tag{3.98c}$$

The numerical calculations are done with the following parameters: $L = 4m$, $K = EI = 2x10^4$ kN.m², and $f_0 = 10$. The Green element equations for a typical element using linear interpolating functions

$$R_{ij}^{(e)}M_j^{(e)} + L_{ij}^{(e)}V_j^{(e)} - T_{ij}^{(e)}f_j^{(e)} = 0 \tag{3.99a}$$

$$EI\left(R_{ij}^{(e)}w_j^{(e)} + L_{ij}^{(e)}\theta_j^{(e)}\right) + T_{ij}^{(e)}M_j^{(e)} = 0 \tag{3.99b}$$

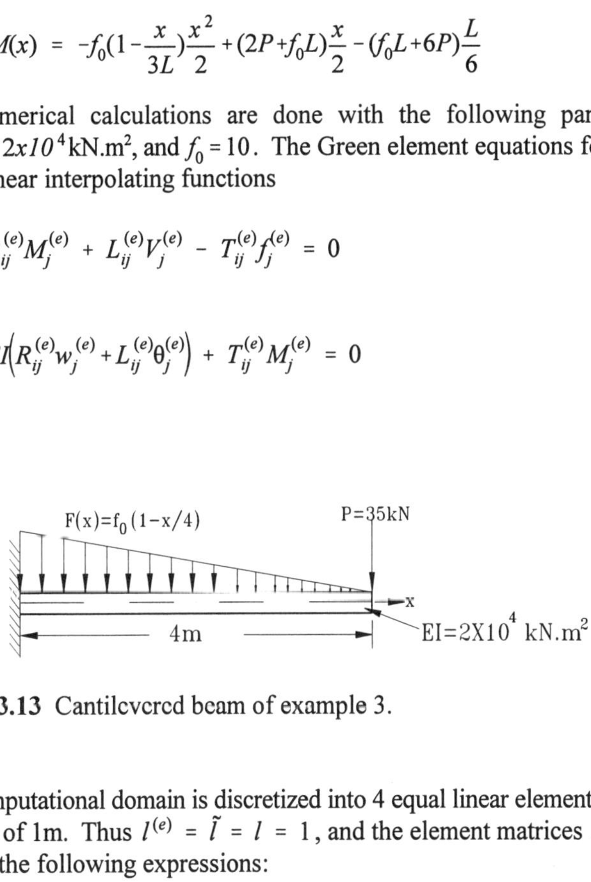

Figure 3.13 Cantilevered beam of example 3.

The computational domain is discretized into 4 equal linear elements each of which has a length of 1m. Thus $l^{(e)} = \tilde{l} = l = 1$, and the element matrices $R_{ij}^{(e)}$, $L_{ij}^{(e)}$, and $T_{ij}^{(e)}$ take on the following expressions:

$$R_{ij}^{(e)} = R_{ij} = (-1)^{i+j-1},\ L_{ij}^{(e)} = L_{ij} = \begin{bmatrix} 1 & -2 \\ 2 & -1 \end{bmatrix},\ \text{and}\ T_{ij}^{(e)} = T_{ij} = \frac{1}{6}\begin{bmatrix} 4 & 5 \\ 5 & 4 \end{bmatrix}.$$

It is necessary to point out that because of the uniform discretization of the computational domain, all the element matrices remain the same for the four elements. The solution provided by eq. (3.99a) can first be done since there exist two boundary data with which to proceed with the solution process. When the element equations of eq. (3.99a) are assembled for all the four elements we have the global matrix equation

$$\begin{bmatrix} -1 & 1 & 1 & -2 & 0 & 0 & 0 & 0 \\ 1 & 2 & -1 & -1 & 0 & 0 & 0 & 0 \\ 0 & 0 & -1 & 1 & 1 & -2 & 0 & 0 \\ 0 & 0 & 1 & 2 & -1 & -1 & 0 & 0 \\ 0 & 0 & 0 & 0 & -1 & 1 & 1 & -2 \\ 0 & 0 & 0 & 0 & 1 & 2 & -1 & -1 \\ 0 & 0 & 0 & 0 & 0 & 0 & -1 & 1 \\ 0 & 0 & 0 & 0 & 0 & 0 & 1 & 2 \end{bmatrix} \begin{Bmatrix} M_1 \\ V_1 \\ M_2 \\ V_2 \\ M_3 \\ V_3 \\ M_4 \\ V_4 \end{Bmatrix} = \frac{1}{6} \begin{Bmatrix} 77.5 \\ 80 \\ 55 \\ 57.5 \\ 32.5 \\ 35 \\ 430 \\ 222.5 \end{Bmatrix}$$

The solution from the above matrix equation is given in Table 3.4. The solution for M, obtained from Table 3.4, is used in solving the second system of discrete equations given by (3.99b). The global matrix equation for eq. (3.99b) for the four elements used in discretizing the computational region is

$$EI. \begin{bmatrix} 1 & -2 & 0 & 0 & 0 & 0 & 0 & 0 \\ -1 & -1 & 0 & 0 & 0 & 0 & 0 & 0 \\ -1 & 1 & 1 & -2 & 0 & 0 & 0 & 0 \\ 1 & 2 & -1 & -1 & 0 & 0 & 0 & 0 \\ 0 & 0 & -1 & 1 & 1 & -2 & 0 & 0 \\ 0 & 0 & 1 & 2 & -1 & -1 & 0 & 0 \\ 0 & 0 & 0 & 0 & -1 & 1 & 1 & -2 \\ 0 & 0 & 0 & 0 & 1 & 2 & -1 & -1 \end{bmatrix} \begin{Bmatrix} w_2 \\ \theta_2 \\ w_3 \\ \theta_3 \\ w_4 \\ \theta_4 \\ w_5 \\ \theta_5 \end{Bmatrix} = \begin{Bmatrix} 207.99 \\ 216.39 \\ 138.89 \\ 146.11 \\ 78.40 \\ 84.72 \\ 23.61 \\ 29.51 \end{Bmatrix}$$

The solution to the global matrix equation is given in Table 3.4, along side the exact solutions which are obtained from eqs. (3.98a) to (3.98c). A computer simulation of the above example was carried using a finer discretization of 10 uniform elements, and the results of that simulation with the exact solutions are presented graphically in Figure 3.14. Good agreement is observed between the numerical and exact solutions.

Table 3.4: **GEM** and exact solutions of example 3

x	GEM solution				Exact Solution			
	$EI.w$ x 10^2	$EI.\theta$ x 10^2	M x 10^2	V	$EI.w$ x 10^2	$EI.\theta$ x 10^2	M x 10^2	V
0			-1.67	55.0			-1.67	55.0
1	-0.75	-1.42	-1.16	46.3	-0.75	-1.41	-1.16	46.3
2	-2.67	-2.36	-0.73	40.0	-2.66	-2.35	-0.73	40.0
3	-5.34	-2.91	-0.35	36.3	-5.31	-2.89	-0.35	36.3
4	-8.36	-3.08			-8.32	-3.07		

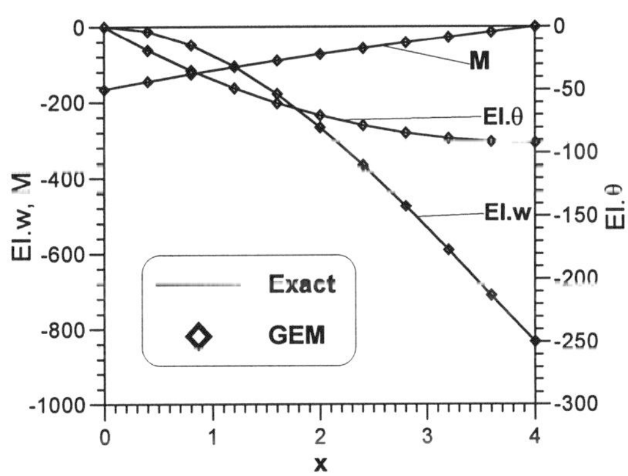

Figure 3.14 **GEM** and exact solutions of $EI.w$, $EI.\theta$, and M for example 3.

3.13.2 *Example 4*

Another example of flexural beam deflection is shown in Figure 3.15; the beam is loaded in a discontinuous fashion, and fixed at one end and on roller support at the other. With the end support arrangement as indicated, the boundary conditions are

$$w(0) = 0, \ \theta(0) = 0, \ w(L) = 0, \ \text{and} \ M(L) = 0 \tag{3.100}$$

The loading on the beam is represented as

$$f(x) = P\delta(x-2) + f_0 H(x-5) \tag{3.101}$$

The exact solutions are readily derived as

$$EIw = -\frac{P}{6}(x-2)^3 H(x-2) - \frac{f_0}{24}(x-5)^4 H(x-5) + \frac{c_1}{6}x^3 + \frac{c_2}{2}x^2 \tag{3.102a}$$

$$EI\theta = -\frac{P}{2}(x-2)^2 H(x-2) - \frac{f_0}{6}(x-5)^3 H(x-5) + \frac{c_1}{2}x^2 + c_2 x \tag{3.102b}$$

$$M = -P(x-2)H(x-2) - \frac{f_0}{2}(x-5)^2 H(x-5) + c_1 x + c_2 \tag{3.102c}$$

in which the constants c_1 and c_2 are given by

$$c_1 = \frac{3}{L^3}\left[\frac{P}{6}(L-2)[3L^2-(L-2)^2] + \frac{f_0}{24}(L-5)^2[6L^2-(L-5)^2]\right] \tag{3.103a}$$

and

$$c_2 = \frac{1}{L^2}\left[\frac{P}{2}(L-2)[(L-2)^2-L^2] + \frac{f_0}{8}(L-5)^2[(L-5)^2-2L^2]\right] \tag{3.103b}$$

where L is the length of the beam which in our present example is 7m, P = 50kN and f_0 = 35kN/m have been used in the calculations. The Green element equations for this example are still eqs. (3.99a) and (3.99b). Because there is only one boundary data for M, (3.99a) cannot be solved by itself; the two discrete equations must have to be solved at the same time. Using a three-element discretization — $l^{(1)}=2$, $l^{(2)}=3$, and $l^{(3)}=2$ — the element matrices are then given by

$$R_{ij}^{(e)} = R_{ij} = (-1)^{i+j-1},\ L_{ij}^{(e)} = \begin{bmatrix} \tilde{l} & -(\tilde{l}+l^{(e)}) \\ l^{(e)}+\tilde{l} & -\tilde{l} \end{bmatrix} \text{ and } T_{ij}^{(e)} = \frac{l^{(e)}}{6}\begin{bmatrix} 3\tilde{l}+l^{(e)} & 3\tilde{l}+2l^{(e)} \\ 3\tilde{l}+2l^{(e)} & 3\tilde{l}+l^{(e)} \end{bmatrix}.$$

Since $\tilde{l}=3$, it implies that $L_{ij}^{(1)} = \begin{bmatrix} 3 & -5 \\ 5 & -3 \end{bmatrix}$, $L_{ij}^{(2)} = \begin{bmatrix} 3 & -6 \\ 6 & -3 \end{bmatrix}$, and $L_{ij}^{(3)} = \begin{bmatrix} 3 & -5 \\ 5 & -3 \end{bmatrix}$.

In addition, $T_{ij}^{(1)} = \frac{1}{3}\begin{bmatrix} 11 & 13 \\ 13 & 11 \end{bmatrix}$, $T_{ij}^{(2)} = \frac{1}{2}\begin{bmatrix} 12 & 15 \\ 15 & 12 \end{bmatrix}$, and $T_{ij}^{(3)} = \frac{1}{3}\begin{bmatrix} 11 & 13 \\ 13 & 11 \end{bmatrix}$.

The reader may now verify that the global matrix arising from the combined solution of eqs. (3.99a) and (3.99b) is given by

$$
\begin{bmatrix}
-1 & -5 & 0 & 0 & 0 & 0 & 0 & 0 & 4.33 & 0 & 3.67 & 0 \\
1 & -3 & 0 & 0 & 0 & 0 & 0 & 0 & 3.67 & 0 & 4.33 & 0 \\
-1 & 3 & 1 & -6 & 0 & 7.5 & 0 & 0 & 6 & 0 & 0 & 0 \\
1 & 6 & -1 & -3 & 0 & 6 & 0 & 0 & 7.5 & 0 & 0 & 0 \\
0 & 0 & -1 & 3. & -5 & 3.67 & 0 & 0 & 0 & 0 & 0 & 0 \\
0 & 0 & 1 & 5 & -3 & 4.33 & 0 & 0 & 0 & 0 & 0 & 0 \\
0 & 0 & 0 & 0 & 0 & 1 & -3 & 5 & 0 & 0 & 0 & 0 \\
0 & 0 & 0 & 0 & 0 & -1 & -5 & 3 & 0 & 0 & 0 & 0 \\
0 & 0 & 0 & 0 & 0 & -1 & 0 & -3 & 1 & 6 & 0 & 0 \\
0 & 0 & 0 & 0 & 0 & 1 & 0 & -6 & -1 & 3 & 0 & 0 \\
0 & 0 & 0 & 0 & 0 & 0 & 0 & 0 & -1 & -3 & 1 & 5 \\
0 & 0 & 0 & 0 & 0 & 0 & 0 & 0 & 1 & -5 & -1 & 3
\end{bmatrix}
\begin{Bmatrix}
EI.w_2 \\ EI.\theta_2 \\ EI.w_3 \\ EI.\theta_3 \\ EI.\theta_4 \\ M_3 \\ V_4 \\ V_3 \\ M_2 \\ V_2 \\ M_1 \\ V_1
\end{Bmatrix}
=
\begin{Bmatrix}
0 \\ 0 \\ 0 \\ 0 \\ 0 \\ 0 \\ 240 \\ 240 \\ 300 \\ 150 \\ 0 \\ 0
\end{Bmatrix}
$$

The solution to the global matrix equation and the exact solution are presented in Table 3.5.

Table 3.5 GEM and exact solutions of example 4

x	**GEM** solution				Exact Solution			
	$EI.w$ x 10^2	$EI.\theta$ x 10	M x 10	V x 10	$EI.w$ x 10^2	$EI.\theta$ x 10	M x 10	V x 10
0			-8.88	5.70			-9.00	5.71
2	-1.02	-6.36	-2.52	5.70	-1.04	-6.57	-2.43	5.71
5	-1.48	4.32 4.61	0.70		-1.60	3.93 4.57	0.71	
7		0.89	-5.30			1.05	-5.29	

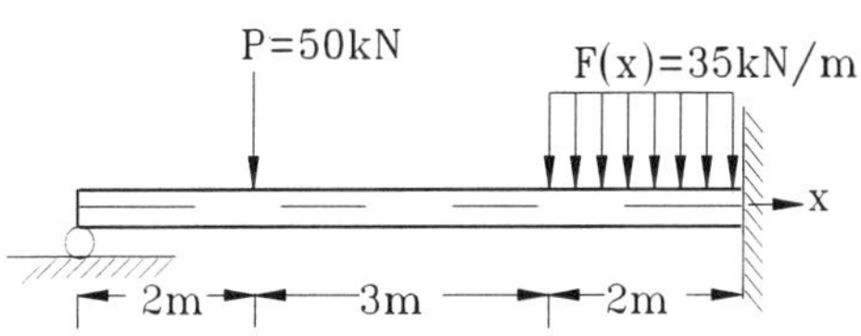

Figure 3.15 Beam of example 4 with fixed and roller supports.

3.14 Remarks

Before now the singular boundary element theory was thought to be not too suitable for the problems addressed in this chapter. It was then thought that the theory was restricted to linear problems, since the free-space Green's function could only be obtained for a linear differential operator. As we have observed that limitation no longer exist, and that has been achieved by using the free-space Green's function of a linear portion of the full differential equation, and treating the nonlinear portion in the same way as the recharge term. That treatment of the nonlinear portion results in a domain integral which before now is considered an unpleasant numerical feature in boundary element methodology. Handling the domain integral in an element-by-element fashion, as we have done, has made its evaluation quite easy and, thus, eliminated the difficulties commonly associated with the evaluation of domain integrals in boundary element applications.

The three Green element models — **GESNH-1**, **GESNH-2**, and **GESNH-3** — represent three levels of approximation of the nonlinear Laplace/Poisson equation for a heterogeneous medium, the latter model being the most sophisticated of the three. The first model, which treats the medium as piecewise homogeneous, yields correct answers for the primary variable but not for its derivative which is discontinuous at the nodes. That certainly is inconsistent with the actual solution. The second and third models produce continuous solutions for the spatial derivative of the primary variable at the nodes. Since the computing effort involved in the third model (**GESNH-3**) is more than that of the second (**GESNH-2**), the latter model is considered most suitable for the nonlinear Laplace/Poisson equation.

3.15 Exercises

3.1 Proceed along the lines of exercise 2.2 in chapter 2 to derive the Green element equations for the nonlinear 1-D. Laplace/Poisson equation so that there is one degree of freedom at each node. Compare the results of your formulation with those of the first and second numerical examples addressed in this chapter.

3.2 The problem governed by

$$\frac{d}{dx}\left(h\frac{dh}{dx}\right) = f(x); \qquad 0 \le x \le L$$

can be considered a linear problem in h^2. Use the Green element formulation of chapter 2 to solve the linear problem in h^2, and compare with the numerical results obtained for example 2 in this chapter. Comment on the results.

3.3 Use any of the Green element models to solve the equation

$$\frac{d}{dx}\left(K(x)\frac{dh}{dx}\right) = f(x); \quad 0 \leq x \leq 1, \; h(0) = 1, \text{ and } \frac{dh(1)}{dx} = \varphi(1) = 1$$

for the following expressions of $K(x)$

(a) $K = 1 + x$

(b) $K = 1 + x^2$

© $K = e^x$

Compare the numerical results with the exact solutions.

3.4 Use an appropriate Green element model to solve the equation

$$\frac{d}{dx}\left([1 + x]\frac{dh}{dx}\right) + f(x) = 0; \quad 0 \leq x \leq 1, \; h(0) = 0, \text{ and } h(1) = 1$$

for the following expressions of the forcing function $f(x)$

(a) $f(x) = 1$

(b) $f(x) = \delta(x - 0.5)$

(c) $f(x) = 1 - x$

(d) $f(x) = \sin(\pi x)$

3.5 Use a Green element model to solve

$$\frac{d}{dx}\left(\frac{1}{1 + h}\frac{dh}{dx}\right) + f(x) = 0; \quad 0 \leq x \leq 1$$

with the boundary conditions: $h(0) = 1$, and $\frac{dh(1)}{dx} = \varphi(1) = 0$.

Carry out your calculations with the following expressions of the forcing function $f(x)$

(a) $\delta(x - 0.5)$

(b) $f(x) = 1 - x$

(c) $f(x) = \sin(\pi x)$

3.6 Solve by the Green element method the deflection problem of a simply supported beam of uniform section which is governed by

$$\frac{d^2}{dx^2}\left[EI\frac{d^2w}{dx^2}\right] + f(x) = 0; \quad 0 \leq x \leq L$$

for the following loading configurations

(a) $f(x) = f_0$ (uniform distributed loading)

(b) $f(x) = f_0 \sin(\pi x/L)$

(c) $f(x) = P\delta(x - L/2)$

Compare your solutions with the exact solutions (choose your parameter values).

3.7 Solve the problem in exercise 3.6 for a beam section whose EI has the following distribution: $0 \leq x \leq L/2 \rightarrow 2EI$; $\quad L/2 < x \leq L \rightarrow EI$.

3.8 Solve the beam deflection problem shown in Figure E3.8 by the Green element method.

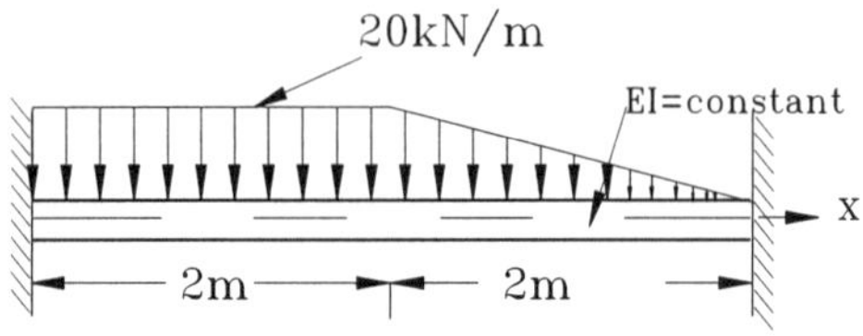

Figure E3.8

3.16 References

1. Cheng, A. H-D. and D. Ouaza, "Groundwater flow," in chapter 8 of *Boundary Element Techniques in Geomechanics*, eds. G.D. Manolis and T.G. Davies, CMP/Elsevier, 1993.

2. Lafe, O.E. and A.H-D. Cheng, "A Perturbation Boundary Element Code for Groundwater flow in Heterogeneous Aquifer," *Water Resources Research*, **23**, pp. 1079-1084, 1987.

3. Partridge, P.W., C.A. Brebbia and L.C. Wrobel, *The Dual Reciprocity Boundary Element method*, CMP/Elsevier, 1991.

4. Cheng, A.H-D., "Darcy's flow with variable Permeability - a Boundary Integral Solution," *Water Resources Research*, **20**, pp. 980-984, 1984.

5. Taigbenu, A.E., "The Green Element Method," *Int. J. for Numerical Methods in Engineering*, **38**, pp 2241-2263, 1995.

6. Taigbenu, A.E., and O.O. Onyejekwe, "Green Element simulations of the Transient Unsaturated Flow Equation," *Applied Mathematical Modelling*, **19**, pp. 675-684, 1995.

7. Strang, G. and G.J. Fix, *An Analysis of the Finite Element Method*, Prentice Hall, Englewood Cliffs, NJ., 1973.

8. Zienkiewicz, O.C. and R.L. Taylor, *The Finite Element Method*, 4th edition, Vol. 1, McGraw-Hill, NY., 1989.

9. Zienkiewicz, O.C., *The Finite Element Method*, 3ed., McGraw-Hill, London, 1977.

10. Reddy, J.N., *An Introduction to the Finite Element Method*, McGraw-Hill, Inc. NY., 1984.

11. Rao, S.S., *The Finite Element Method in Engineering*, 2ed. Pergamon Press plc, Oxford, 1989.

12. Ottosen, N.S. and H. Petersson, *Introduction to the Finite Element Method*, Prentice Hall Int. Ltd. UK., 1992.

13. Hornbeck, R.W., *Numerical Methods*, Quantum Pub. Inc., U.S.A., 1975.

4 HELMHOLTZ EQUATION

4.1 Introduction

As we begin this fourth chapter we do suppose that you have begun to observe a pattern with the formulation of the Green element equations. Here we shall follow the same procedure that we had employed in the two previous chapters, and that is to seek a Fredholm integral equation of the second kind with a singular kernel for the differential equation, discretize the integral equation over the computational domain, and then approximate the distributions of functional quantities (known and unknown) with appropriate element shape functions. The outcome of these steps is a system of discrete equations that completely describes the solution behavior of the system being modeled. Before embarking on developing the system of discrete equations for the *Helmholtz equation*, we choose a physical problem to which it applies. We do this so as to elicit a good engineering appreciation of the differential equation being addressed.

4.2 Derivation of Helmholtz Equation

A problem similar to the example of confined flow in a porous medium of chapter 2 is used to derive the Helmholtz equation. Here the medium is assumed infinitely extensive in the other lateral dimension to x and has a leaky stratum of constant thickness b_L overlying the confined medium. It is assumed that an unconfined aquifer overlies the leaky stratum which is maintained at a uniform head h_0 (See Figure 4.1). The statement of mass conservation for an infinitesimal control volume of length dx under steady conditions remains as earlier derived in eq. (2.4) and restated here

$$\frac{dq}{dx} = N(x) \tag{4.1}$$

However, the recharge per unit length of aquifer $N(x)$ consists of recharge through the leaky stratum which we denote as $N_L(x)$ and that from distributed and point sources/sinks which we denote as $N_R(x)$. By Darcy law

$$N_L(x) = K_L \frac{h - h_0}{b_L} \tag{4.2}$$

in which K_L is the hydraulic conductivity of the leaky stratum. Introducing eq. (4.2) into eq. (4.1), and expressing q in terms of the gradient of the piezometric head in the confined aquifer we have

$$Kb\frac{d^2h}{dx^2} = -N_R(x) - \frac{K_L}{b_L}(h - h_0)$$

or

$$\boxed{\frac{d^2h}{dx^2} + \omega^2 h = f(x)} \tag{4.3}$$

in which

$$\omega^2 = \frac{K_L}{Kbb_L} \quad \text{and} \quad f(x) = \frac{1}{Kb}\left(\frac{K_L}{b_L}h_0 - N_R\right) \tag{4.4}$$

Eq. (4.3) is the linear form of the Helmholtz equation, and in the groundwater literature the reciprocal of ω is known as the *leakage factor*. It should be noted that in deriving eq. (4.3) we have assumed that the medium is homogeneous. If that is not the case then the heterogeneous form of eq. (4.3) becomes (the reader should verify this)

$$\frac{d}{dx}\left(K\frac{dh}{dx}\right) + \omega^2 h = f(x) \tag{4.5}$$

in which

$$\omega^2 = \frac{K_L}{bb_L} \quad \text{and} \quad f(x) = \frac{1}{b}\left(\frac{K_L}{b_L}h_0 - N_R\right) \tag{4.6}$$

Either eq. (4.3) or (4.5) can readily be solved by the Green element method. However, in this text we shall only address eq. (4.3). (Exercise 4.4 at the end of this chapter offers the reader the opportunity to derive the Green element equations of eq. (4.5).) In the subsequent section, two Green element models of eq. (4.3) are presented; one uses the free-space Green's function of the diffusive term d^2h/dx^2, while the other uses that for the full equation $d^2h/dx^2 \pm \omega^2 h$. The Helmholtz problem is also referred to as the *eigenvalue* problem, where the parameter ω^2 is the *eigenvalue* or *characteristic parameter* associated with the differential equation, and h is the corresponding *eigenvector*. There are many other areas of engineering applications where eq. (4.3) is encountered. These applications include those of determining the natural vibration or buckling or frequency characteristics of slender columns, elastic beams, etc., and the 1-D. propagation characteristics of electromagnetic waves in a wave guide, or acoustic waves in a closed chamber. The importance of these engineering problems underscores achieving a suitable method of solution of the Helmholtz equation.

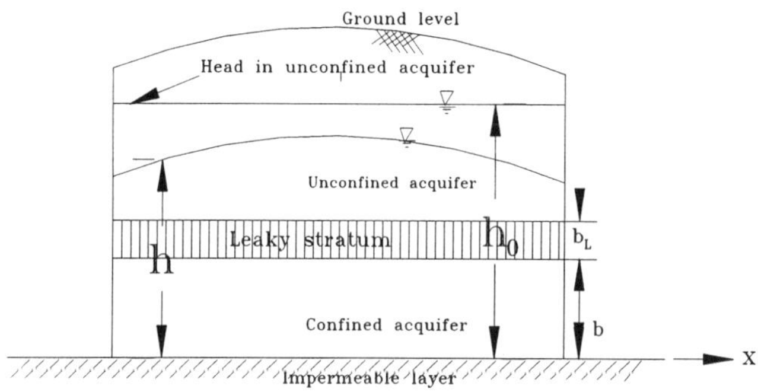

Figure 4.1 Leaky confined flow used in deriving Helmholtz equation.

4.3 Two Green Element Models

4.3.1 *Model 1*

The first Green element model for the Helmholtz equation uses the fundamental solution of the differential equation (2.20) which is given by eq. (2.30). Based on that fundamental solution, an integral equation of the Helmholtz equation is obtained by invoking Green's identity which is given by eq. (2.25). Within a typical element $[x_1^{(e)},x_2^{(e)}]$, the integral equation is

$$\int_{x_1^{(e)}}^{x_2^{(e)}} [h\delta(x-x_i) - G(f(x) - \omega^2 h)]dx = h\frac{dG}{dx}\bigg|_{x_1^{(e)}}^{x_2^{(e)}} - G\frac{dh}{dx}\bigg|_{x_1^{(e)}}^{x_2^{(e)}} \tag{4.7}$$

which simplifies to

$$\begin{aligned}&-\lambda^{(e)}h_i^{(e)} - G^*(x_1^{(e)},x_i^{(e)})h_1^{(e)} + G^*(x_2^{(e)},x_i^{(e)})h_2^{(e)} + G(x_1^{(e)},x_i^{(e)})\varphi_1^{(e)} - \\ &G(x_2^{(e)},x_i^{(e)})\varphi_2^{(e)} + \int_{x_1^{(e)}}^{x_2^{(e)}} G(x,x_i^{(e)})\,[f(x) - \omega^2 h(x)]dx = 0; \quad i=1,2\end{aligned} \tag{4.8}$$

in which, for instance,

$$G^*(x_1^{(e)},x_i^{(e)}) = \left.\frac{dG(x,x_i^{(e)})}{dx}\right|_{x=x_1^{(e)}} \text{ and } \varphi_1^{(e)} = \left.\frac{dh(x)}{dx}\right|_{x=x_1^{(e)}} \tag{4.9}$$

To arrive at eq. (4.8) we have essentially treated the term $\omega^2 h$ as a part of the recharge function $f(x)$. To proceed further from eq. (4.8) we need to prescribe some functional distribution within the element for the known function $f(x)$ and the unknown quantity $h(x)$ in terms of their nodal values. We elect to use linear interpolation for both quantities, and that is obtained using the element shape functions introduced in section 3.6. That is

$$f(x) \approx \Omega_j^{(e)} f_j^{(e)} \text{ and } h(x) \approx \Omega_j^{(e)} h_j^{(e)} \tag{4.10}$$

in which $\Omega_j^{(e)}$'s are the linear element shape functions defined by eq. (3.27). Introducing eq. (4.10) into eq. (4.8) yields

$$\begin{aligned} &-\lambda^{(e)} h_i^{(e)} - G^*(x_1^{(e)},x_i^{(e)}) h_1^{(e)} + G^*(x_2^{(e)},x_i^{(e)}) h_2^{(e)} + \\ &G(x_1^{(e)},x_i^{(e)})\varphi_1^{(e)} - G(x_2^{(e)},x_i^{(e)})\varphi_2^{(e)} + \\ &\int_{x_1^{(e)}}^{x_2^{(e)}} G(x,x_i^{(e)})\Omega_j^{(e)}(f_j^{(e)} - \omega^2 h_j^{(e)})\,dx = 0; \quad i,j = 1,2 \end{aligned} \tag{4.11}$$

Recall from chapter 2 that we set the arbitrary constant k in the expression for the free-space Green's function to the longest spatial element size $\tilde{l}$ used to discretize the computational domain. With that

$$G = \frac{|x-x_i| + \tilde{l}}{2} \text{ and } \frac{dG}{dx} \equiv G^* = \frac{1}{2}[H(x-x_i) - H(x_i - x)]$$

Then,

$$G(x_1^{(e)},x_1^{(e)}) = G(x_2^{(e)},x_2^{(e)}) = \frac{\tilde{l}}{2}, \; G(x_2^{(e)},x_1^{(e)}) = G(x_1^{(e)},x_2^{(e)}) = \frac{\tilde{l}+l^{(e)}}{2} \tag{4.12a}$$

$$G^*(x_1^{(e)},x_1^{(e)}) = G^*(x_2^{(e)},x_2^{(e)}) = 0, \; G^*(x_1^{(e)},x_2^{(e)}) = -G^*(x_2^{(e)},x_1^{(e)}) = -\frac{1}{2} \tag{4.12b}$$

Introducing the above relations into eq. (4.11) gives

$$-h_1^{(e)} + h_2^{(e)} + \tilde{l}\varphi_1^{(e)} - (\tilde{l}+l^{(e)})\varphi_2^{(e)} + l^{(e)}\int_0^1 (l^{(e)}\zeta+\tilde{l})\Omega_j^{(e)}(f_j^{(e)}-\omega^2 h_j^{(e)})d\zeta = 0 \quad (4.13a)$$

for $i=1$, and

$$h_1^{(e)} - h_2^{(e)} + (\tilde{l}+l^{(e)})\varphi_1^{(e)} - \tilde{l}\varphi_2^{(e)} + l^{(e)}\int_0^1 [l^{(e)}(1-\zeta)+\tilde{l}]\Omega_j^{(e)}(f_j^{(e)}-\omega^2 h_j^{(e)})d\zeta = 0 \quad (4.13b)$$

for $i=2$.
Eqs. (4.13a) and (4.13b) are expressed in matrix form as

$$\boxed{R_{ij}^{(e)}h_j^{(e)} + L_{ij}^{(e)}\varphi_j^{(e)} + T_{ij}^{(e)}(f_j^{(e)} - \omega^2 h_j^{(e)}) = 0, \quad i,j=1,2} \quad (4.14)$$

where the element matrices $R_{ij}^{(e)}$ and $L_{ij}^{(e)}$ are given by eqs. (2.42a) and (2.42b), respectively, and $T_{ij}^{(e)}$ had earlier been encountered in chapter 3. It is obtained from eq. (3.33) and restated below

$$T_{ij}^{(e)} = \frac{l^{(e)}}{6}\begin{bmatrix} 3\tilde{l}+l^{(e)} & 3\tilde{l}+2l^{(e)} \\ 3\tilde{l}+2l^{(e)} & 3\tilde{l}+l^{(e)} \end{bmatrix} \quad (4.15)$$

Eq. (4.14) is the system of discrete element equations based on one Green element replication of the Helmholtz equation. It provides all the information to evaluate the values of h and φ at the nodes. The global matrix equation is obtained by assembling the element equations for all the elements used to discretize the computational region.

4.3.2 *Model 2*

Our second Green element model of the Helmholtz equation (4.3) derives its free-space Green's function from the complementary equation

$$\frac{d^2G}{dx^2} + \omega^2 G = \delta(x-x_i); \quad -\infty \le x \le \infty \quad (4.16)$$

To obtain the integral equation of eq. (4.3) within an element, we invoke Green's identity which is given by eq. (2.25). Within that typical element $[x_1^{(e)},x_2^{(e)}]$, the integral equation is

$$\int_{x_1^{(e)}}^{x_2^{(e)}} \{h\delta(x-x_i) - Gf(x)\}\,dx = h\frac{dG}{dx}\bigg|_{x_1^{(e)}}^{x_2^{(e)}} - G\frac{dh}{dx}\bigg|_{x_1^{(e)}}^{x_2^{(e)}} \tag{4.17}$$

which, using the property of the Dirac delta function derived in chapter 2, simplifies to

$$\begin{aligned}&-\lambda^{(e)}h_i^{(e)} - G^*(x_1^{(e)},x_i^{(e)})h_1^{(e)} + G^*(x_2^{(e)},x_i^{(e)})h_2^{(e)} + G(x_1^{(e)},x_i^{(e)})\varphi_1^{(e)} - \\ &G(x_2^{(e)},x_i^{(e)})\varphi_2^{(e)} + \int_{x_1^{(e)}}^{x_2^{(e)}} G(x,x_i^{(e)})f(x)dx = 0; \quad i=1,2\end{aligned} \tag{4.18}$$

The line integral of the recharge term can be simplified by using the approach adopted in the formulation of model 1, and that is to prescribe some distribution within the element for the known function $f(x)$ in terms of its nodal values. We use a linear interpolation function to approximate $f(x)$ within the element.

$$\begin{aligned}&-\lambda^{(e)}h_i^{(e)} - G^*(x_1^{(e)},x_i^{(e)})h_1^{(e)} + G^*(x_2^{(e)},x_i^{(e)})h_2^{(e)} + G(x_1^{(e)},x_i^{(e)})\varphi_1^{(e)} - \\ &G(x_2^{(e)},x_i^{(e)})\varphi_2^{(e)} + \int_{x_1^{(e)}}^{x_2^{(e)}} G(x,x_i^{(e)})\Omega_j^{(e)}f_j^{(e)}dx = 0; \quad i,j=1,2\end{aligned} \tag{4.19}$$

At this stage we can substitute the expressions for the free-space Green's function $G(x,x_i)$ and its spatial derivative into eq. (4.19). Those expressions are:

$$\boxed{\begin{aligned}G(x,x_i) &= \frac{\sin(\omega|x-x_i|)}{2\omega} \\ \frac{dG}{dx} &= G^*(x,x_i) = \frac{\cos(\omega|x-x_i|)}{2}[H(x-x_i)-H(x_i-x)]\end{aligned}} \tag{4.20}$$

The expression for G is graphically presented in Figure 4.2 for selected values of ω. If we substitute those expressions into eq. (4.19) we run into some difficulties because $G(x_1^{(e)},x_1^{(e)}) = 0$ and $G(x_2^{(e)},x_2^{(e)}) = 0$, and that makes the element matrix equation singular. How do we resolve this? One is tempted to quickly suggest that we introduce an arbitrary constant into the expression for the free-space Green's function, as we did for the linear 1-D. Laplace operator, but that is fraught with the problem of violating the original differential equation from which the Green's function was obtained. But do we have an alternative to introducing a constant into the Green's function to overcome this problem of dealing with a singular matrix? We do not really have an alternative, but that constant cannot be an arbitrary one in the true sense of the word, but rather a *penalty parameter* which should be chosen such that we can solve the matrix equation to obtain meaningful results and at the same time minimally violate the differential operator. That penalty parameter is denoted as γ_p in order to distinguish it from the arbitrary constant k that was used in chapter 2. Later in the text we shall carry out some numerical experiments to give suitable range of values for this parameter. It should be pointed out that the penalty parameter used here is conceptually different from that which is employed in *penalty finite element formulations* of fluid flow problems. The only similarity between them is that the introduction of the penalty parameter, though artificial, allows meaningful solutions to be obtained from the numerical scheme. When the penalty parameter is introduced, the free-space Green's function becomes

$$\boxed{G(x,x_i) = \frac{\sin(\omega|x-x_i|)+\gamma_p}{2\omega}} \tag{4.21}$$

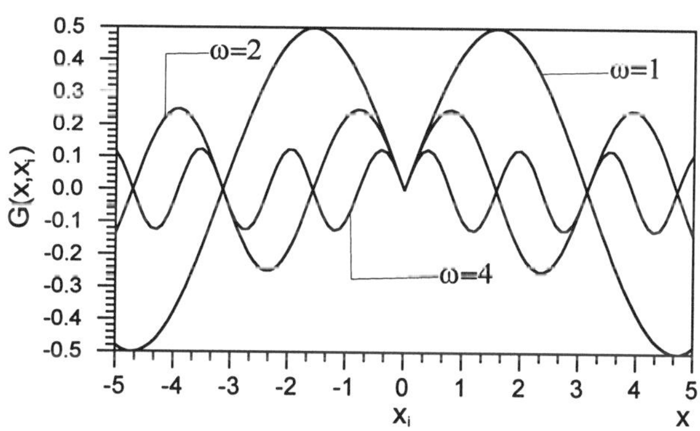

Figure 4.2 Plot of the fundamental solution to $d^2G/dx^2+\omega^2G=\delta(x-x_i)$.

And the following relations hold

$$\begin{aligned} G(x_1^{(e)},x_1^{(e)}) &= G(x_2^{(e)},x_2^{(e)}) = \frac{\gamma_p}{2\omega} \\ G(x_2^{(e)},x_1^{(e)}) &= G(x_1^{(e)},x_2^{(e)}) = \frac{\gamma_p+\sin(\omega l^{(e)})}{2\omega} \end{aligned} \tag{4.22a}$$

$$\begin{aligned} G^*(x_1^{(e)},x_1^{(e)}) &= G^*(x_2^{(e)},x_2^{(e)}) = 0 \\ G^*(x_2^{(e)},x_1^{(e)}) &= -G^*(x_1^{(e)},x_2^{(e)}) = \frac{\cos(\omega l^{(e)})}{2} \end{aligned} \tag{4.22b}$$

Introducing the above relations into eq. (4.20), when the source point $x_i^{(e)}$ is located at $x_1^{(e)}$, gives

$$-h_1^{(e)} + \cos(\omega l^{(e)}) h_2^{(e)} + \frac{\gamma_p}{\omega}\varphi_1^{(e)} - \frac{\gamma_p + \sin(\omega l^{(e)})}{\omega}\varphi_2^{(e)} + \frac{l^{(e)}}{\omega}\int_0^1 [\gamma_p + \sin(\omega l^{(e)}\zeta)]\Omega_j^{(e)} f_j^{(e)} d\zeta = 0; \quad j=1,2 \tag{4.23a}$$

and when the source point $x_i^{(e)}$ is at $x_2^{(e)}$

$$-h_2^{(e)} + \cos(\omega l^{(e)}) h_1^{(e)} - \frac{\gamma_p}{\omega}\varphi_2^{(e)} + \frac{\gamma_p + \sin(\omega l^{(e)})}{\omega}\varphi_1^{(e)} + \frac{l^{(e)}}{\omega}\int_0^1 [\gamma_p + \sin(\omega l^{(e)}(1-\zeta))][\Omega_1^{(e)} f_1^{(e)} + \Omega_2^{(e)} f_2^{(e)}] d\zeta = 0 \tag{4.23b}$$

Eqs. (4.23a) and (4.23b) are expressed in matrix form as

$$\boxed{R_{ij}^{(e)} h_j^{(e)} + L_{ij}^{(e)} \varphi_j^{(e)} + T_{ij}^{(e)} f_j^{(e)} = 0, \quad i,j=1,2} \tag{4.24}$$

in which the element matrices are given as

$$R_{ij}^{(e)} = \begin{bmatrix} -1 & \cos(\omega l^{(e)}) \\ \cos(\omega l^{(e)}) & -1 \end{bmatrix} \tag{4.25a}$$

$$L_{ij}^{(e)} = \frac{1}{\omega}\begin{bmatrix} \gamma_p & -[\gamma_p + \sin(\omega l^{(e)})] \\ [\gamma_p + \sin(\omega l^{(e)})] & -\gamma_p \end{bmatrix} \tag{4.25b}$$

$$\begin{aligned} T_{11}^{(e)} &= \frac{l^{(e)}}{\omega}\int_0^1 (1-\zeta)[\gamma_p + \sin(\omega l^{(e)}\zeta)] d\zeta = \\ T_{22}^{(e)} &= \frac{l^{(e)}}{\omega}\int_0^1 \zeta[\gamma_p + \sin(\omega l^{(e)}(1-\zeta))] d\zeta = \frac{l^{(e)}}{\omega}\left[\frac{\gamma_p}{2} + \frac{1}{\omega l^{(e)}}\left(1 - \frac{\sin(\omega l^{(e)})}{\omega l^{(e)}}\right)\right] \end{aligned} \tag{4.25c}$$

$$
\begin{aligned}
T_{12}^{(e)} &= \frac{l^{(e)}}{\omega}\int_0^1 \zeta[\gamma_p+\sin(\omega l^{(e)}\zeta)]d\zeta = \\
T_{21}^{(e)} &= \frac{l^{(e)}}{\omega}\int_0^1 (1-\zeta)[\gamma_p+\sin(\omega l^{(e)}(1-\zeta))]d\zeta = \\
&\qquad \frac{l^{(e)}}{\omega}\left[\frac{\gamma_p}{2}+\frac{1}{\omega l^{(e)}}\left(\frac{\sin(\omega l^{(e)})}{\omega l^{(e)}}-\cos(\omega l^{(e)})\right)\right]
\end{aligned}
\tag{4.25d}
$$

Eq. (4.24) is the system of discrete element equations based on a second Green element replication of the Helmholtz equation. It provides all the information to evaluate the values of h and φ at the nodes. The global matrix equation is obtained by assembling the element equations for all the elements. From the system of discrete element equations it appears that the value of the penalty parameter has to be chosen in relation to the magnitude $\sin(\omega l^{(e)})$. Its most critical value will be determined by the smallest spatial element size $l_{\min}^{(e)}$. In subsequent numerical experiments we shall examine the solution accuracy of the model for various values of

$$
\boxed{\kappa = \frac{\gamma_p}{\sin(\omega l_{\min}^{(e)})}}
\tag{4.26}
$$

Helmholtz equation could take another form given by

$$
\frac{d^2h}{dx^2} - \omega^2 h = f(x)
\tag{4.27}
$$

The difference between (4.27) and (4.3) is the sign of the second term on the left-hand-side. Modeling of eq. (4.27) by the first Green element formulation is essentially the same as modeling eq. (4.3) (the sign change is simply effected in the last term of the element equation (4.14)), but that is not the case with the second formulation. For the second model, whereas the integral equation for the element remains eq. (4.19), the expression for the free space Green's function and its spatial derivative are no longer eq. (4.20) but

$$
\boxed{G(x,x_i)=\frac{\sinh(\omega|x-x_i|)}{2\omega}} \quad \text{and} \quad \boxed{G^*=\frac{\cosh(\omega|x-x_i|)}{2}[H(x-x_i)-H(x_i-x)]}
\tag{4.28}
$$

The expression for $G(x,x_i)$ is graphically presented in Figure 4.3. If we employ this expression for the free-space Green's function we run into the same problem of dealing with a singular system of element equations. This difficulty is overcome in a similar manner by introducing a penalty parameter γ_p, that is

$$G(x,x_i) = \frac{\sinh(\omega|x-x_i|)+\gamma_p}{2\omega} \tag{4.29}$$

With this expression for G in eq. (4.29), some of the quantities in the integral equation (4.19) now simplify to

$$\begin{aligned} G(x_1^{(e)},x_1^{(e)}) &= G(x_2^{(e)},x_2^{(e)}) = \frac{\gamma_p}{2\omega} \\ G(x_2^{(e)},x_1^{(e)}) &= G(x_1^{(e)},x_2^{(e)}) = \frac{\gamma_p+\sinh(\omega l^{(e)})}{2\omega} \end{aligned} \tag{4.30a}$$

$$\begin{aligned} G^*(x_1^{(e)},x_1^{(e)}) &= G^*(x_2^{(e)},x_2^{(e)}) = 0 \\ G^*(x_2^{(e)},x_1^{(e)}) &= -G^*(x_1^{(e)},x_2^{(e)}) = \frac{\cosh(\omega l^{(e)})}{2} \end{aligned} \tag{4.30b}$$

Using the above relations, eq. (4.19), when the source point $x_i^{(e)}$ is located at $x_1^{(e)}$, yields

$$\begin{aligned} &-h_1^{(e)} + \cosh(\omega l^{(e)})h_2^{(e)} + \frac{\gamma_p}{\omega}\varphi_1^{(e)} - \frac{\gamma_p+\sinh(\omega l^{(e)})}{\omega}\varphi_2^{(e)} + \\ &\frac{l^{(e)}}{\omega}\int_0^1 [\gamma_p+\sinh(\omega l^{(e)}\zeta)]\Omega_j^{(e)} f_j^{(e)} d\zeta = 0; \quad j=1,2 \end{aligned} \tag{4.31a}$$

and when the source point $x_i^{(e)}$ is at $x_2^{(e)}$

$$\begin{aligned} &-h_2^{(e)} + \cosh(\omega l^{(e)})h_1^{(e)} - \frac{\gamma_p}{\omega}\varphi_2^{(e)} + \frac{\gamma_p+\sinh(\omega l^{(e)})}{\omega}\varphi_1^{(e)} + \\ &\frac{l^{(e)}}{\omega}\int_0^1 [\gamma_p+\sinh(\omega l^{(e)}(1-\zeta))]\Omega_j^{(e)} f_j^{(e)} d\zeta = 0; \quad j=1,2 \end{aligned} \tag{4.31b}$$

Combining eqs. (4.31a) and (4.31b) gives the element matrix equation (4.24), but the element matrices have the following expressions:

$$R_{ij}^{(e)} = \begin{bmatrix} -1 & \cosh(\omega l^{(e)}) \\ \cosh(\omega l^{(e)}) & -1 \end{bmatrix} \tag{4.32a}$$

$$L_{ij}^{(e)} = \frac{1}{\omega}\begin{bmatrix} \gamma_p & -[\gamma_p + \sinh(\omega l^{(e)})] \\ [\gamma_p + \sinh(\omega l^{(e)})] & -\gamma_p \end{bmatrix} \tag{4.32b}$$

$$T_{11}^{(e)} = \frac{l^{(e)}}{\omega}\int_0^1 (1-\zeta)[\gamma_p + \sinh(\omega l^{(e)}\zeta)]d\zeta =$$

$$T_{22}^{(e)} = \frac{l^{(e)}}{\omega}\int_0^1 \zeta[\gamma_p + \sinh(\omega l^{(e)}(1-\zeta))]d\zeta = \frac{l^{(e)}}{\omega}\left[\frac{\gamma_p}{2} - \frac{1}{\omega l^{(e)}}\left(1 - \frac{\sinh(\omega l^{(e)})}{\omega l^{(e)}}\right)\right] \tag{4.32c}$$

$$T_{12}^{(e)} = \frac{l^{(e)}}{\omega}\int_0^1 \zeta[\gamma_p + \sinh(\omega l^{(e)}\zeta)]d\zeta =$$

$$T_{21} = \frac{l^{(e)}}{\omega}\int_0^1 (1-\zeta)[\gamma_p + \sinh(\omega l^{(e)}(1-\zeta))]d\zeta = \tag{4.32d}$$

$$\frac{l^{(e)}}{\omega}\left[\frac{\gamma_p}{2} - \frac{1}{\omega l^{(e)}}\left(\frac{\sinh(\omega l^{(e)})}{\omega l^{(e)}} - \cosh(\omega l^{(e)})\right)\right]$$

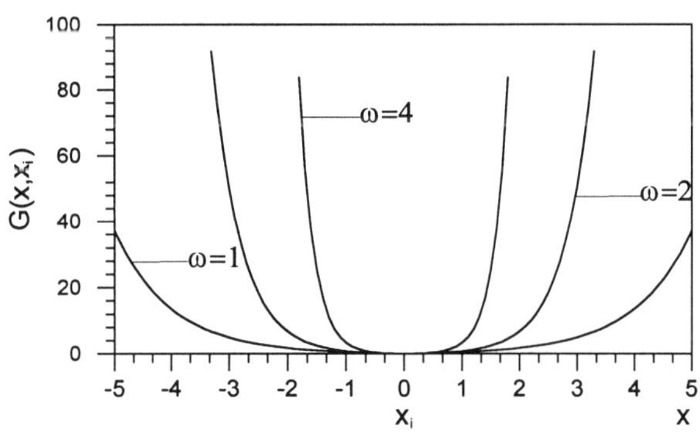

Figure 4.3 Plot of the fundamental solution to $d^2G/dx^2 - \omega^2 G = \delta(x - x_i)$.

From the system of discrete element equations the value of the penalty parameter depends on its magnitude relative to $\sinh(\omega l^{(e)})$. Its most critical value is determined by the smallest spatial element size $l_{\min}^{(e)}$ or the relationship

$$\boxed{\kappa = \frac{\gamma_p}{\sinh(\omega l_{\min}^{(e)})}} \tag{4.33}$$

4.4 Numerical Experiments to Establish Suitable Range of Values for the Penalty Parameter

Three examples of the 1-D. Helmholtz equation are used to evaluate the range of values of the penalty parameter that offers acceptable solutions from the second model of the Green element method. These examples are designed in such a way that their exact solutions can be derived and employed as bench mark results for carrying out accuracy assessment. The three examples and their exact solutions are given below:

Example 1

The problem statement for this example is given as

$$\frac{d^2h}{dx^2} + \omega^2 h = 0, \quad h(0) = 1, \text{ and } \varphi(1) = 1 \tag{4.34a}$$

The exact solution can readily be verified as

$$h(x) = \cos(\omega x) + \frac{[1 + \omega\sin(\omega)]\sin(\omega x)}{\omega\cos(\omega)} \tag{4.34b}$$

Example 2

The differential equation and boundary data for this example are

$$\frac{d^2h}{dx^2} + \omega^2 h = x^2, \quad h(0) = 0, \text{ and } \varphi(1) = 1 \tag{4.35a}$$

The exact solution is

$$h(x) = \frac{1}{\omega^4}\left[\frac{2\cos[\omega(1-x)] - \omega(\omega^2-2)\sin(\omega x)}{\cos(\omega)} + \omega^2 x^2 - 2\right] \tag{4.35b}$$

Example 3

Our third example is governed by the following differential equation and boundary data

$$\frac{d^2h}{dx^2} - \omega^2 h = 0, \quad h(0) = 1, \text{ and } \varphi(1) = 1 \tag{4.36a}$$

The exact solution is

$$h(x) = \frac{1 + \omega e^{-\omega}\sinh(\omega x)}{\omega\cosh(\omega)} + e^{-\omega x} \tag{4.36b}$$

In all three examples, the Green element calculations, based on our second model, are carried out using a 10-element uniform discretization of the computational domain. A value of $\omega = 2$ is used for examples 1 and 3, and unity for example 2. The solutions are obtained from a computer program based on the second model formulation and are presented in Table 4.1 in terms of the L_2 error norm which is defined as

$$L_2 = [\sum_i (h_i^N - h_i^E)^2]^{0.5} \tag{4.37}$$

where the superscripts N and E stand for the numerical and exact solutions, respectively. From the results in Table 4.1, the value of the penalty factor could be chosen in such a way that κ lies between 10^{-4} and 10^{-5} in order to achieve optimal results.

Table 4.1 L_2 error values of solutions from the second Green element model for various penalty parameter values.

κ	Solution for h	Solution for φ
	Example 1	
10^{-3}	4.09×10^{-2}	6.45×10^{-2}
10^{-4}	4.14×10^{-2}	1.09×10^{-2}
10^{-5}	6.28×10^{-4}	1.04×10^{-2}
10^{-6}	5.45×10^{-6}	3.80×10^{-1}
	Example 2	
10^{-3}	6.44×10^{-3}	1.02×10^{-2}
10^{-4}	3.70×10^{-3}	5.67×10^{-3}
10^{-5}	3.42×10^{-3}	1.05×10^{-2}
10^{-6}	3.40×10^{-3}	3.11×10^{-2}
	Example 3	
10^{-3}	8.44×10^{-4}	3.88×10^{-2}
10^{-4}	1.04×10^{-2}	4.27×10^{-2}
10^{-5}	1.06×10^{-2}	8.22×10^{-2}
10^{-6}	1.06×10^{-2}	2.08×10^{-1}

4.5 Hand Computations

A coarse spatial discretization of the computational domain for the second numerical example is done so that hand computations of the first and second models of the Green element formulation can be carried out. A value of unity is used for the parameter ω in the computations. As we had earlier pointed out, the purpose for presenting these hand computations is to enable the reader have a good feel for the method, and to appreciate the fact that the method transcends a bunch of mathematical abstractions.

The discrete element equations of the first Green element model for this example are obtained from eq. (4.14), and stated below. It is noted that $f(x) = x^2$

$$R_{ij}^{(e)} h_j^{(e)} + L_{ij}^{(e)} \varphi_j^{(e)} + T_{ij}^{(e)} (x_j^{(e)2} - \omega^2 h_j^{(e)}) = 0, \quad i,j = 1,2$$

The computational domain [0, 1] is discretized into 5 equal elements, implying that the spatial element size $l^{(e)} = l = \tilde{l} = 0.2$. The element matrices $R_{ij}^{(e)}$, $L_{ij}^{(e)}$, and $T_{ij}^{(e)}$ will be the same for each element because of the uniform discretization. From eqs. (2.42a), (2.42b), and (4.15), we obtain the expressions for these element matrices as

$$R_{ij}^{(e)} = R_{ij} = (-1)^{i+j-1}, \; L_{ij}^{(e)} = L_{ij} = \frac{1}{5}\begin{bmatrix} 1 & -2 \\ 2 & -1 \end{bmatrix} \text{ and } T_{ij}^{(e)} = T_{ij} = \frac{1}{150}\begin{bmatrix} 4 & 5 \\ 5 & 4 \end{bmatrix}$$

In addition, $x_1^{(1)2} = 0$, $x_2^{(1)2} = x_1^{(2)2} = 1/25$, $x_2^{(2)2} = x_1^{(3)2} = 4/25$, $x_2^{(3)2} = x_1^{(4)2} = 9/25$, $x_2^{(4)2} = x_1^{(5)2} = 16/25$, and $x_2^{(5)2} = 1$. With the information above, the global coefficient matrix equation for this example becomes

$$\begin{bmatrix}
1 & \frac{29}{6} & -2 & 0 & 0 & 0 & 0 & 0 & 0 & 0 \\
2 & -\frac{154}{30} & -1 & 0 & 0 & 0 & 0 & 0 & 0 & 0 \\
0 & -\frac{154}{30} & 1 & \frac{29}{6} & -2 & 0 & 0 & 0 & 0 & 0 \\
0 & \frac{29}{6} & 2 & -\frac{154}{30} & -1 & 0 & 0 & 0 & 0 & 0 \\
0 & 0 & 0 & -\frac{154}{30} & 1 & \frac{29}{6} & -2 & 0 & 0 & 0 \\
0 & 0 & 0 & \frac{29}{6} & 2 & -\frac{154}{30} & -1 & 0 & 0 & 0 \\
0 & 0 & 0 & 0 & 0 & -\frac{154}{30} & 1 & \frac{29}{6} & -2 & 0 \\
0 & 0 & 0 & 0 & 0 & \frac{29}{6} & 2 & -\frac{154}{30} & -1 & 0 \\
0 & 0 & 0 & 0 & 0 & 0 & 0 & -\frac{154}{30} & 1 & \frac{29}{6} \\
0 & 0 & 0 & 0 & 0 & 0 & 0 & \frac{29}{6} & 2 & -\frac{154}{30}
\end{bmatrix}
\begin{Bmatrix} \varphi_1 \\ h_2 \\ \varphi_2 \\ h_3 \\ \varphi_3 \\ h_4 \\ \varphi_4 \\ h_5 \\ \varphi_5 \\ h_6 \end{Bmatrix}
= \frac{-1}{750}
\begin{Bmatrix} 5 \\ 4 \\ 24 \\ 21 \\ 61 \\ 56 \\ 116 \\ 109 \\ -1311 \\ -570 \end{Bmatrix}$$

Its solution, obtained by a matrix solver, is presented in Table 4.2. For the second model, the element equations are obtained from eq. (4.24)

$$R_{ij}^{(e)} h_j^{(e)} + L_{ij}^{(e)} \varphi_j^{(e)} + T_{ij}^{(e)} x_j^{(e)2} = 0, \quad i,j=1,2$$

whose element matrices, for this 5-element spatial discretization, are (Note that $\omega = 1$ and $l^{(e)} = 0.2$, and we have chosen to use a value for $\gamma_p = 10^{-5}$.)

$$R_{ij}^{(e)} = R_{ij} = \begin{bmatrix} -1 & \cos(0.2) \\ \cos(0.2) & -1 \end{bmatrix} = \begin{bmatrix} -1 & 0.98 \\ 0.98 & -1 \end{bmatrix}$$

$$L_{ij}^{(e)} = L_{ij} = \begin{bmatrix} \gamma_p & -\gamma_p - \sin(0.2) \\ \gamma_p + \sin(0.2) & -\gamma_p \end{bmatrix} = \begin{bmatrix} 10^{-5} & -0.199 \\ 0.199 & -10^{-5} \end{bmatrix}$$

$$T_{ij}^{(e)} = T_{ij} = \frac{1}{5}\begin{bmatrix} \frac{\gamma_p}{2} + 0.0333 & \frac{\gamma_p}{2} + 0.0664 \\ \frac{\gamma_p}{2} + 0.0664 & \frac{\gamma_p}{2} + 0.0333 \end{bmatrix} = \frac{1}{5}\begin{bmatrix} 0.0333 & 0.0664 \\ 0.0664 & 0.0333 \end{bmatrix}$$

The element equations are assembled to form the global matrix equation

$$\begin{bmatrix} 10^{-5} & .98 & -.199 & 0 & 0 & 0 & 0 & 0 & 0 & 0 \\ .199 & -1. & -10^{-5} & 0 & 0 & 0 & 0 & 0 & 0 & 0 \\ 0 & -1 & 10^{-5} & .98 & -.199 & 0 & 0 & 0 & 0 & 0 \\ 0 & .98 & .199 & -1 & -10^{-5} & 0 & 0 & 0 & 0 & 0 \\ 0 & 0 & 0 & -1 & 10^{-5} & .98 & -.199 & 0 & 0 & 0 \\ 0 & 0 & 0 & .98 & .199 & -1 & -10^{-5} & 0 & 0 & 0 \\ 0 & 0 & 0 & 0 & 0 & -1 & 10^{-5} & .98 & -.199 & 0 \\ 0 & 0 & 0 & 0 & 0 & .98 & .199 & -1 & -10^{-5} & 0 \\ 0 & 0 & 0 & 0 & 0 & 0 & 0 & -1 & 10^{-5} & .98 \\ 0 & 0 & 0 & 0 & 0 & 0 & 0 & .98 & .199 & -1 \end{bmatrix} \begin{Bmatrix} \varphi_1 \\ h_2 \\ \varphi_2 \\ h_3 \\ \varphi_3 \\ h_4 \\ \varphi_4 \\ h_5 \\ \varphi_5 \\ h_6 \end{Bmatrix} = \frac{-1}{150} \begin{Bmatrix} 0.066 \\ 0.033 \\ 0.299 \\ 0.199 \\ 0.731 \\ 0.565 \\ 1.362 \\ 1.130 \\ -22.642 \\ 1.893 \end{Bmatrix}$$

whose solution is presented in Table 4.2. The solutions from the two models are virtually the same, and they closely reproduce the exact solution.

4.6 Computer Simulations of the Three Examples

The three numerical examples of the Helmholtz equation presented in section 4.4 are simulated with the two Green element models derived in this chapter for the equation. In all the examples, the computational region is discretized into 10 uniform linear elements, and a value of the penalty parameter used in the second model is chosen such that $\kappa = 10^{-5}$. The numerical and exact solutions for the three examples are presented in Figures 4.4 through 4.6. Since we had earlier presented the L_2 norm of the error of the numerical solution from the second model in Table 4.1, we do the same for the first model in Table 4.3 for completeness. The results from both are virtually the same; model 2 has an advantage in accuracy over model 1 for example 1, and *vice versa* for examples 2 and 3.

Table 4.2 Numerical solutions from hand computations of two Green element models of the Helmholtz example 2.

Position	Exact		Model 1		Model 2	
x	h	φ	h	φ	h	φ
0		1.264		1.252		1.254
0.2	0.251	1.241	0.249	1.231	0.249	1.233
0.4	0.494	1.185	0.490	1.177	0.491	1.178
0.6	0.724	1.114	0.719	1.109	0.720	1.109
0.8	0.940	1.046	0.934	1.043	0.935	1.043
1	1.144		1.138		1.138	

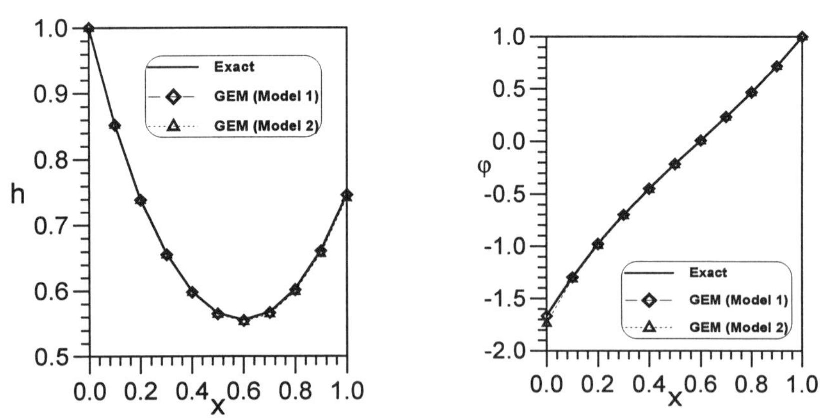

(a) (b)

Figure 4.4 **GEM** and exact solutions of example 1 — **(a)** solution for h, **(b)** solution for φ

Table 4.3 L_2 error of solutions from the first Green element model for the three Helmholtz examples

Example	Numerical solution for h	Numerical solution for φ
1	6.87×10^{-2}	1.09×10^{-1}
2	3.87×10^{-3}	6.20×10^{-3}
3	3.85×10^{-3}	6.93×10^{-3}

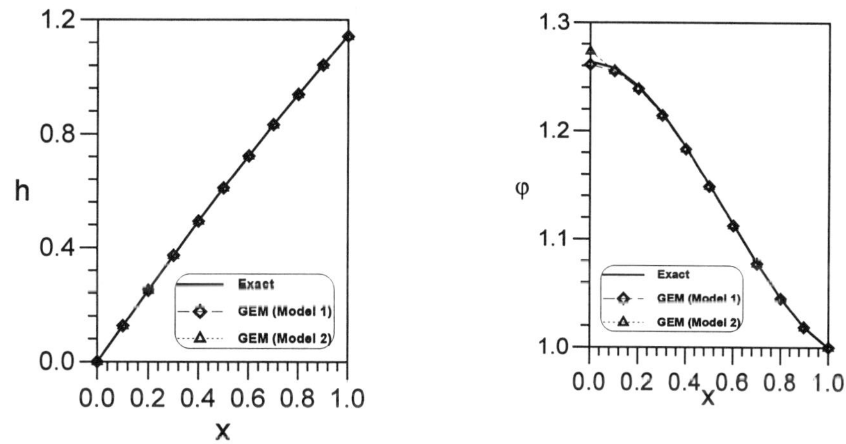

(a) (b)

Figure 4.5 GEM and exact solutions of example 2 — **(a)** solution for h, **(b)** solution for φ

4.7 Remarks

The Helmholtz equation is addressed in this chapter because of its extensive areas of applications in engineering. Two Green element models are proposed for its solution; the first uses the free-space Green's function of the 1-D. Laplace differential operator, while the second employs the free-space Green's function of the Helmholtz differential operator. To overcome encountering a singular element matrix for the flux vector in the second model, a penalty parameter is introduced into the expression for the free-space Green's function, and its optimum range of values is obtained from numerical experiments to be such that κ lies between 10^{-4} and 10^{-5}. The solutions from both Green element models are identical, but the first model has the advantage of being more versatile, having been employed for all the problems addressed in the earlier two chapters.

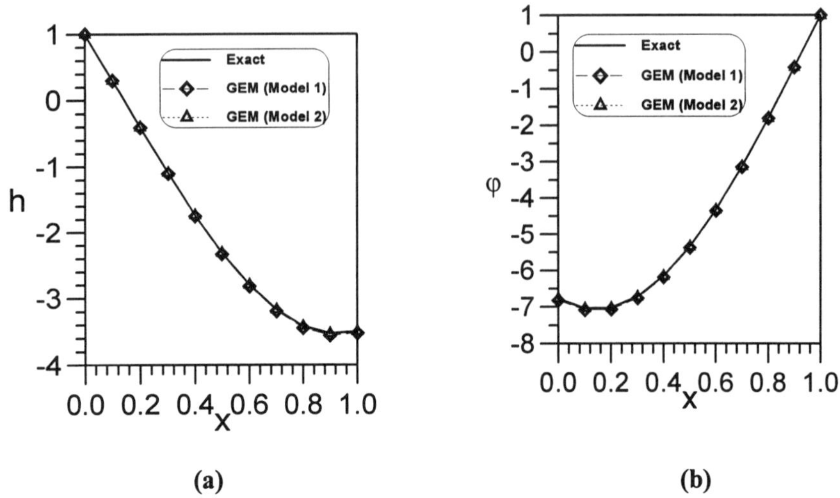

Figure 4.6 GEM and exact solutions of example 3 — **(a)** solution for h, **(b)** solution for φ

4.8 Exercises

4.1 Reformulate the two Green element models derived in this chapter for the linear Helmholtz equation so that there is only one degree of freedom at each node. Apply these new formulations to the three numerical examples presented in this chapter, and compare with those presented in the text.

4.2 Apply the two Green element models derived in this chapter to solve

$$\frac{d^2h}{dx^2} \pm h + f(x) = 0; \quad 0 \le x \le 1, \ h(0) = 1, \text{and } h(1) = 0$$

for the following expressions of $f(x)$

(a) $f(x) = \delta(x - 0.5)$
(b) $f(x) = 1 - x$
(c) $f(x) = \sin(\pi x)$
(d) $f(x) = \cos(\pi x)$

4.3 Repeat exercise 2 with the following boundary conditions:

$$\frac{dh(0)}{dx} \equiv \varphi(0) = 1 \ \text{ and } \ h(1) = 1$$

4.4 Formulate an appropriate Green element model for the heterogeneous form of the equation given by eq. (4.5).

4.5 Use the Green element model formulated in Question 4.4 to solve the following equation

$$\frac{d}{dx}\left[K(x)\frac{dh}{dx}\right] + h = 0, \quad 0 \le x \le 1, \; h(0) = 0 \text{ and } h(1) = 1$$

for the expressions of $K(x)$

(a) $K(x) = 1 + x$

(b) $K(x) = 1 + x^2$

(c) $K = e^x$

5 TRANSIENT DIFFUSION

5.1 Introduction

Before now, all the problems that have been addressed were steady. In other words neither the primary variable nor the flux had any time history. Steady state solutions may be viewed as states of equilibrium which only become altered when the dynamic forces ensuring equilibrium are altered. With this understanding it means that transient problems could be considered as going through quasi-steady states. Although steady state solutions have their value, especially in evaluating the performance of a system under equilibrium conditions, the real world is dynamic in nature. States of equilibrium are rarely maintained for too long, so that the time history of the performance of the system has to be closely monitored in order to capture its complete picture.

5.2 Derivation of the Transient Diffusion Equation

One of the simplest transient second-order partial differential equations is the *diffusion* or *heat conduction equation.* It is encountered in many fields of engineering — heat conduction through a bar, confined and unconfined aquifer flow, soil consolidation, diffusion of a contaminant, and many others. It can be a linear or nonlinear equation, depending on medium parameter characteristics. A conducting bar subject to heat flux is used to derive the 1-D. transient diffusion equation (Figure 5.1a). The length of the rod or bar offers the predominant direction of heat transport; variations in temperature in the other directions are assumed negligible. The bar is allowed to receive influx of heat from an external source by an amount $F(x,t)$ [J/(m.s)]. Our control volume in which the statement of conservation of heat is examined is an infinitesimal length of the bar shown in Figure 5.1b. The principle of conservation of heat states: heat flux into the control volume minus heat flux out of the control volume equals temporal rate of change in internal energy within the control volume. The heat influx is

$$Q(x,t) \; + \; F(x,t)dx \tag{5.1}$$

Heat flux out of the element

$$Q(x+dx,t) \; = \; Q(x,t) + \frac{\partial Q}{\partial x}dx \tag{5.2}$$

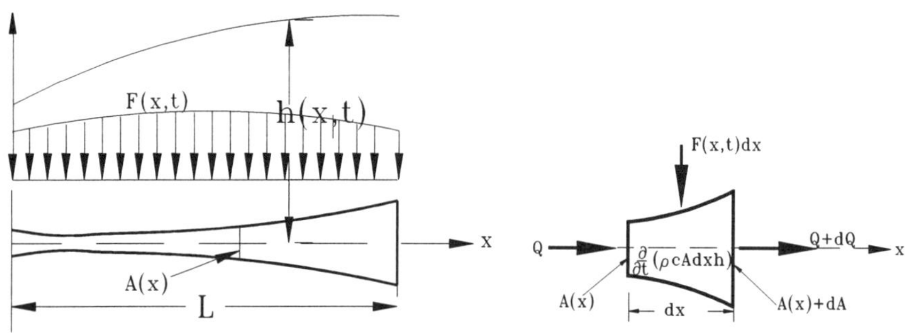

(a) **(b)**

Figure 5.1 Definition sketch for unsteady heat transfer — **(a)** a conducting bar; **(b)** an infinitesimal element.

The net flux of heat across the control surface is obtained by subtracting eq. (5.2) from (5.1), that is

$$-\frac{\partial Q}{\partial x}dx + F(x,t)dx \tag{5.3}$$

By Fourier law of heat conduction stated in chapter 3 (see eq. (3.3))

$$Q(x,t) \; = \; -kA\frac{\partial h(x,t)}{\partial x} \tag{5.4}$$

in which h is the temperature, A is the area of the rod, and k is the heat conduction coefficient. Introducing (5.4) into (5.3) yields

$$\frac{\partial}{\partial x}\left(kA(x)\frac{\partial h}{\partial x}\right) dx \; + \; F(x,t)dx \tag{5.5}$$

The temporal change in internal energy within the infinitesimal control volume is the temporal derivative of the product of the mass of element and heat storage per unit mass, and it is expressed as

$$\frac{\partial}{\partial t}(\rho c A dx h) = \rho c A dx \frac{\partial h}{\partial t} \tag{5.6}$$

in which ρ is the *density* of the *rod* [kg/m^3], c is the *specific heat* [J/(kg.K)], and the product of the quantities ρc is referred to as the *heat capacity* [J/(m^3.K)], and it is a measure of the ability of the rod to store heat. Eqs. (5.5) and (5.6) are combined to give the statement of conservation of heat within the control volume

$$\frac{\partial}{\partial x}\left(kA(x)\frac{\partial h}{\partial x}\right) dx + F(x,t)dx = \rho c A dx \frac{\partial h}{\partial t}$$

or

$$\frac{\partial}{\partial x}\left(K\frac{\partial h}{\partial x}\right) = \frac{\partial h}{\partial t} + f(x,t) \tag{5.7}$$

in which $K = k/(\rho c)$ is the *diffusivity*, and $f(x,t) = -F(x,t)/(\rho c A)$ is referred to as the *heat source density*.

The history of application of the boundary element method to the transient diffusion equation is a fairly long one. We touched on some of the landmark developments in the introductory remarks in chapter 1. Much of the effort was devoted at preserving the boundary only character of the method. To preserve the excellent performance of the boundary element theory for elliptic-type differential operators, the transient diffusion equation was transformed to a steady one by Laplace transform. While the boundary-only character is retained for the differential equation in the transformed plane, the restoration of the solution to the real x-t plane is not straightforward. Although most of the applications of the Laplace transform formulation have been for 2-D. spatial problems, the character of the formulation is essentially the same for 1- and 2-dimensions. In recent years Cheng and his co-workers have devoted much effort to efficient inversion of the numerical solution from the transformed plane into the real plane, and they have reported substantial success, although it is difficult to assess how general such inversion procedures are (Cheng and Ou [1]).

We certainly will draw issues to unusually prolonged lengths if we present existing **BEM** formulations side-by-side **GEM** formulations for transient diffusion and then, as we did in chapter 2 for the steady diffusion equation, highlight the advantages and disadvantages of both methods. (The reader may wish to consult the following references on **BE** applications to the transient diffusion equation — Banerjee and Butterfield [2], and Liggett and Liu [3].) Here we shall focus only on Green element applications to transient diffusion by presenting two of its models.

5.3 Boundary and Initial Conditions

To obtain unique solutions to the flow equation it is important to specify conditions for h and/or $K\partial h/\partial x \equiv K\varphi$ at the end points of the computational region at all times, and those data throughout the flow length at the initial time t_0. Without these conditions, there are infinitely many distributions of h which could satisfy the differential equation. The boundary conditions are, in general, of three types. The 1st-type boundary condition or Dirichlet or essential condition specifies the primary variable h at a boundary

$$h(x=x_0,t) = h_0(t) \quad \text{and} \quad h(x=x_L,t) = h_L(t) \tag{5.8}$$

The 2nd-type or Neumann or natural condition specifies the flux (fluid flux or heat flux) across an area, and it is expressed as

$$-K\frac{\partial h(x,t)}{\partial x}\bigg]_{x=x_0} = q_0(t) \quad \text{and} \quad -K\frac{\partial h(x,t)}{\partial x}\bigg]_{x=x_L} = q_L(t) \tag{5.9}$$

A 3rd-type or Cauchy condition prescribes a linear combination of h and its flux at a boundary. That is,

$$\left[a_0 h + b_0 K\frac{\partial h}{\partial x}\right]_{x=x_0} = v_0(t) \quad \text{and} \quad \left[a_L h + b_L K\frac{\partial h}{\partial x}\right]_{x=x_L} = v_L(t) \tag{5.10}$$

in which a_0, b_0, a_L, and b_L are known coefficients. An additional condition has to be imposed for a unique solution to be achieved. That condition, referred to as the *initial condition*, describes an earlier attained equilibrium state of the system at time t_0. The initial condition specifies h everywhere on the flow domain

$$h(x,t=t_0) = p_0(x), \quad on \quad x_0 \le x \le x_L \tag{5.11}$$

5.4 Quasi-Steady Green Element Model (QSGE)

The first Green element model for transient diffusion follows that which we employed for steady problems in the previous chapters. Our complimentary differential equation to eq. (5.7) is again eq. (2.20)

$$\frac{d^2G}{dx^2} = \delta(x-x_i), \quad on \quad -\infty \le x \le \infty \tag{2.20}$$

Applying Green's identity to eqs. (5.7) and (2.20) yields the integral representation for the element $[x_1^{(e)},x_2^{(e)}]$ (For now we shall only examine the linear diffusion equation, that is K is a constant.)

$$
\begin{aligned}
&-2\lambda^{(e)}h_i^{(e)} + [H(x_2^{(e)}-x_i^{(e)}) - H(x_i^{(e)}-x_2^{(e)})]h_2^{(e)} - \\
&[H(x_1^{(e)}-x_i^{(e)}) - H(x_i^{(e)}-x_1^{(e)})]h_1^{(e)} - (|x_2^{(e)}-x_i^{(e)}| + \tilde{l})\varphi_2^{(e)} + \\
&(|x_1^{(e)}-x_i^{(e)}| + \tilde{l})\varphi_1^{(e)} + \frac{1}{K}\int_{x_1^{(e)}}^{x_2^{(e)}} (|x-x_i^{(e)}| + \tilde{l})\left(\frac{\partial h}{\partial t} + f(x,t)\right) dx \;=\; 0, \quad i=1,2
\end{aligned}
\tag{5.12}
$$

(Note that we have set the arbitrary constant in the solution to eq. (2.20) to the longest spatial element size $\tilde{l}$ as prescribed in chapter 2.) Eq. (5.12) is similar to eq. (2.38) of the steady problem in the sense that the term with the temporal derivative has been treated as the external recharge term. From our knowledge of interpolation in an element, we elect to approximate in space the distribution of $f(x,t)$ and $h(x,t)$ by linear interpolating functions

$$
h(x,t) \;\approx\; \Omega_j^{(e)}(\zeta)h_j^{(e)}(t) \quad \text{and} \quad f(x,t) \;\approx\; \Omega_j^{(e)}(\zeta)f_j^{(e)}(t) \tag{5.13}
$$

in which $\Omega_j^{(e)}(\zeta)$ represents the linear element interpolating functions, and $h_2^{(e)}(t)$, for instance, means the value of h at node $x_2^{(e)}$ at time t. From eq. (5.13) the temporal derivative of h is given by

$$
\frac{\partial h(x,t)}{\partial t} \;\approx\; \Omega_j^{(e)}(\zeta)\frac{dh_j^{(e)}(t)}{dt} \tag{5.14}
$$

Eqs. (5.13) and (5.14) are substituted into eq. (5.12) to give

$$
\begin{aligned}
&-2\lambda^{(e)}h_i^{(e)} + [H(x_2^{(e)}-x_i^{(e)}) - H(x_i^{(e)}-x_2^{(e)})]h_2^{(e)} - \\
&[H(x_1^{(e)}-x_i^{(e)}) - H(x_i^{(e)}-x_1^{(e)})]h_1^{(e)} - (|x_2^{(e)}-x_i^{(e)}| + \tilde{l})\varphi_2^{(e)} + \\
&(|x_1^{(e)}-x_i^{(e)}| + \tilde{l})\varphi_1^{(e)} + \frac{1}{K}\int_{x_1^{(e)}}^{x_2^{(e)}} (|x-x_i^{(e)}| + \tilde{l})\Omega_j^{(e)}\left(\frac{dh_j^{(e)}}{dt} + f_j^{(e)}\right) dx = 0, \quad i,j=1,2
\end{aligned}
\tag{5.15}
$$

Eq. (5.15) is split into two equations by evaluating it with $x_i^{(e)} \;=\; x_1^{(e)}$

$$-h_1^{(e)} + h_2^{(e)} + \tilde{l}\varphi_1^{(e)} - (\tilde{l}+l^{(e)})\varphi_2^{(e)} + \frac{l^{(e)}}{K}\int_0^1 \Omega_j^{(e)}\left(\frac{dh_j^{(e)}}{dt} + f_j^{(e)}\right)(\tilde{l}+l^{(e)}\zeta)d\zeta = 0, \quad j=1,2 \tag{5.16a}$$

and with $x_i^{(e)} = x_2^{(e)}$

$$h_1^{(e)} - h_2^{(e)} + (\tilde{l}+l^{(e)})\varphi_1^{(e)} - \tilde{l}\varphi_2^{(e)} + \frac{l^{(e)}}{K}\int_0^1 \Omega_j\left(\frac{dh_j^{(e)}}{dt} + f_j^{(e)}\right)[\tilde{l}+l^{(e)}(1-\zeta)]d\zeta = 0, \quad j=1,2 \tag{5.16b}$$

Eqs. (5.16a) and (5.16b) are combined to form the element matrix equations of one of the Green element formulations of the transient linear diffusion equation.

$$\boxed{R_{ij}^{(e)}h_j^{(e)} + L_{ij}^{(e)}\varphi_j^{(e)} + \frac{T_{ij}^{(e)}}{K}\left(\frac{dh_j^{(e)}}{dt} + f_j^{(e)}\right) = 0, \quad i,j=1,2} \tag{5.17}$$

in which $R_{ij}^{(e)}$ and $L_{ij}^{(e)}$ retain the expressions in eqs. (2.42a) and (2.42b), and

$$T_{1j}^{(e)} = l^{(e)}\int_0^1 \Omega_j^{(e)}(\zeta)(\tilde{l}+l^{(e)}\zeta)d\zeta, \quad j=1,2 \tag{5.18a}$$

$$T_{2j}^{(e)} = l^{(e)}\int_0^1 \Omega_j^{(e)}(\zeta)[\tilde{l}+l^{(e)}(1-\zeta)]d\zeta, \quad j=1,2 \tag{5.18b}$$

We had encountered $T_{ij}^{(e)}$ in chapters 3 and 4, and its expression is re-stated from eqs. (3.33) and (4.15)

$$T_{ij}^{(e)} = \frac{l^{(e)}}{6}\begin{bmatrix} (3\tilde{l}+l^{(e)}) & (3\tilde{l}+2l^{(e)}) \\ (3\tilde{l}+2l^{(e)}) & (3\tilde{l}+l^{(e)}) \end{bmatrix} \tag{5.19}$$

Eq. (5.17) is a system of discrete first-order differential equations in time. There are a number of solution avenues available to us to simplify the temporal derivative. Since eq. (5.17) is a system of initial-value differential equations, *Euler* and *Runge-Kutta* methods should be quite appropriate for the solution of the system of equations. To use a difference approximation for the temporal derivative, the time history is also divided into elements with a typical element, denoted s, having two nodes or time levels: t_1

being the previous time level at which solution information are completely specified, and t_2 being the current time level at which solution information are sought (see Figure 5.2). That difference approximation is given by

$$\left.\frac{dh_j^{(e)}}{dt}\right|_{t=t_1+\alpha\Delta t} \approx \frac{h_j^{(e)}(t_2) - h_j^{(e)}(t_1)}{\Delta t} = \frac{h_j^{(e,2)} - h_j^{(e,1)}}{\Delta t}, \quad 0 \le \alpha \le 1 \tag{5.20}$$

Eq (5.20) indicates that temporal derivative is evaluated at a time $t_1 + \alpha\Delta t$, where α is a *difference weighting factor* whose value varies between 0 and 1. The value of α positions the time level at which the temporal derivative is evaluated. Using familiar nomenclature in finite difference and finite element circles, a scheme with $\alpha = 0$ is said to be the *fully explicit scheme*, that with $\alpha = 0.5$ is the *Crank-Nicholson scheme*, that with $\alpha = 0.67$ the *Galerkin scheme*, while that with $\alpha = 1.0$ is usually referred to as the *fully implicit scheme*. Since the temporal derivative has been evaluated at $t_1 + \alpha\Delta t$, it is reasonable to evaluate the other terms of eq. (5.17) at that time level using a weighted average of the form

$$\alpha\left[R_{ij}^{(e)}h_j^{(e)} + L_{ij}^{(e)}\varphi_j^{(e)} + \frac{T_{ij}^{(e)}}{K}f_j^{(e)}\right]_{t=t_2} + \omega\left[R_{ij}^{(e)}h_j^{(e)} + L_{ij}^{(e)}\varphi_j^{(e)} + \frac{T_{ij}^{(e)}}{K}f_j^{(e)}\right]_{t=t_1}$$
$$+ \left.\frac{T_{ij}^{(e)}}{K}\frac{dh_j^{(e)}}{dt}\right|_{t=t_1+\alpha\Delta t} = 0, \quad i,j=1,2, \quad 0 \le \alpha \le 1 \tag{5.21}$$

in which $\omega = 1 - \alpha$. We thus far have used the superscript (e) on variables and parameters to indicate that the equations are derived specifically for a typical element. We wish to omit this superscript from henceforth, but may only revert to it when confusion is envisaged between the element equations and the global equations. Eq. (5.20) is now introduced into eq. (5.21) to yield

$$\boxed{\begin{aligned}&\alpha\left[R_{ij}h_j^{(2)} + L_{ij}\varphi_j^{(2)} + \frac{T_{ij}}{K}f_j^{(2)}\right] + \omega\left[R_{ij}h_j^{(1)} + L_{ij}\varphi_j^{(1)} + \frac{T_{ij}}{K}f_j^{(1)}\right] + \\ &\frac{T_{ij}}{K}\left(\frac{h_j^{(2)} - h_j^{(1)}}{\Delta t}\right) = 0, \quad i,j=1,2, \quad 0 \le \alpha \le 1\end{aligned}} \tag{5.22}$$

The bracketed superscripts 2 and 1 indicate the current and previous time levels, respectively. In other words, $h_j^{(2)}$ is the current value of h at node j. Eq. (5.22) represents the element equations of the first Green element model of the diffusion equation, and it is herein referred to as the Quasi-steady Green element (**QSGE**) formulation. It completely characterizes the solution within a spatial element and a temporal element. We are now in a position to assemble eq. (5.22) for all the spatial elements in the same way as we did for the 1-D. Laplace/Poisson problem of chapter 2. The matrix equation is of the form

$$[A]\begin{Bmatrix} h^{(2)} \\ \varphi^{(2)} \end{Bmatrix} = \{S^{(1)}\} \tag{5.23}$$

The global coefficient matrix $[A]$ has the same size as that for the 1-D. Laplace/Poisson equation with row dimension of $2M$ and half bandwidth of 2. The unknown vector consists of h and φ at the current time for nodes where they are not specified by the boundary condition, and the known vector $\{S^{(1)}\}$ has entries from the boundary, initial and externally imposed recharge data. The global coefficient matrix equation can be solved to given the solution information at t_2, which then becomes the initial time for a subsequent time. Thus, the solution procedure is a *time marching scheme.*

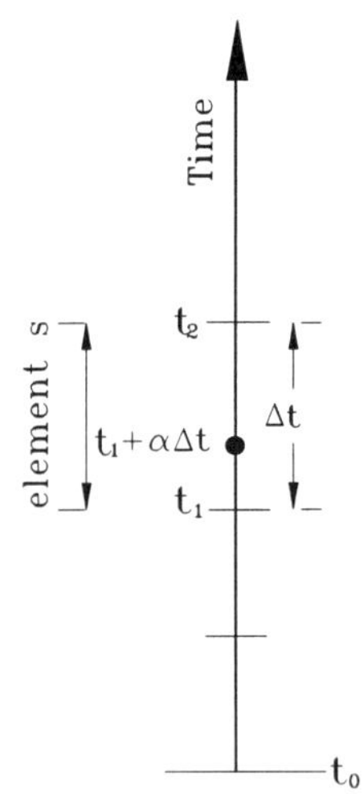

Figure 5.2 Discretization of time domain.

5.5 Transient Green Element Model (TGE)

Our second **GEM** formulation for the transient diffusion comes from the use of the free-space Green's function to the linear transient diffusion equation. Earlier use of this free-space Green's function along the lines of conventional boundary element method had not produced, as the linear steady diffusion equation, a strictly boundary method.

The reason for this is because the initial data has to be integrated over the entire domain. That formulation is herein adopted in a Green element sense.

The current formulation uses the fundamental solution to the time-dependent complementary differential equation

$$\frac{\partial^2 G}{\partial x^2} + \frac{1}{K}\frac{\partial G}{\partial t} = \delta(x-x_i)\delta(t_2-t), \quad -\infty \le x \le \infty, \ t_2 \ge t \tag{5.24}$$

in which t_2 is the source point in the time domain. Here is the first time we are encountering a time variable in the complementary differential equation, and this is also reflected in the expression of the free-space Green's function. The solution to eq. (5.24), which is the free-space Green's function, is given by

$$G(x,t;x_i,t_2) = \frac{H(t_2-t)}{\sqrt{[4K\pi(t_2-t)]}}\exp\left[-\frac{(x-x_i)^2}{4K(t_2-t)}\right] \tag{5.25}$$

To derive the integral representation of the linear form of eq. (5.7), Green's identity of eq. (2.25) is invoked. However the identity has to be slightly modified to account for the fact the h and G are now functions of two variables x and t instead of just one. This slight modification of the identity simply involves integration in time of eq. (2.25) within the temporal element. That is

$$\int_{t_1}^{t_2}\int_{x_1}^{x_2}\left(h\frac{\partial^2 G}{\partial x^2} - G\frac{\partial^2 h}{\partial x^2}\right)dx\,dt = \int_{t_1}^{t_2}\left\{h\frac{\partial G}{\partial x}\bigg|_{x_1}^{x_2} - G\frac{\partial h}{\partial x}\bigg|_{x_1}^{x_2}\right\}dt \tag{5.26}$$

The linear diffusion equation and eq. (5.24) are introduced into eq. (5.26) to give

$$\int_{t_1}^{t_2}\int_{x_1}^{x_2}\left[h\left(\delta(x-x_i)\delta(t-\tau) - \frac{1}{K}\frac{\partial G}{\partial t}\right) - \frac{G}{K}\left(\frac{\partial h}{\partial t} + f(x,t)\right)\right]dx\,dt = \int_{t_1}^{t_2}\left[h\frac{\partial G}{\partial x} - G\varphi\right]_{x_1}^{x_2}dt \tag{5.27}$$

Eq. (5.27) is simplified after establishing a number of relationships. By the property of the Dirac delta function discussed in chapter 2

$$\int_{t_1}^{t_2}\int_{x_1}^{x_2} h(x,t)\delta(x-x_i)\delta(t_2-t)\,dx\,dt = \lambda h(x_i,t_2) = \lambda h_i^{(2)} \tag{5.28a}$$

in which $\lambda = 1$ if x_i is within the interval $[x_1,x_2]$, and $\lambda = 0.5$ if x_i is at x_1 or x_2. In addition

$$-\frac{1}{K}\int_{t_1}^{t_2}\int_{x_1}^{x_2}\left(h\frac{\partial G}{\partial t}+G\frac{\partial h}{\partial t}\right)dxdt = -\frac{1}{K}\int_{t_1}^{t_2}\int_{x_1}^{x_2}\frac{\partial}{\partial t}(hG)dxdt$$
$$= -\frac{1}{K}\int_{x_1}^{x_2}(hG)\Big|_{t_1}^{t_2}dx = \frac{1}{K}\int_{x_1}^{x_2}G(x,t_1;x_i,t_2)h(x,t_1)dx \tag{5.28b}$$

In the derivation of eq. (5.28b) we have employed one of the characteristics of the Green's function and that is $G(x,t_2;x_i,t_2) = 0$. Introducing the relationships provided by eqs. (5.28a) and (5.28b) into eq. (5.27) yields

$$-\lambda h_i^{(2)} + \int_{t_1}^{t_2}\left[h(x_2,t)G^*(x_2,t;x_i,t_2) - h(x_1,t)G^*(x_1,t;x_i,t_2)\right]dt -$$
$$\int_{t_1}^{t_2}\left[G(x_2,t;x_i,t_2)\varphi(x_2,t) - G(x_1,t;x_i,t_2)\varphi(x_1,t)\right]dt - \tag{5.29}$$
$$\frac{1}{K}\int_{x_1}^{x_2}G(x,t_1;x_i,t_2)h(x,t_1)dx + \frac{1}{K}\int_{t_1}^{t_2}\int_{x_1}^{x_2}G(x,t;x_i,t_2)f(x,t)dxdt = 0$$

where

$$G^*(x,t;x_i,t_2) = \frac{\partial G}{\partial x} = -\frac{(x-x_i)}{2K(t_2-t)}G(x,t;x_i,t_2) \tag{5.30}$$

In the numerical procedure of implementing eq. (5.29) it is considered that the solution at t_1 is known, and the solution at t_2 is desired. Furthermore, there is the need to prescribe a distribution for $h(x,t)$ and $\varphi(x,t)$ within the spatial element $[x_1,x_2]$ and the temporal element $[t_1,t_2]$. A linear interpolation function in both space and time is proposed for the quantities, namely

$$h(x,t) \approx \Omega^{(m)}(\tau)\Omega_j(\zeta)h_j^{(m)}$$
$$= \Omega^{(1)}(\tau)[\Omega_1(\zeta)h_1^{(1)}+\Omega_2(\zeta)h_2^{(1)}] + \Omega^{(2)}(\tau)[\Omega_1(\zeta)h_1^{(2)}+\Omega_2(\zeta)h_2^{(2)}] \tag{5.31}$$

in which $\tau = (t-t_1)/\Delta t$ is a new coordinate system in the time element s whose origin is at t_1, and we shall refer to it as the *local temporal coordinate*, and the Ω's are the linear element shape functions in time and space which were earlier defined in the space dimension (see eq. (3.27)). We re-state them for the sake of completeness

$$\Omega_1(\zeta) = 1-\zeta;\quad \Omega_2(\zeta) = \zeta,\quad \Omega^{(1)}(\tau) = 1-\tau,\quad \Omega^{(2)}(\tau) = \tau \tag{5.32}$$

It should be noted that the Einstein summation convention has been applied to the two repeated indices (including the bracketed ones) in eq. (5.31). Using the definition of the local temporal coordinate, the following quantities in eq. (5.29) can now be more explicitly stated

$$G(x,t;x_i,t_2) = G(l(\zeta-\zeta_i),\Delta t(1-\tau);0,0) \tag{5.33a}$$

$$G^*(x,t;x_i,t_2) = -\frac{l(\zeta-\zeta_i)}{2K\Delta t(1-\tau)}G(l(\zeta-\zeta_i),\Delta t(1-\tau);0,0) \tag{5.33b}$$

$$G(x,t_1;x_i,t_2) = G(l(\zeta-\zeta_i),\Delta t;0,0) \tag{5.33c}$$

Using the relationships supplied by eqs. (5.30 - 5.33), eq (5.29) can be simplified further to give

$$\begin{aligned}
&-\lambda h_i^{(2)} + \Delta t\int_0^1 \Omega^{(m)}(\tau)h_2^{(m)}G^*(x_2,\Delta t(1-\tau);x_i,0)d\tau \; - \\
&\Delta t\int_0^1 \Omega^{(m)}(\tau)h_1^{(m)}G^*(x_1,\Delta t(1-\tau);x_i,0)]d\tau \; - \\
&\Delta t\int_0^1 \Omega^{(m)}(\tau)\varphi_2^{(m)}G(x_2,\Delta t(1-\tau);x_i,0)d\tau \; + \\
&\Delta t\int_0^1 \Omega^{(m)}(\tau)\varphi_1^{(m)}G(x_1,\Delta t(1-\tau);x_i,0)d\tau \; - \\
&-\frac{l}{K}\int_0^1 \Omega_j(\zeta)h_j^{(1)}G(l(\zeta-\zeta_i),\Delta t;0,0)d\zeta \; + \\
&\frac{l\Delta t}{K}\int_0^1\int_0^1 G(l(\zeta-\zeta_i),\Delta t(1-\tau);0,0)\Omega^{(m)}(\tau)\Omega_j(\zeta)f_j^{(m)}d\zeta d\tau = 0
\end{aligned} \tag{5.34}$$

Eq. (5.34) is split into two equations by considering the source point x_i located at x_1

$$\begin{aligned}
&-\frac{1}{2}h_1^{(2)} + \Delta t\int_0^1 \Omega^{(m)}(\tau)h_2^{(m)}G^*(l,\Delta t(1-\tau);0,0)d\tau \; - \\
&\Delta t\int_0^1 \Omega^{(m)}(\tau)\varphi_2^{(m)}G(l,\Delta t(1-\tau);0,0)d\tau \; + \\
&\Delta t\int_0^1 \Omega^{(m)}(\tau)\varphi_1^{(m)}G(0,\Delta t(1-\tau);0,0)d\tau \; - \\
&\frac{l}{K}\int_0^1 \Omega_j(\zeta)h_j^{(1)}G(l\zeta,\Delta t;0,0)d\zeta \; + \\
&\frac{l\Delta t}{K}\int_0^1\int_0^1 G(l\zeta,\Delta t(1-\tau);0,0)\Omega^{(m)}(\tau)\Omega_j(\zeta)f_j^{(m)}d\zeta d\tau = 0
\end{aligned} \tag{5.35a}$$

and then when the source point x_i is located at x_2

$$
\begin{aligned}
&-\frac{1}{2}h_2^{(2)} - \Delta t\int_0^1 \Omega^{(m)}(\tau)h_1^{(m)}G^*(-l,\Delta t(1-\tau);0,0)d\tau - \\
&\Delta t\int_0^1 \Omega^{(m)}(\tau)\varphi_2^{(m)}G(0,\Delta t(1-\tau);0,0)d\tau + \\
&\Delta t\int_0^1 \Omega^{(m)}(\tau)\varphi_1^{(m)}G(-l,\Delta t(1-\tau);0,0)d\tau - \\
&-\frac{l}{K}\int_0^1 \Omega_j(\zeta)h_j^{(1)}G(l(\zeta-1),\Delta t;0,0)d\zeta \\
&+\frac{l\Delta t}{K}\int_0^1\int_0^1 G(l(\zeta-1),\Delta t(1-\tau);0,0)\Omega^{(m)}(\tau)\Omega_j(\zeta)f_j^{(m)}d\zeta d\tau = 0
\end{aligned}
\tag{5.35b}
$$

where $h_2^{(2)}$, for instance, stands for the value of h at x_2 and at the current time. Eqs. (5.35a) and (5.35b) are combined in a matrix of the form

$$
\boxed{R_{ij}^{m}h_j^{(m)} + L_{ij}^{(m)}\varphi_j^{(m)} + T_{ij}^{(m)}f_j^{(m)} = 0}
\tag{5.36}
$$

in which

$$
R_{11}^{(2)} = R_{22}^{(2)} = -0.5 \tag{5.37a}
$$

$$
\begin{aligned}
R_{12}^{(2)} = R_{21}^{(2)} &= \Delta t\int_0^1 \Omega^{(2)}(\tau)G^*(l,\Delta t(1-\tau);0,0)d\tau \\
&= \frac{\sqrt{\gamma}}{2\sqrt{\pi}}\int_0^1 \frac{\tau}{(1-\tau)^{3/2}}\exp\left[\frac{-\gamma}{1-\tau}\right]d\tau = \frac{1}{\sqrt{\pi}}\left[\sqrt{\pi}(\frac{1}{2}+\gamma)erfc(\sqrt{\gamma})-\sqrt{\gamma}\exp(-\gamma)\right]
\end{aligned}
\tag{5.37b}
$$

$$
\begin{aligned}
L_{11}^{(2)} = -L_{22}^{(2)} &= \Delta t\int_0^1 \Omega^{(2)}(\tau)G(0,\Delta t(1-\tau);0,0)d\tau \\
&= \frac{l}{4\sqrt{\pi\gamma}}\int_0^1 \frac{\tau}{(1-\tau)^{1/2}}d\tau = -\frac{l}{3\sqrt{\pi\gamma}}
\end{aligned}
\tag{5.37c}
$$

$$L_{12}^{(2)} = -L_{21}^{(2)} = \Delta t\int_0^1 \Omega^{(2)}(\tau)G(l,\Delta t(1-\tau);0,0)d\tau$$
$$= \frac{l}{4\sqrt{\pi\gamma}}\int_0^1 \frac{\tau}{(1-\tau)^{1/2}}\exp\left[\frac{-\gamma}{1-\tau}\right]d\tau \tag{5.37d}$$
$$= \frac{l}{2\sqrt{\pi\gamma}}\left[-\sqrt{\pi\gamma}(1+\frac{2}{3}\gamma)erfc(\sqrt{\gamma}) + \frac{2}{3}(1+\gamma)\exp(-\gamma)\right]$$

$$R_{11}^{(1)} = R_{22}^{(1)} = -l\int_0^1 \Omega_1(\zeta)G(l\zeta,\Delta t;0,0)d\zeta = \frac{\sqrt{\gamma}}{\sqrt{\pi}}\int_0^1 (1-\zeta)\exp[-\gamma\zeta^2]d\zeta$$
$$= \frac{1}{2\sqrt{\pi}}\left\{\sqrt{\pi}erf(\sqrt{\gamma}) + \frac{1}{\sqrt{\gamma}}[\exp(-\gamma)-1]\right\} \tag{5.37e}$$

$$R_{12}^{(1)} = R_{21}^{(1)} = \Delta t\int_0^1 \Omega^{(1)}(\tau)G^*(l,\Delta t(1-\tau);0,0)d\tau - l\int_0^1 \Omega_2(\zeta)G(l\zeta,\Delta t;0,0)d\zeta$$
$$= \frac{\sqrt{\gamma}}{2\sqrt{\pi}}\left\{\int_0^1 \frac{1}{(1-\tau)^{1/2}}\exp\left[\frac{-\gamma}{1-\tau}\right]d\tau + 2\int_0^1 \zeta\exp[-\gamma\zeta^2]d\zeta\right\} \tag{5.37f}$$
$$= \frac{1}{\sqrt{\pi\gamma}}\left[\frac{1}{2} + (\gamma - \frac{1}{2})\exp(-\gamma)\right] - \gamma erfc(\sqrt{\gamma})$$

$$L_{11}^{(1)} = -L_{22}^{(1)} = \Delta t\int_0^1 \Omega^{(1)}(\tau)G(0,\Delta t(1-\tau);0,0)d\tau$$
$$= \frac{l}{4\sqrt{\pi\gamma}}\int_0^1 (1-\tau)^{1/2}d\tau = -\frac{l}{6\sqrt{\pi\gamma}} \tag{5.37g}$$

$$L_{12}^{(1)} = -L_{21}^{(1)} = \Delta t\int_0^1 \Omega^{(1)}(\tau)G(l,\Delta t(1-\tau);0,0)d\tau$$
$$= \frac{l}{4\sqrt{\pi\gamma}}\int_0^1 (1-\tau)^{1/2}\exp\left[\frac{-\gamma}{1-\tau}\right]d\tau \tag{5.37h}$$
$$= \frac{l\gamma^{3/2}}{6\sqrt{\pi\gamma}}\left[2\sqrt{\pi}erfc(\sqrt{\gamma}) + \frac{(1-2\gamma)}{\gamma^{3/2}}\exp(-\gamma)\right]$$

$$
\begin{aligned}
T_{11}^{(2)} = T_{22}^{(2)} &= l\Delta t \int_0^1\int_0^1 \Omega^{(2)}(\tau)\Omega_1(\zeta)G(l\zeta,\Delta t(1-\tau);0,0)d\tau d\zeta \\
&= -\frac{\Delta t\sqrt{\gamma}}{\sqrt{\pi}}\int_0^1\int_0^1 \frac{(1-\zeta)\tau}{(1-\tau)^{1/2}}\exp\left[\frac{-\gamma\zeta^2}{1-\tau}\right]d\tau d\zeta = \frac{\Delta t}{2\sqrt{\pi}}\left\{\sqrt{\pi}\int_1^0 (1-\tau)erf\left(\frac{\sqrt{\gamma}}{\sqrt{\tau}}\right)d\tau + \right. \\
&\left. \frac{2\gamma}{15}\left[\frac{2\gamma^2+4\gamma-1}{\gamma^{3/2}}\exp(-\gamma) - \sqrt{\pi}(5+2\gamma)erfc(\sqrt{\gamma})\right]\right\}
\end{aligned}
\tag{5.37i}
$$

$$
\begin{aligned}
T_{12}^{(2)} = T_{21}^{(2)} &= l\Delta t \int_0^1\int_0^1 \Omega^{(2)}(\tau)\Omega_2(\zeta)G(l\zeta,\Delta t(1-\tau);0,0)d\tau d\zeta \\
&= -\frac{\Delta t\sqrt{\gamma}}{\sqrt{\pi}}\int_0^1\int_0^1 \frac{\zeta\tau}{(1-\tau)^{1/2}}\exp\left[\frac{-\gamma\zeta^2}{1-\tau}\right]d\tau d\zeta \\
&= \frac{2\Delta t}{15\sqrt{\pi\gamma}}\left[-1+\gamma^{3/2}\left(\frac{(1-2\gamma^2-4\gamma)}{\gamma^{3/2}}\exp(-\gamma)+\sqrt{\pi}(5+2\gamma)erfc(\sqrt{\gamma})\right)\right]
\end{aligned}
\tag{5.37j}
$$

$$
\begin{aligned}
T_{11}^{(1)} = T_{22}^{(1)} &= l\Delta t \int_0^1\int_0^1 \Omega^{(1)}(\tau)\Omega_1(\zeta)G(l\zeta,\Delta t(1-\tau);0,0)d\tau d\zeta \\
&= -\frac{\Delta t\sqrt{\gamma}}{\sqrt{\pi}}\int_0^1\int_0^1 (1-\zeta)(1-\tau)^{1/2}\exp\left[\frac{-\gamma\zeta^2}{1-\tau}\right]d\tau d\zeta \\
&= \frac{\Delta t}{2\sqrt{\pi}}\left\{\sqrt{\pi}\int_1^0 \tau erf\left(\frac{\sqrt{\gamma}}{\sqrt{\tau}}\right)d\tau + \frac{2}{5\sqrt{\gamma}} + \right. \\
&\left. \frac{2}{15}\left[\frac{(2\gamma-4\gamma^2-3)}{\sqrt{\gamma}}\exp(-\gamma)+4\sqrt{\pi}\gamma^2 erfc(\sqrt{\gamma})\right]\right\}
\end{aligned}
\tag{5.37k}
$$

$$
\begin{aligned}
T_{12}^{(1)} = T_{21}^{(1)} &= l\Delta t \int_0^1\int_0^1 \Omega^{(1)}(\tau)\Omega_2(\zeta)G(l\zeta,\Delta t(1-\tau);0,0)d\tau d\zeta \\
&= -\frac{\Delta t\sqrt{\gamma}}{\sqrt{\pi}}\int_0^1\int_0^1 \zeta(1-\tau)^{1/2}\exp\left[\frac{-\gamma\zeta^2}{1-\tau}\right]d\tau d\zeta \\
&= \frac{\Delta t}{5\sqrt{\pi\gamma}}\left\{-1+\frac{1}{3}\left[(3+4\gamma^2-2\gamma)\exp(-\gamma)-4\sqrt{\pi\gamma^5}erfc(\sqrt{\gamma})\right]\right\}
\end{aligned}
\tag{5.37l}
$$

where

$$\gamma = \frac{l^2}{(4K\Delta t)} \tag{5.38a}$$

and

$$erf(x) = \frac{2}{\sqrt{\pi}}\int_0^x \exp(-s^2)ds; \quad erfc(x) = 1 - erf(x) \tag{5.38b}$$

are, respectively, the *error function* and the *complementary error function.* The line integrals in eqs. (5.37i) and (5.37k) are evaluated by standard Gauss quadrature. (Here is the first time we have encountered an integral that cannot be evaluated analytically, and that is a reflection of the complexity of the free-space Green's function that has been employed in this formulation.) We restate that eq. (5.36) is written for a spatial element with end nodal coordinates x_1 and x_2 and for a temporal interval between t_1 and t_2. That equation is the discrete analog of the differential equation. It provides the essential information required to solve the temporal diffusion equation by the Green element formulation which uses the fundamental solution of the full diffusion equation. This current formulation is referred to as the *transient Green element* (**TGE**) formulation.

It is worth noting that the coefficients of the elemental matrices are more complicated to evaluate than those of the quasi-steady Green element **QSGE** model and the reason for this is simply because of the complexity of the free-space Green's function. Without carrying out any actual CPU comparison of the **QSGE** and **TGE** models, it is obvious that the latter model will be less efficient than the former. The only advantage then of the **TGE** model may be in terms of accuracy. The next section on numerical examples will certainly confirm whether or not **TGE** is a more accurate formulation than **QSGE**.

5.6 Stability Characteristics of GEM for the Diffusion Equation

In the previous sections we derived two time-marching formulations of the Green element method for the transient linear diffusion equation. Essentially the numerical process is one in which the known data at the beginning of the time step are updated to new values at the end of the time step according to the replication of the differential equation. For a well-posed initial-value problem, the solution at any time level is expected to be bounded, and that can only be achieved by a numerical scheme which does not allow the solution and accumulated numerical errors to grow out of bounds.

The method employed to analyze the stability characteristics of the two Green element formulations (**QSGE** and **TGE**) is that proposed by Stone and Brian [4] which was used for the evaluation of their finite difference formulation, and has also been employed by Leendertse [5,6], Gray and Pinder [7] and Gray and Lynch [8] for assessing the stability characteristics of various numerical schemes for surface and subsurface flow problems. In this approach, the general solution of the diffusion process is represented as an infinite series of Fourier wave components, and the stability characteristics of the numerical scheme are assessed by comparing the amounts by which the amplitude and phase angle of each wave component are modified after each time step in relation to their theoretical values. The ideal situation is for the numerical scheme to modify each Fourier wave component in amplitude and phase angle by the same amount as their corresponding analytic values.

We may consider the distribution of the dependent variable at the initial time as comprising the true solution (which is usually not known *apriori*) and numerical errors that have accumulated up to that time level. Speaking of numerical errors, their sources are round off errors whose magnitudes depend on the precision of the computing machine, errors from approximating the differential equation by a system of discrete equations, and errors which the analyst introduces by not properly programming the numerical scheme. Whatever is the functional distribution of the dependent variable at initial time, it should have an upper bound, and as such one view of numerical stability stems from the ability of the scheme to keep that initial distribution within bounds as numerical calculations progress from one time level to another. A numerical scheme is thus said to be stable if the ratio of the absolute value of the amplitude of each wave component at the end of a time interval to its correspond value at the beginning of the time step is less than unity. However, assessing stability from only this angle does not provide a complete picture of the stability characteristics of a numerical method.

We are also interested in knowing how correctly the numerical scheme propagates each Fourier wave component in speed and amplitude in comparison to its theoretical propagation values. A scheme is said to be *dissipative* if it is unable to modify the amplitude of each Fourier wave component after each time step by the same theoretical value. Every numerical scheme faces its stiffest challenge when the solution profile has large gradients or steep fronts that are predominately characterized by the short Fourier wavelengths. If numerical dissipation is large for these short wavelengths, then additional artificial diffusion is introduced by the numerical scheme so that these steep gradients become smeared. The effects of the inability of a numerical scheme to correctly modify the phase angle of each wave component after each time step are referred to as *dispersive effects*. Dispersive effects cause the small wave lengths to be out phase with the longer wave lengths, giving rise to artificial oscillation of the numerical solution.

We begin our analysis by restating the differential operator of the diffusion process with no external forcing function

$$K\frac{\partial^2 h}{\partial x^2} - \frac{\partial h}{\partial t} = 0 \tag{5.45}$$

We propose a general solution in terms of an infinite series of *Fourier-wave* components of the form

$$h = \sum_{n=-\infty}^{\infty} \Phi_n \exp[\hat{i}(\sigma_n x + \beta_n t)] \tag{5.46}$$

where for each n-th Fourier wave component, $\beta_n = 2\pi/T_n$ is the *temporal frequency*, T_n is the *period*, $\sigma_n = 2\pi/L_n$ is the *spatial frequency* or *wave number*, and L_n is the wave length, and $\hat{i}$ is the imaginary representation of a complex variable. Because eq. (5.45) is linear, we can take each term of the equation one at a time. Considering only one term of the series solution eq. (5.46) can be expressed as

$$h = h_t \sim \Phi \exp[\hat{i}(\sigma x + \beta t)] \tag{5.47}$$

Note that we have dropped the subscript n from the parameters of eq. (5.46) since we are considering only one term of the series. The relationship between the temporal frequency β and the wave number σ can be obtained by introducing eq. (5.47) into the differential equation (5.45). When that is done the following dispersion relationship is obtained

$$\beta = \hat{i}K\sigma^2 \tag{5.48}$$

Eq. (5.48) is now substituted into the general solution of eq. (5.47) to give

$$h = h_t \sim \Phi \exp(\hat{i}\sigma x)\exp(-K\sigma^2 t) \tag{5.49}$$

The expression given by eq. (5.49) indicates that each Fourier wave component is a *standing* or *stationary wave* whose amplitude is modified in time by the second exponential function. In other words, after a time step Δt each Fourier wave retains its shape or has no phase change but has an amplitude of $\exp(-K\sigma^2\Delta t)$ less than its original value. Eq. (5.49) gives the solution at the beginning of the time level t, while after a time step Δt is

$$h_{t+\Delta t} \sim \Phi \exp(\hat{i}\sigma x)\exp[-K\sigma^2(t+\Delta t)] \tag{5.50}$$

From eq. (5.50), it means that the modulus of the *theoretical amplification* of each Fourier wave after one time step is given by

$$|g_T| = \frac{|h_{t+\Delta t}|}{|h_t|} = \exp(-K\sigma^2\Delta t) \tag{5.51}$$

In addition, the *theoretical phase change* of each Fourier component, which is denoted by θ_T, is given by

$$\exp(\hat{i}\theta_T) = \frac{g_T}{|g_T|} \tag{5.52}$$

The *theoretical amplitude modification factor* or theoretical amplification g_T has no imaginary component so that $\theta_T = 0$. Our next step is to examine the characteristics of each Fourier wave component when the two Green element formulations approximate the differential equation by their discrete analogs. To achieve this, we shall consider a flow domain $[x_0,x_L]$ that is discretized into uniform linear elements each of which has a length l. The numerical replication of the differential equation is written for an internal node x_i sandwiched between two other nodes at x_{i-1} and x_{i+1} (see Figure 5.3). We shall first derive the *numerical amplitude modification factor* or *numerical amplification* g_N, and the *numerical phase change* θ_N for the Quasi-steady Green element (**QSGE**) formulation of eq. (5.45). To do this, we recognize that two quantities h and φ are computed at each node by the numerical scheme, so that there is the need to generate two discrete equations at x_i. The first discrete equation is obtained by considering the segment $[x_{i-1},x_i]$ and the second by considering the segment $[x_i,x_{i+1}]$. The discrete equation at any source node for the **QSGE** is taken from eq. (5.22) and restated below without any external forcing function $f(x,t)$

$$\alpha\left[R_{ij}h_j^{(2)} + L_{ij}\varphi_j^{(2)}\right] + \omega\left[R_{ij}h_j^{(1)} + L_{ij}\varphi_j^{(1)}\right] + \frac{T_{ij}}{K}\left(\frac{h_j^{(2)} - h_j^{(1)}}{\Delta t}\right) = 0, \quad i,j=1,2, \quad 0 \le \alpha \le 1 \tag{5.53}$$

The expressions for the element matrices for a uniformly discretized domain are

$$R_{ij} = (-1)^{(i+j-1)}, \; L_{ij} = l\begin{bmatrix} 1 & -2 \\ 2 & -1 \end{bmatrix}, \text{ and } \; T_{ij} = \frac{l^2}{6}\begin{bmatrix} 4 & 5 \\ 5 & 4 \end{bmatrix} \tag{5.54}$$

We rearrange eq. (5.53) so that coefficients of similar quantities are grouped together. That results in

$$[\alpha R_{ij} + \frac{T_{ij}}{K\Delta t}]h_j^{(2)} + \alpha L_{ij}\varphi_j^{(2)} + [\omega R_{ij} - \frac{T_{ij}}{K\Delta t}]h_j^{(1)} + \omega L_{ij}\varphi_j^{(1)} = 0, \; i,j=1,2$$

or

$$P_{ij}^{(m)}h_j^{(m)} + N_{ij}^{(m)}\varphi_j^{(m)} = 0, \quad i,j,m=1,2 \tag{5.55}$$

where

$$P_{ij}^{(2)} = \alpha R_{ij} + \frac{T_{ij}}{K\Delta t}, \quad P_{ij}^{(1)} = \omega R_{ij} - \frac{T_{ij}}{K\Delta t} \tag{5.56a}$$

$$N_{ij}^{(2)} = \alpha L_{ij}, \quad N_{ij}^{(1)} = \omega L_{ij} \tag{5.56b}$$

The first discrete equation obtained for the element $[x_{i-1},x_i]$ with the source node at x_i is

$$P_{22}^{(m)}h_i^{(m)} + P_{21}^{(m)}h_{i-1}^{(m)} + N_{22}^{(m)}\varphi_i^{(m)} + N_{21}^{(m)}\varphi_{i-1}^{(m)} = 0 \tag{5.57a}$$

and the second discrete equation comes from evaluating the integral equation within the element $[x_i,x_{i+1}]$ with the source node at x_i

$$P_{11}^{(m)}h_i^{(m)} + P_{12}^{(m)}h_{i+1}^{(m)} + N_{11}^{(m)}\varphi_i^{(m)} + N_{12}^{(m)}\varphi_{i+1}^{(m)} = 0 \tag{5.57b}$$

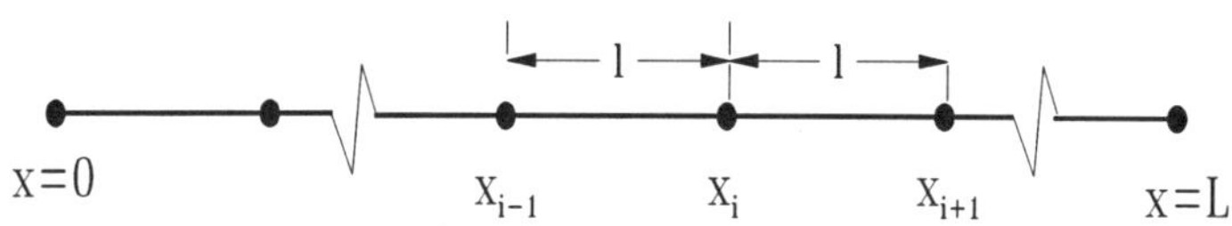

Figure 5.3 Spatial uniform discretization for stability analysis.

It should be noted that the spatial derivative of h, that is φ, also satisfies the differential eq. (5.45) so that a general solution like that of eq (5.49) can be written for φ, that is

$$\varphi_t \sim X\exp(\hat{i}\sigma x)\exp(-K\sigma^2 t) \tag{5.58}$$

From the prescribed general solutions of eqs. (5.49) and (5.58), the following relations can be deduced

$$h_{i+1}^{(2)} = h_i^{(2)}\exp(\hat{i}\sigma l) = g_N h_i^{(1)}\exp(\hat{i}\sigma l) \tag{5.59a}$$

$$h_{i-1}^{(2)} = h_i^{(2)}\exp(-\hat{i}\sigma l) = g_N h_i^{(1)}\exp(-\hat{i}\sigma l) \tag{5.59b}$$

$$\varphi_{i+1}^{(2)} = \varphi_i^{(2)}\exp(\hat{i}\sigma l) = g_N\varphi_i^{(1)}\exp(\hat{i}\sigma l) \tag{5.59c}$$

$$\varphi_{i-1}^{(2)} = \varphi_i^{(2)}\exp(-\hat{i}\sigma l) = g_N\varphi_i^{(1)}\exp(-\hat{i}\sigma l) \tag{5.59d}$$

Introducing the above relations into the system of discrete element equations of (5.57a) and (5.57b), we obtain the numerical stability matrix equation which is given by

$$\begin{bmatrix} a_{11}g_N+b_{11} & a_{12}g_N+b_{12} \\ a_{21}g_N+b_{21} & a_{22}g_N+b_{22} \end{bmatrix} \begin{Bmatrix} h_i^{(1)} \\ \varphi_i^{(1)} \end{Bmatrix} = \begin{Bmatrix} 0 \\ 0 \end{Bmatrix} \tag{5.60}$$

and where, for the **QSGE** model, the elements of the matrix are

$$\begin{aligned} a_{11} &= a_{11}^r + \hat{i}a_{11}^i = P_{22}^{(2)} + P_{21}^{(2)}\cos(\sigma l) - \hat{i}P_{21}^{(2)}\sin(\sigma l) \\ &= -\alpha Cr^2 + \frac{2}{3} + [\alpha Cr^2 + \frac{5}{6}]\cos(\sigma l) \; - \; \hat{i}[\alpha Cr^2 + \frac{5}{6}]\sin(\sigma l) \end{aligned} \tag{5.61a}$$

$$\begin{aligned} a_{12} &= a_{12}^r + \hat{i}a_{12}^i = N_{22}^{(2)} + N_{21}^{(2)}\cos(\sigma l) - \hat{i}N_{21}^{(2)}\sin(\sigma l) \\ &= l\left[\alpha Cr^2[-1 + 2\cos(\sigma l)] - \hat{i}2\alpha Cr^2\sin(\sigma l)\right] \end{aligned} \tag{5.61b}$$

$$a_{21} = a_{21}^r + \hat{i}a_{21}^i = P_{11}^{(2)} + P_{12}^{(2)}\cos(\sigma l) + \hat{i}P_{12}^{(2)}\sin(\sigma l) = a_{11}^r - \hat{i}a_{11}^i \tag{5.61c}$$

$$a_{22} = a_{22}^r + \hat{i}a_{22}^i = N_{11}^{(2)} + N_{12}^{(2)}\cos(\sigma l) + \hat{i}N_{12}^{(2)}\sin(\sigma l) = -a_{12}^r + \hat{i}a_{12}^i \tag{5.61d}$$

$$\begin{aligned} b_{11} &= b_{11}^r + \hat{i}b_{11}^i = P_{22}^{(1)} + P_{21}^{(1)}\cos(\sigma l) - \hat{i}P_{21}^{(1)}\sin(\sigma l) \\ &= -\omega Cr^2 - \frac{2}{3} + [\omega Cr^2 - \frac{5}{6}]\cos(\sigma l) - \hat{i}[\omega Cr^2 - \frac{5}{6}]\sin(\sigma l) \end{aligned} \tag{5.61e}$$

$$\begin{aligned} b_{12} &= b_{12}^r + \hat{i}b_{12}^i = N_{22}^{(1)} + N_{21}^{(1)}\cos(\sigma l) - \hat{i}N_{21}^{*}\sin(\sigma l) \\ &= l\left[\omega Cr^2[-1 + 2\cos(\sigma l)] - \hat{i}2\omega Cr^2\sin(\sigma l)\right] \end{aligned} \tag{5.61f}$$

$$b_{21} = b_{21}^r + \hat{i}b_{21}^i = P_{11}^{(1)} + P_{12}^{(1)}\cos(\sigma l) + \hat{i}P_{12}^{(1)}\sin(\sigma l) = b_{11}^r - \hat{i}b_{11}^i \tag{5.61g}$$

$$b_{22} = b_{22}^r + \hat{i}b_{22}^i = N_{11}^{(1)} + N_{12}^{(1)}\cos(\sigma l) + \hat{i}N_{12}^{(1)}\sin(\sigma l) = -b_{12}^r + \hat{i}b_{12}^i \tag{5.61h}$$

The superscripts r and i of the matrix elements in eqs. (5.61a - h) denote the real and imaginary components of those elements, and the dimensionless parameter Cr stands for the *Courant number* which has the expression

$$Cr = \frac{\sqrt{K\Delta t}}{l} \tag{5.62}$$

The Courant number, for a given value of K, gives the relative magnitude of the temporal and spatial discretizations. We can now eliminate $h_i^{(1)}$ and $\varphi_i^{(1)}$ from eq. (5.60) to obtain an expression for the numerical amplification g_N. It is given by the quadratic equation

$$Ag_N^2 + Bg_N + C = 0 \tag{5.63}$$

where

$$A = a_{22}a_{11} - a_{12}a_{21} \tag{5.64a}$$

$$B = a_{22}b_{11} + a_{11}b_{22} - a_{12}b_{21} - a_{21}b_{12} \tag{5.64b}$$

$$C = b_{11}b_{22} - b_{12}b_{21} \tag{5.64c}$$

The expression for g_N is readily obtained from

$$g_N = \frac{-B \pm \sqrt{B^2 - 4AC}}{2A} \tag{5.65}$$

Two values of g_N are obtained from eq. (5.65). One of them is inadmissible because it is either unacceptably too large or too small, and, for most values of Courant number, has uniform values for all Fourier wave components. For the **QSGE** formulation it is the value of g_N with positive square root that is inadmissible. The numerical phase change after a time step Δt is given as

$$\exp(\hat{i}\theta_N) = \frac{g_N}{|g_N|} \tag{5.66}$$

The reader needs to confirm that g_N has no imaginary component so that the phase change for all Fourier wave components produced by the numerical scheme is the same as the theoretical phase change. That is $\theta_N = \theta_T = 0$. A numerical scheme is said to be stable if

$$|g_N| < 1 \tag{5.67}$$

The computed values of g_N from eq. (5.65) with values of the temporal weighting factor $\alpha = 0.5, 0.67, 0.75$ and 1 are presented in Figures 5.4 and 5.5 for two values of $Cr = 0.1$ and $Cr = 1$. These figures show that the **QSGE** formulation is unconditionally stable when $\alpha \geq 0.5$. In other words, the formulation will not permit the numerical solution to grow in an unbounded fashion for $\alpha \geq 0.5$. Though the formulation is stable for the given values of α and Cr, we still need to compare the numerical amplitude modification g_N to its theoretical value g_T. The ratio of the moduli of both parameters gives the *amplification factor* T

$$T = \frac{|g_N|}{|g_T|} = \frac{|g_N|}{\exp(-K\sigma^2\Delta t)} = \frac{|g_N|}{\exp\left[-4\pi^2 Cr^2\left(\frac{l}{L}\right)^2\right]} \tag{5.68}$$

When $T > 1$, it does not mean that the numerical scheme is unstable, but that it amplifies the wave amplitude by a larger amount than the theoretical value, and conversely when $T < 1$. Ideally T should have unit value. Figure 5.6 shows the plot of T versus L/l for $Cr = 0.1$, and the results show that T is about unity for all Fourier wave components, and its value is the same for all values of $\alpha \geq 0.5$. Thus for small values of Courant number the numerical scheme correctly modifies the amplitude of all harmonics of Fourier wave components. However, when $Cr = 1$, T depends significantly on the value of α, and the plot of T versus L/l in Figure 5.7 shows that none of the **QSGE** schemes correctly modifies the amplitudes of the harmonics with small wave lengths. The **QSGE** scheme with $\alpha = 0.5$ correctly modifies the harmonics with wave length greater than 6.5 times the spatial element size ($L > 6.5l$), while the scheme with $\alpha = 0.67$ correctly modifies the wave amplitude when L is greater than 4.5 times the element spatial size l, that with $\alpha = 0.75$ reproduces the theoretical amplitude modification when $L > 4l$, and that with $\alpha = 1$ when $L > 7l$. These results do indicate that when the Courant number is large or a large time step is adopted in the numerical computations, then any steep gradient of the solution at any time may not be damped or diffused throughout the flow region as quickly as its theoretical rate of diffusion, and as such could maintain its form much longer than the physical process allows. To avoid this, a smaller time step could be used. Although all the schemes are dissipative with respect to the harmonics of small wavelengths, the **QSGE** model with $\alpha = 0.67$ and 0.75 provide the least dissipative effects.

We now turn our attention to the **TGE** formulation, and examine its stability characteristics. The procedure for establishing this is similar to that adopted for the **QSGE** formulation, except that we do not have to consider various temporal difference approximations. The system of discrete element equations for the **TGE** formulation without external recharge comes from eq. (5.36). That is

$$R_{ij}^{m} h_j^{(m)} + L_{ij}^{(m)} \varphi_j^{(m)} = 0 \tag{5.69}$$

Comparing eq. (5.69) and eq. (5.55) we observe that they are of the same form. That means that the expression for g_N for the **TGE** formulation can be obtained from the solution of a quadratic equation that is similar to eq. (5.63). Two roots of g_N are obtained; the one with the negative square root is inadmissible. We should be somewhat curious about the controlling parameter on which depends the value of g_N

for the **TGE** formulation. A clue comes from examining the expressions of $R_{ij}^{(m)}$ and $L_{ij}^{(m)}$ in eqs. (5.37a) through (5.37h). There we observe that the element matrices depend on γ whose expression is given by eq. (5.38a) and is related to the Courant number of eq. (5.62) in the following manner

$$\gamma = \frac{l^2}{4K\Delta t} = \frac{1}{4Cr^2} \tag{5.70}$$

Thus, the numerical amplitude modification factor g_N for the **TGE** formulation depends only on the Courant number. Figures 5.4 through 5.7 show plots of g_N versus L/l and T versus L/l for Courant values of 0.1 and 1. The formulation is unconditionally stable. However, when $Cr = 0.1$, it damps the amplitudes of the small wavelength harmonics much more than that of the **QSGE** formulation, and when $Cr = 1$ it is only able to reproduce the correct Fourier wave amplitudes with wavelengths $L \geq 5l$. This implies that, as with the **QSGE** formulation, solution profiles with steep gradients may pose some computational difficulties if large time steps or large values of the Courant number are used in the numerical calculations since the small wavelength harmonics, that dominate those profiles, may not be dissipated as quickly as theoretically expected.

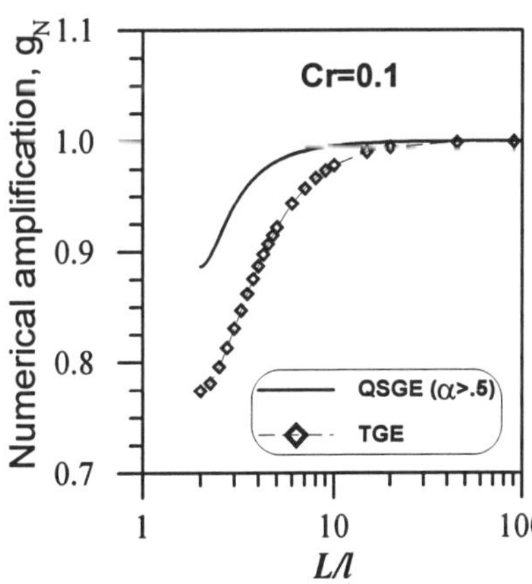

Figure 5.4 Variations of numerical amplification when $Cr = 0.1$

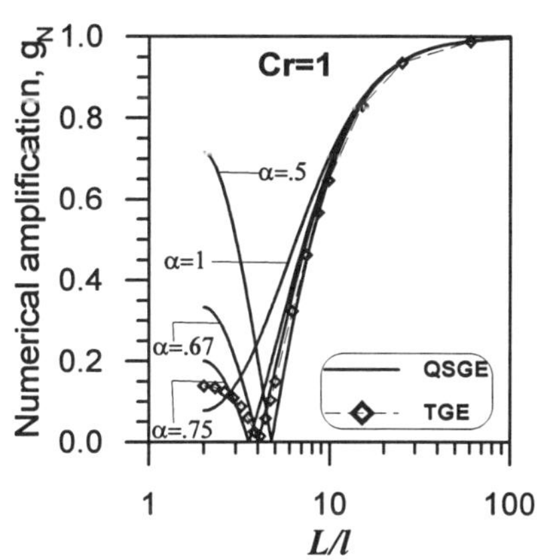

Figure 5.5 Variations of numerical amplification when $Cr = 1.0$

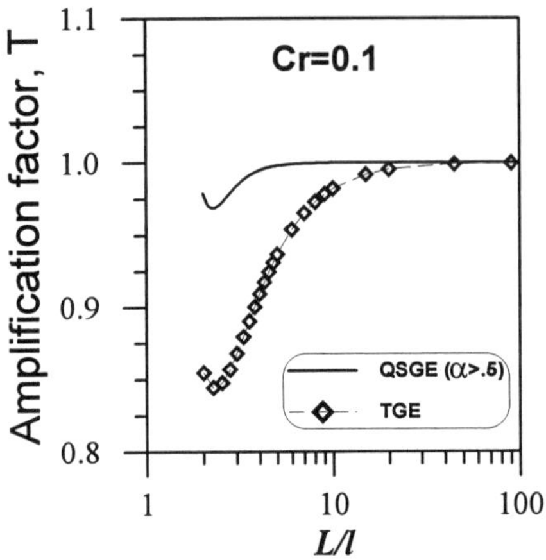

Figure 5.6: Variations of amplification factor for $Cr = 0.1$

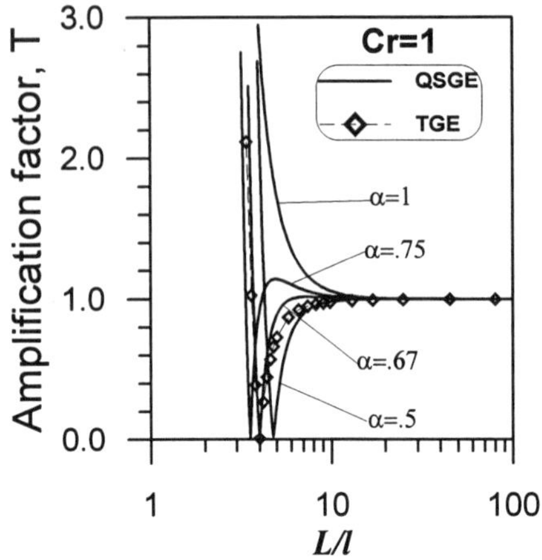

Figure 5.7 Variations of amplification factor for $Cr = 1.0$

5.7 Numerical Examples of Transient Linear Diffusion

As we had done in previous chapters, the performance of the two Green element models proposed in this chapter is examined using some numerical examples. Three numerical examples of transient linear diffusion for which analytical solutions are available are used to demonstrate the capabilities of the models. Their exact solutions serve as bench mark results in assessing the accuracy of the numerical solutions.

Example 1

Our first example is one of heat conduction in a bar of unit length in which the temperatures at both ends are maintained at certain values (steady Dirichlet boundary conditions). The problem is described by

$$\frac{\partial^2 h}{\partial x^2} = \frac{\partial h}{\partial t}, \quad h(0,t) = 1, \; h(1,t) = 0, \; t>0 \text{ and } h(x,0) = 0 \tag{5.71}$$

Its exact solution is derived by method of Laplace transform, and given by

$$h(x,t) = 1 - x + \frac{2}{\pi}\sum_{n=1}^{\infty} \frac{(-1)^n}{n} \exp(-n^2\pi^2 t)\sin[n\pi(1-x)] \tag{5.72}$$

Example 2

Our second example is similar to that of the first, except that unit magnitude of flux of heat is maintained at the ends of the bar (Neumann boundary conditions). The problem is thus described by

$$\frac{\partial^2 h}{\partial x^2} = \frac{\partial h}{\partial t}, \quad \varphi(0,t) = \varphi(1,t) = 1, \quad t>0 \text{ and } h(x,0) = 0 \tag{5.73}$$

Its exact solution, also obtained by the method of Laplace transform, is given by

$$h(x,t) = -x+\frac{1}{2} - \frac{4}{\pi^2}\sum_{n=1}^{\infty} \frac{1}{(2n-1)^2}\exp[-(2n-1)^2\pi^2 t]\cos[(2n-1)\pi x] \tag{5.74}$$

Example 3

In this example one end of the unit length of the bar is subjected to a sinusoidal temperature variation (transient Dirichlet boundary condition), while the other is maintained at zero temperature. Our problem is described by

$$\frac{\partial^2 h}{\partial x^2} = \frac{\partial h}{\partial t}, \; h(0,t) = \sin(2\pi t), \; h(1,t) = 0, \; t>0 \text{ and } h(x,0) = 0 \tag{5.75}$$

Its exact solution is derived by method of Laplace transform, and given by

$$\begin{aligned} h(x,t) = \frac{1}{\mu_3}&\Big[\sin(2\pi t)\big(\sinh\mu_1\cos\mu_1\sinh\mu_2\cos\mu_2 + \sin\mu_1\cosh\mu_1\sin\mu_2\cosh\mu_2\big) + \\ &\cos(2\pi t)\big(\sin\mu_1\cosh\mu_1\sinh\mu_2\cos\mu_2 - \sinh\mu_1\cos\mu_1\sin\mu_2\cosh\mu_2\big)\Big] - \\ &4\sum_{n=1}^{\infty}(-1)^n\frac{n}{n^4\pi^2+4}\exp(-n^2\pi^2 t)\sin[n\pi(1-x)] \end{aligned} \tag{5.76}$$

in which

$$\mu_1 = \sqrt{\pi}(1-x), \; \mu_2 = \sqrt{\pi}, \; \mu_3 = \sinh^2\mu_2\cos^2\mu_2 + \sin^2\mu_2\cosh^2\mu_2 \tag{5.77}$$

The Green element calculations of the three transient linear diffusion examples are carried out on a 10-element uniformly discretized domain so that the spatial element size is 0.1, and a uniform time step of 0.05 is adopted for the first two examples, and 0.025 for the third example. Since the performance of the **QSGE** formulation also depends on the value of the temporal weighting factor α, our first set of numerical experiments is designed to assess the value of α that gives the most accurate results. We evaluate the accuracy of the numerical solutions from the **QSGE** model using values of $\alpha = 0.5$, 0.67, 0.75, and 1.0. (Because of the oscillations of the numerical solutions of example 1 when $\alpha = 0.5$ (Crank-Nicholson scheme), we did not examine the accuracy of that scheme further.) To assess the solutions for h and φ at any time level, the root-mean-square-deviations between the numerical and exact solutions are computed, and these are represented as the error in the accompanying plots. The error plots for these schemes are presented in Figures 5.8a and 5.8b for example 1, Figures 5.9a and 5.9b for example 2, and in Figures 5.10a and 5.10b for example 3. An overall assessment of these plots indicates that the scheme with $\alpha = 0.67$ gives optimal results. A comparison of the **QSGE** model with $\alpha = 0.67$ is now done against that of the **TGE** model using those same plots of Figs 5.8a, b; 5.9a, b; and 5.10a, b. The **TGE** model has an advantage in terms of accuracy over the **QSGE** model.

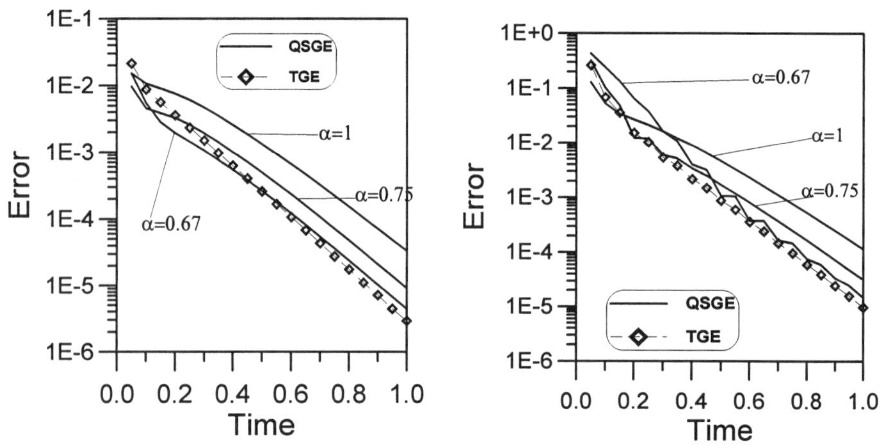

(a) **(b)**

Figure 5.8 Error of **GEM** solutions to example 1 — **(a)** for $h(x,t)$, **(b)** for $\varphi(x,t)$.

5.8 Nonlinear Diffusion

So far we have only considered the case where the diffusivity or hydraulic conductivity K is a constant. However, there are many engineering applications where the K depends also on the primary variable h as in flow in unsaturated aquifers (which is addressed in greater detail in chapter 8), heat conduction in nonlinear materials, *etc*. Under such conditions application of the boundary element method is completely ruled out, but we are still in a position to apply the Green element method and this is one of the strengths of the method.

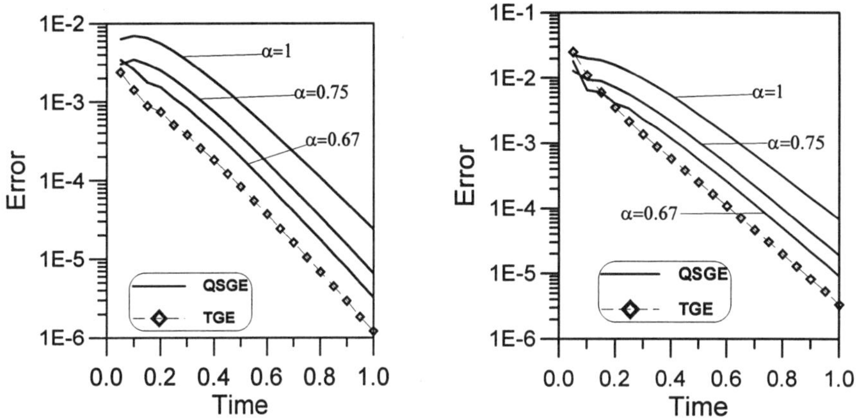

(a) **(b)**

Figure 5.9 Error of **GEM** solutions to example 2 — **(a)** for $h(x,t)$, **(b)** for $\varphi(x,t)$.

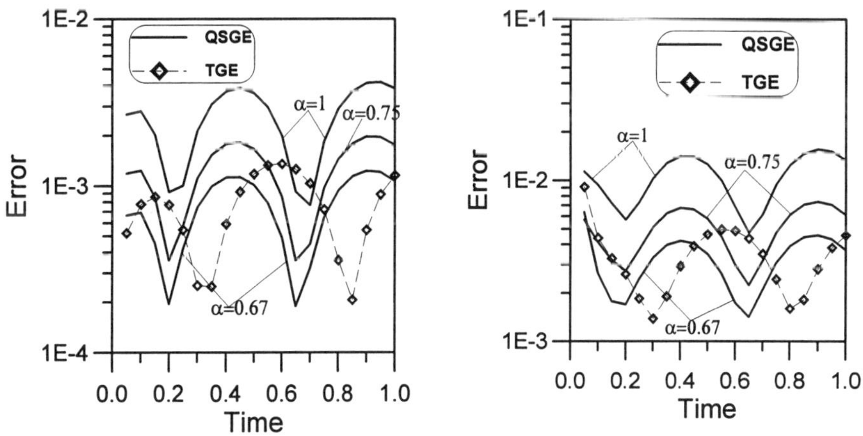

(a) **(b)**

Figure 5.10 Error of **GEM** solutions to example 3 — **(a)** for $h(x,t)$, **(b)** for $\varphi(x,t)$.

We restate our differential equation

$$\frac{\partial}{\partial x}\left(K(h)\frac{\partial h}{\partial x} \right) = \frac{\partial h}{\partial t} + f(x,t) \tag{5.78}$$

The relationship between K and h is provided by an equation commonly referred to as the *constitutive relations*. You recall that when we applied the Green element method to the stationary version of eq. (5.78) in chapter 3 we came up with three formulations or models which were a reflection of various levels of approximation of the differential equation by the method. There is nothing to suggest that those same three formulations cannot be applicable to the transient equation. Because the Green element equations are derived from the integral representation of the boundary element theory within a typical element, medium parameters like K are approximated in the same way as the primary variable. It is for this reason that even when K is dependent on the primary variable, it does not pose any computational difficulty to **GEM**.

Instead of pursuing the derivation of three similar Green element models for eq. (5.78) along the lines of **GESNH-1**, **GESNH-2**, and **GESNH-3** of chapter 3, we shall only adopt one of them. We elect to proceed with the formulation which is similar to the **GESNH-2** model, and leave the derivation of the other two models as an exercise to the reader.

Eq. (5.78) is expressed in another form as

$$\frac{\partial^2 h}{\partial x^2} = \frac{1}{K}\left[-\frac{\partial K}{\partial x}\frac{\partial h}{\partial x} + \frac{\partial h}{\partial t} + f(x)\right] \tag{5.79}$$

We use the complementary differential equation (2.20) and Green's identity to transform the differential equation (5.79) to an integral one which applies to a typical element $[x_1,x_2]$. That integral representation is

$$\begin{aligned}&-\lambda h(x_i,t) + G^*(x_2,x_i)h(x_2,t) - G^*(x_1,x_i)h(x_1,t)\\&- G(x_2,x_i)\varphi(x_2,x_i) + G(x_1,x_i)\varphi(x_1,t) +\\&\int_{x_1}^{x_2}\frac{G(x,x_i)}{K}\left[-\frac{\partial K}{\partial x}.\varphi(x,t) + \frac{\partial h(x,t)}{\partial t} + f(x,t)\right]dx = 0\end{aligned} \tag{5.80}$$

We approximate the functional quantities ln$K$ or Θ, $1/K$ or Ψ, $\partial h/\partial x$ or φ, and $f(x)$ by linear element shape functions, namely

$$\ln K \equiv \Theta \approx \Omega_j(\zeta)\Theta_j(t) \tag{5.81a}$$

$$\frac{1}{K} \equiv \Psi \approx \Omega_j(\zeta)\Psi_j(t) \tag{5.81b}$$

$$\frac{\partial h}{\partial x} \equiv \varphi \approx \Omega_j(\zeta)\varphi_j(t) \tag{5.81c}$$

$$f(x,t) \equiv f(\zeta,t) \approx \Omega_j(\zeta)f_j(t) \tag{5.81d}$$

Furthermore

$$\frac{\partial \Theta}{\partial x} = \frac{d\zeta}{dx}\frac{\partial \Theta}{\partial \zeta} \approx \frac{\Theta_2 - \Theta_1}{l} \text{ and } \frac{\partial h}{\partial t} \approx \Omega_j \frac{dh_j}{dt} \tag{5.82}$$

Substituting the expressions for $G(x,x_i)$ and $G^*(x,x_i)$, and those in eqs. (5.81) and (5.82) into eq. (5.80) yields two equations; the first comes from the source point at x_1

$$\begin{aligned} &-h_1 + h_2 + \tilde{l}\varphi_1 - (\tilde{l}+l)\varphi_2 + \\ &l\int_0^1 \Omega_j \left[\frac{d\Omega_n}{l d\zeta}\Theta_n\varphi_j + \Omega_n\Psi_n\left(\frac{dh_j}{dt} + f_j\right)\right](\tilde{l} + l\zeta)d\zeta = 0, \quad j,n=1,2 \end{aligned} \tag{5.83a}$$

and the second from the source point at x_2

$$\begin{aligned} &h_1 - h_2 + (\tilde{l}+l)\varphi_1 - \tilde{l}\varphi_2 + \\ &l\int_0^1 \Omega_j \left[\frac{d\Omega_n}{l d\zeta}\Theta_n\varphi_j + \Omega_n\Psi_n\left(\frac{dh_j}{dt} + f_j\right)\right][\tilde{l} + l(1-\zeta)]d\zeta = 0, \quad j,n=1,2 \end{aligned} \tag{5.83b}$$

A matrix representation of eqs. (5.83a) and (5.83b), which gives the system of discrete element equations, is

$$R_{ij}h_j + (L_{ij} - V_{inj}\Theta_n)\varphi_j + U_{ijn}\Psi_n\left(\frac{dh_j}{dt} + f_j\right) = 0, \qquad i,j,n=1,2 \tag{5.84}$$

where the element matrices in eq. (5.84) have been encountered in earlier chapters. The expressions for R_{ij} and L_{ij} are found in eqs. (2.42a) and (2.42b), respectively, and those for V_{ijn} and U_{ijn} can be obtained from eqs. (3.42c) and (3.45c). Our next step is to simplify the temporal derivative. The generalized 2-level time discretization scheme employed in section 5.4 will certainly be useful in achieving this simplification. The temporal derivative is approximated by a difference expression at $t = t_{1+\alpha} = t_1 + \alpha\Delta t$ (see Figure 5.2) which is given by eq. (5.20), so that the system of discrete element equations (5.84) becomes

$$\begin{aligned} &\alpha\left(R_{ij}h_j^{(2)} + [L_{ij} - V_{inj}\Theta_n^{(2)}]\varphi_j^{(2)}\right) + \\ &\omega\left(R_{ij}h_j^{(1)} + [L_{ij} - V_{inj}\Theta_n^{(1)}]\varphi_j^{(1)}\right) + U_{ijn}[\alpha\Psi_n^{(2)} + \omega\Psi_n^{(1)}]. \\ &\left(\frac{h_j^{(2)} - h_j^{(1)}}{\Delta t} + \alpha f_j^{(2)} + \omega f_j^{(1)}\right) \equiv g_i = 0; \quad i,j,n=1,2, \ 0 \le \alpha \le 1 \end{aligned}$$

or

$$\left[\alpha R_{ij} + \frac{U_{ijn}[\alpha\Psi_n^{(2)} + \omega\Psi_n^{(1)}]}{\Delta t}\right] h_j^{(2)} + \alpha\left[L_{ij} - V_{inj}\Theta_n^{(2)}\right]\varphi_j^{(2)} +$$
$$\left[\omega R_{ij} - \frac{U_{ijn}[\alpha\Psi_n^{(2)} + \omega\Psi_n^{(1)}]}{\Delta t}\right] h_j^{(1)} + \omega\left[L_{ij} - V_{inj}\Theta_n^{(1)}\right]\varphi_j^{(1)} + \qquad (5.85)$$
$$U_{ijn}[\alpha\Psi_n^{(2)} + \omega\Psi_n^{(1)}][\alpha f_j^{(2)} + \omega f_j^{(1)}] \equiv g_i = 0 \qquad i,j,n=1,2,\ 0\le\alpha\le 1$$

in which α represents the time weighting factor which we had encountered in section 5.4, and $\omega = 1-\alpha$. By the second equivalent expression in eq. (5.85) we have grouped together coefficients of similar quantities. Also it is noted that bracketed superscripts 2 and 1 denote the current and previous time levels, respectively, and $\Delta t = t_2 - t_1$ is the time step. Since eq. (5.85) is a system of discrete nonlinear equations, it can only be solved for the unknown nodal quantities h and φ by an appropriate nonlinear solution algorithm. We elect to use the Newton-Raphson algorithm. We must point out that the Picard algorithm can also be employed. As we indicated in chapter 3, the major step in correctly implementing the Newton-Raphson algorithm is deriving the Jacobian matrix. It is given by

$$J_{ij}^{(2,k)} = \begin{cases} \left.\dfrac{\partial g_i}{\partial h_j}\right|_{h_j=h_j^{(2,k)}} = \begin{array}{l} \alpha R_{ij} + \dfrac{U_{ilj}[\alpha\Psi_l^{(2,k)} + \omega\Psi_l^{(1)}]}{\Delta t} + \alpha U_{iln}\dfrac{[h_n^{(2,k)} - h_n^{(1)}]}{\Delta t}\dfrac{d\Psi_l}{dh_j} \\ -\alpha V_{iln}\varphi_n^{(2,k)}\dfrac{d\Theta_l}{dh_j} + \alpha U_{inl}[\alpha f_l^{(2)} + \omega f_l^{(1)}]\dfrac{d\Psi_n}{dh_j} \end{array} \\ \left.\dfrac{\partial g_i}{\partial \varphi_j}\right|_{\varphi_j=\varphi_j^{(2,k)}} = \alpha[L_{ij} - V_{ilj}\Theta_l^{(2,k)}] \end{cases} \qquad (5.86)$$

in which the superscript k stands for the iteration level.

5.9 Numerical Example on Transient Nonlinear Diffusion

One rare transient nonlinear diffusion problem with an exact solution is that governed by

$$\frac{\partial}{\partial x}\left(h\frac{\partial h}{\partial x}\right) = \frac{\partial h}{\partial t}; \quad 0\le x\le 1; \quad t>0 \qquad (5.87)$$

with the boundary and initial data

$$h(0,t) = 0, \; \varphi(1,t) = 0, \text{ and } h(x,0) = \mathrm{X}(x) \tag{5.88}$$

in which the function $\mathrm{X}(x)$ is defined implicitly by the integral equation

$$x = \frac{1}{B(1/2,1/3)} \int_0^{\mathrm{X}(x)} (1-u)^{-1/2} u^{-1/3}\, du = 0.3865\; B_x(1/2,1/3) \tag{5.89}$$

where $B(1/2\;1/3)$ stands for the *complete Beta function*, and $B_x(1/2,1/3)$ the *incomplete Beta function*. The exact solution to the problem is given by Boussinesq [9] using similarity considerations. The solution is given by

$$h(x,t) = \frac{\mathrm{X}(x)}{[1.1155t+1]} \tag{5.90}$$

while the solution for the flux at $x = 0$ is given by

$$q = -\left[h\frac{\partial h}{\partial x}\right]_{x=0} = \frac{0.8624}{[1.1155t+1]^2} \tag{5.91}$$

We certainly will experience some computational difficulties if we solve the above example with the Green element formulation which we have just proposed for transient nonlinear diffusion problems because of the physically unrealistic situation at $x = 0$ which suggests that the parameter $K = h$ in our example has zero value there. (All along we have always indicated the parameter K is greater than zero, and that is the case in practice.) This example represents unconfined groundwater flow, and the actual condition which is practically observed at $x = 0$ is a *seepage face* in which the head h equals the elevation of the *water table* because of the existence of atmospheric pressure on that face. Because of the special nature of this example, we shall derive a more appropriate Green element formulation which can handle the situation posed by that boundary condition. It begins by transforming the original differential equation (5.87) to

$$\frac{\partial^2 h^2}{\partial x^2} = 2\frac{\partial h}{\partial t}; \quad 0 \le x \le 1 \tag{5.92}$$

The quasi-steady Green element formulation is now applied to eq. (5.92) which yields the discrete element equations

$$R_{ij}h_j^2 + 2L_{ij}q_j + 2T_{ij}\frac{dh_j}{dt} = 0 \tag{5.93}$$

(We leave the derivation of eq. (5.93) to the reader.) It should be noted that $q_j = [h\partial h/\partial x]_j$ in eq. (5.93), and it denotes the flux. Applying the generalized 2-level temporal discetization scheme to eq. (5.93) yields

$$\begin{aligned} &\alpha R_{ij} h_j^{(2)2} + \frac{2T_{ij}}{\Delta t} h_j^{(2)} + 2\alpha L_{ij} q_j^{(2)} + \\ &\omega R_{ij} h_j^{(1)2} - \frac{2T_{ij}}{\Delta t} h_j^{(1)} + 2\omega L_{ij} q_j^{(1)} \equiv g_i = 0 \end{aligned} \tag{5.94}$$

The Jacobian matrix, when the Newton-Raphson algorithm is applied to eq. (5.94), is

$$J_{ij}^{(2,k)} = \begin{cases} \left[\dfrac{\partial g_i}{\partial h_j}\right]_{h_j = h_j^{(2,k)}} = \alpha R_{il} \dfrac{dh_l^2}{dh_j} + 2\dfrac{T_{ij}}{\Delta t} \\ \left[\dfrac{\partial g_i}{\partial q_j}\right]_{q_j = q_j^{(2,k)}} = 2\alpha L_{ij} \end{cases} \tag{5.95}$$

We solve this example with the above Green element model using a 10-element uniform discretization of the computational region, a uniform time step of 0.05, and a temporal weighting factor $\alpha = 0.67$. The calculations are done with 4 iterations being allowed at each time step. The results of **GEM** and the exact solution for h at various time levels, and for the flux at $x = 0$ are presented in Figures 5.11a and 5.11b. We observe good agreement between both solutions.

Thus far we have made no mention of the possibility of applying the **TGE** formulation to the transient nonlinear diffusion equation. It can be done, though with some limitations. To achieve this, we will need to approximate the parameter $K(h)$ within each element by a uniform value $\bar{K}$ (somewhat similar to the approximation used for the same parameter in the **GESNH-1** model in chapter 3). By so doing the computational domain is treated as piecewise homogeneous or multiply zoned. Although the parameter K, being a function of the primary variable h, is a function of space and time, $\bar{K}$ has to be a constant within an element and within any time step in order to apply the **TGE** formulation. The real challenge is how to choose $\bar{K}$. We suggest that within an element $[x_1, x_2]$, $\bar{K}$ be chosen by the approximation

$$K[(h(x,t)] \approx \bar{K} = \bar{K}^{(1)} = K\left[\frac{1}{2}\left(h_1^{(1)} + h_2^{(1)}\right)\right] \tag{5.96}$$

In other words, by eq. (5.96), $\bar{K}$ is chosen using the mean of the primary variable at the previous time level t_1. It is also necessary to point out that with this approach, the coefficients of the element matrices whose expressions are given in eqs. (5.37a) through (5.37l) will constantly have to be recomputed at every time step. Considering the complicated nature of these coefficients, it could be quite taxing to evaluate them at every time step.

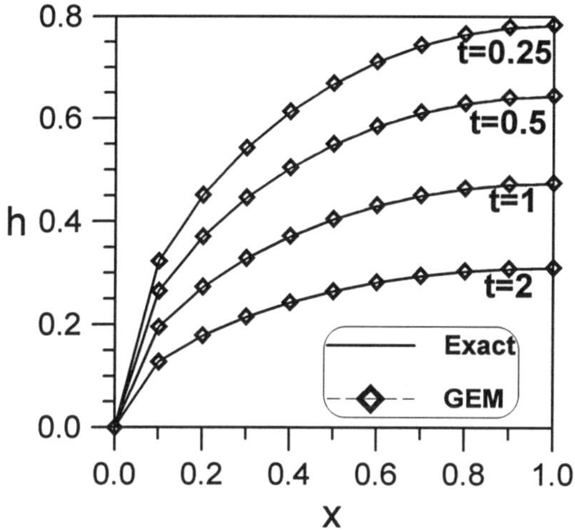

(a)

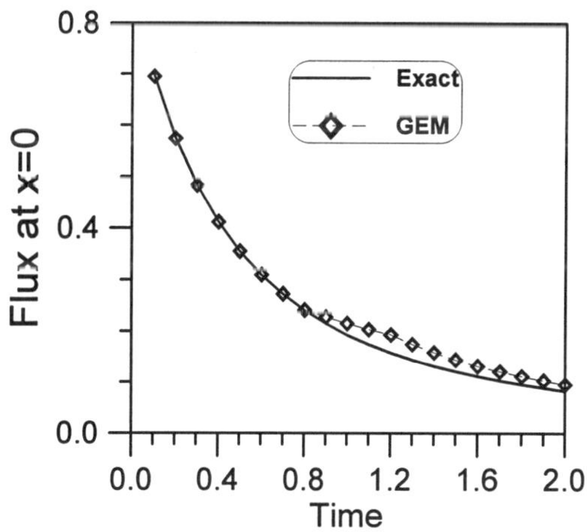

(b)

Figure 5.11 Exact and **GEM** solutions of nonlinear diffusion example — **(a)** for $h(x,t)$, **(b)** for flux at $x=0$

5.10 Remarks

Two Green element models are proposed for the transient diffusion equation in this chapter. The first, denoted as the **QSGE** model, employs the free-space Green's function of the 1-D. Laplace differential operator to derive its integral representation, and simplifies the temporal derivative by finite difference approximation. Various schemes of the **QSGE** model are achieved depending on the time level within the time step that the difference approximation is carried out. The various schemes are represented by different values of the temporal weighting factor α which could assume any value between 0 and 1. A value of $\alpha = 0.67$ is, by numerical experiments, found to give optimal results. Numerical solutions which are based on the Crank Nicholson scheme may give equally good results for problems without significant spatial gradients of the primary variable.

The second Green element model (**TGE** model) proposed for the transient diffusion equation uses the free-space Green's function to the transient linear diffusion differential operator. Because of the complicated nature of the free-space Green's function, the expressions of its element matrices are not only more complicated, but require more computing resources to evaluate. However, the model consistently gave more accurate results than the **QSGE** model for the three numerical examples examined.

Numerical stability assessment of the two models show that the **QSGE** model with $\alpha \geq 0.5$ and the **TGE** model are always stable for values of Courant number less than unity. However, as the Courant number approaches unity, the harmonics with small wavelengths, which are dominant when the solution has steep gradients, are not correctly represented in amplitude.

Both Green element models can be applied to transient nonlinear diffusion, but the **QSGE** is more amenable to such problems. The main reason for this is tied to the simple nature of its free-space Green's function.

5.11 Exercises

5.1 Derive the Green element equations for the transient linear diffusion equation so that there is only one degree of freedom at each node. Use these new models to solve the examples presented in this chapter.

5.2 Examine the performance of the **QSGE** model which uses the relationship below to approximate the temporal derivative

$$\left.\frac{dh_j}{dt}\right]_{t=t_2} = \frac{\alpha}{\Delta t}\left(h_j^{(2)} - h_j^{(1)}\right) + (1-\alpha)\left.\frac{dh_j}{dt}\right]_{t=t_1} ; \quad 1 \leq \alpha \leq 2$$

Compare its performance to that of the generalized 2-level time discretization scheme presented in this chapter.

5.3 Use the Green element method to solve the equation

$$\frac{\partial^2 h}{\partial x^2} - \frac{\partial h}{\partial t} + x^2 = 0; \quad 0 \le x \le 1, \quad t > 0$$

with the initial condition $h(x,t{=}0) = 0$, and the following sets of boundary conditions:

(a) $h(0,t) = 1,\ h(1,t) = 0$

(b) $\dfrac{\partial h(0,t)}{\partial x} \equiv \varphi(0,t) = 1,\ \dfrac{\partial \varphi(1,t)}{\partial x} \equiv \varphi(1,t) = 1$

(c) $h(0,t) = \cos(\pi t),\ \dfrac{\partial h(1,t)}{\partial x} \equiv \varphi(1,t) = 0$

(d) $h(0,t) = e^{-t},\ \dfrac{\partial h(1,t)}{\partial x} \equiv \varphi(1,t) = 0$

5.4 Employ any appropriate Green element model to solve the equation

$$\frac{\partial}{\partial x}\left[K(x)\frac{\partial h}{\partial x}\right] - \frac{\partial h}{\partial t} = 0; \quad 0 \le x \le 1, \quad t > 0$$

using the initial and boundary conditions in exercise 5.3, and for the following expressions for $K(x)$

(a) $K(x) = 1 + x$

(b) $K(x) = 1 + x^2$

(c) $K(x) = e^{x}$

5.5 Apply the Green element method to solve the equation

$$\frac{\partial}{\partial x}\left[\frac{1}{1+h}\frac{\partial h}{\partial x}\right] - \frac{\partial h}{\partial t} = 0; \quad 0 \le x \le 1, \quad t > 0$$

$$h(x,t{=}0) = 0, \quad h(0,t) = 1 \text{ and } \frac{\partial h(1,t)}{\partial x} = \varphi(1,t) = 0$$

5.12 References

1. Cheng, A. H-D. and K. Ou, "An efficient Laplace transform Solution for Multiaquifer Systems," *Water Resources Research*, **25**(4), pp. 742-748, 1989.

2. Banerjee, P.K. and R. Butterfield, *Boundary Element Methods in Engineering Science*, McGraw-Hill, London, U.K., 1981.

3. Liggett, J.A. and P. L-F. Liu, *The Boundary Integral Equation Method for Porous Media Flow*, George Allen & Unwin, London, 1983.

4. Stone, H.L. and P.L.T. Brian , "Numerical Solution of Convective Transport Problems", *A.I. Ch.E. J.*, **9**(5), pp. 681-688, 1963.

5. Leendertse, J.J., "Aspects of a Computational model for long-period water-wave propagation", *Rep. RM-5294-PR*, 165pp., The Rand Corp., Santa Monica, Calif. U.S.A., 1967.

6. Leendertse, J.J., "A Water-quality Simulation Model for well-mixed Estuaries and Coastal Seas, Vol. 1, Principles of Computation", *Rep. RM-6230-RC*, 71pp., The Rand Corp., Santa Monica, Calif, U.S.A., 1970.

7. Gray, W.G. and G.F. Pinder, "An Analysis of the Numerical Solution of the Transport Equation," *Water Resources research*, **12**, pp. 547-555, 1976.

8. Gray, W.G. and D.R. Lynch, "Time-Stepping Schemes for Finite Element Tidal Model Computations," in *Surface Flow*, ed. W.G. Gray, pp. 1-14, CML Pub. Ltd., Southampton, U.K., 1984.

9. Boussinesq, J., "Recherches Theoretiques sur l'ecolement des Nappes d'eau infiltrees dans sol et sur le debit des sources", *J. de Mathematiques pures et Appliquees*, **10**(5th series), pp. 5-78, 363-394, 1904.

6 TRANSPORT EQUATION

6.1 Introduction

Having gone through the earlier chapters in which derivations of the Green element method were done for linear and nonlinear Laplace/Poisson problems in homogeneous and heterogeneous media, and for the transient linear and nonlinear diffusion equations, you should by now have a good feel of the way the method is derived for a differential operator. For each of these problems, we sought a system of discrete element equations and then assembled it for all the elements to obtain the global matrix equation from which the nodal solutions were computed. The core of the Green element formulation, therefore, lies in one's ability to correctly derive these element equations for the differential operator. When that is done, everything else can be considered straightforward. Our attention in this chapter focuses on the derivation of the Green element equations for the contaminant transport problem. Under conditions of transport in a homogenous domain with uniform ambient flow, three distinct Green element formulations are derived. However, the amount of effort involved in formulating the element equations escalates appreciably as we progress from the first formulation to the third one. At the end of the exercise of going through these three formulations, coupled with the understanding gained from the two formulations for the diffusion equation, we will be in a position to assess the relationship between computational efficiency and versatility of **GEM**, and the nature of the free-space Green's function employed. Furthermore, as you must have come to observe, domain integrations do not pose any unusual computational difficulties in Green element formulations. There is a very simple reason for this. In Green element formulations, the source and field nodes are always on the same element so that domain integrations are much easier to evaluate.

The transient one-dimensional transport or contaminant or diffusion-advection equation can be derived by examining the hydrodynamics of the spread of a substance or pollutant as it is advected in fluid with an ambient velocity $u(x,t)$. Our first formulation uses the free-space Green's function of the 1-D. Laplace differential operator, making it similar to the **QSGE** formulation of chapter 5, and the second formulation uses the free-space Green's function of the 1-D. transient diffusion equation so that it is essentially the **TGE** formulation of Chapter 5. The third formulation employs the free-space Green's function of the linear diffusion-advection differential operator, so that it is applicable only to homogeneous transport with uniform ambient velocity of flow.

The partial differential equation that describes one-dimensional transport with first-order decay in a homogeneous medium under transient fluid flow conditions is given by

$$D\frac{\partial^2 h(x,t)}{\partial x^2} - u(x,t)\frac{\partial h(x,t)}{\partial x} - \frac{\partial h(x,t)}{\partial t} - \mu h = f(x,t), \quad on \ x_0 \leq x \leq x_L \tag{6.1}$$

in which $h(x,t)$ is the dependent variable which represents the *concentration* of the pollutant, D is the *hydrodynamic dispersion coefficient* [L^2/T], $u = u(x,t)$ is the velocity in the x-direction [L/T], μ is the rate constant of first-order decay for a nonconservative pollutant [T^{-1}], $f(x,t)$ represents an external contaminant source, and $L = x_L - x_0$ is considered the length of the flow domain. Eq. (6.1) continues to receive enormous interest in numerical circles because of its extensive applications in the areas of agriculture, chemical, environmental, and water resources engineering. The solution of eq. (6.1) requires information on the boundary conditions at x_0 and x_L, and the data on $h(x,t)$ at the initial time t_0, and those conditions described in section 5.2 of chapter 5 could be applied.

6.2 Model 1: Quasi-Steady Green Element Model (QSGE)

Our current Green element formulation is based on the Fredholm singular integral theory which employs the free-space Green's function of the Laplace differential operator which we have encountered right from chapter 2. That formulation should now be quite familiar to the reader to enable us skip a number of steps. As we had earlier pointed out, the central issue in formulating a Green element replication of a differential equation is to derive its system of discrete element equations. Those equations should provide all the information required to characterize the solution, and should be applicable to the spatial and temporal elements. The current model uses the complementary differential equation

$$\frac{d^2G}{dx^2} = \delta(x - x_i), \qquad -\infty \leq x \leq \infty \tag{6.2}$$

which is essentially eq. (2.20). When we invoke Green's identity on eqs. (6.1) and (6.2), and utilize the properties of the Dirac delta function and the solution to eq. (6.2), the outcome is the integral equation which applies to a typical element $[x_1, x_2]$

$$
\begin{aligned}
&D\Big[-\lambda h(x_i,t) + G^*(x_2,x_i)h(x_2,t) - G^*(x_1,x_i)h(x_1,t) \\
&\quad - G(x_2,x_i)\varphi(x_2,t) + G(x_1,x_i)\varphi(x_1,t)\Big] + \\
&\int_{x_1}^{x_2} G(x,x_i)\left[u(x,t)\varphi(x,t) + \frac{\partial h(x,t)}{\partial t} + \mu h + f(x,t)\right]dx = 0
\end{aligned}
\tag{6.3}
$$

To evaluate the line integral over a typical element, there is the need to prescribe a distribution for $h(x,t)$, $u(x,t)$, $\varphi(x,t)$, and $f(x,t)$ over each element. We elect to use linear element shape functions to approximate these quantities

$$
\begin{aligned}
&h(x,t) \approx \Omega_j(\zeta)h_j(t), \qquad \varphi(x,t) \approx \Omega_j(\zeta)\varphi_j(t) \\
&u(x,t) \approx \Omega_j(\zeta)u_j(t), \quad f(x,t) \approx \Omega_j(\zeta)f_j(t)
\end{aligned}
\tag{6.4}
$$

Introducing eq. (6.4) into eq. (6.3) yields

$$
\begin{aligned}
&D\Big[-\lambda h_i + G^*(x_2,x_i)h_2 - G^*(x_1,x_i)h_1 - G(x_2,x_i)\varphi_2 + G(x_1,x_i)\varphi_1\Big] + \\
&\int_{x_1}^{x_2} G(x,x_i)\Omega_j\left\{u_j\Omega_n\varphi_n + \frac{dh_j}{dt} + \mu h_j + f_j\right\}dx = 0; \quad i,j,n = 1,2
\end{aligned}
\tag{6.5}
$$

We recall the expressions for $G(x,x_i)$ and $G^*(x,x_i)$ from eqs. (2.30) and (2.31) in chapter 2, and set the arbitrary constant k in the expression of the free-space Green's function to the longest spatial element size $\tilde{l}$ that is used to discretize the computational domain. With that

$$
G(x_1,x_1) = G(x_2,x_2) = \frac{\tilde{l}}{2}, \quad G(x_2,x_1) = G(x_1,x_2) = \frac{\tilde{l}+l}{2}
$$

$$
G^*(x_1,x_1) = G^*(x_2,x_2) = 0, \quad G^*(x_2,x_1) = -G^*(x_1,x_2) = \frac{1}{2}
$$

Introducing the above relations into eq. (6.5) gives (with the source point x_i at x_1)

$$
\begin{aligned}
&D\Big[-h_1 + h_2 + \tilde{l}\varphi_1 - (\tilde{l}+l)\varphi_2\Big] + \\
&l\int_0^1 \Omega_j\left(\Omega_n u_n\varphi_j + \frac{dh_j}{dt} + \mu h_j + f_j\right)(\tilde{l}+l\zeta)d\zeta = 0, \quad j,n = 1,2
\end{aligned}
\tag{6.6a}
$$

and with x_i at x_2

$$D\left[h_1 - h_2 + (\tilde{l}+l)\varphi_1 - \tilde{l}\varphi_2\right] + \\ l\int_0^1 \Omega_j\left(\Omega_n u_n \varphi_j + \frac{dh_j}{dt} + \mu h_j + f_j\right)[\tilde{l} + l(1-\zeta)]\,d\zeta = 0, \qquad j,n=1,2 \tag{6.6b}$$

Eqs. (6.6a) and (6.6b) are combined to form the element matrix equations of the first Green element model of the transient transport equation under nonuniform ambient flow conditions.

$$\boxed{DR_{ij}h_j + (DL_{ij} + U_{ikj}u_k)\varphi_j + T_{ij}\left(\frac{dh_j}{dt} + \mu h_j + f_j\right) = 0, \qquad i,j,k=1,2} \tag{6.7}$$

This Green element model is also referred to as the quasi-steady formulation and denoted **QSGE** as that in chapter 5 since they are based on the same fundamental solution. The element matrices in eq. (6.7) have been encountered in one form or another in earlier chapters; they are simply summarized below for ease of reference

$$R_{ij} = \begin{bmatrix} -1 & 1 \\ 1 & -1 \end{bmatrix} = (-1)^{(i+j-1)}, \quad L_{ij} = \begin{bmatrix} \tilde{l} & -(\tilde{l}+l) \\ (\tilde{l}+l) & -\tilde{l} \end{bmatrix} \tag{6.8a}$$

$$T_{ij} = l\int_0^1 \Omega_j G(\zeta,\zeta_i)\,d\zeta = \frac{l}{6}\begin{bmatrix} 3\tilde{l}+l & 3\tilde{l}+2l \\ 3\tilde{l}+2l & 3\tilde{l}+2l \end{bmatrix} \tag{6.8b}$$

$$U_{1ij} = l\int_0^1 \Omega_i\Omega_j G(\zeta,\zeta_1)\,d\zeta = \frac{l}{12}\begin{bmatrix} 4\tilde{l}+l & 2\tilde{l}+l \\ 2\tilde{l}+l & 4\tilde{l}+3l \end{bmatrix} \tag{6.8c}$$

$$U_{2ij} = l\int_0^1 \Omega_i\Omega_j G(\zeta,\zeta_2)\,d\zeta = \frac{l}{12}\begin{bmatrix} 4\tilde{l}+3l & 2\tilde{l}+l \\ 2\tilde{l}+l & 4\tilde{l}+l \end{bmatrix} \tag{6.8d}$$

Eq. (6.7) provides all the information required to solve for the nodal unknowns h and φ. The system of element equations is still a system of first-order ordinary differential equations in time which can be solved for h and φ at the nodes by employing an appropriate approximation for the temporal derivative. We elect to use the 2-level time discretization scheme that was employed in chapter 5. The scheme approximates the temporal derivative at $t = t_{1+\alpha} = t_1 + \alpha\Delta t$ (see Figure 5.2) by a difference expression that is given by eq. (5.20). When this approximation is incorporated into the system of discrete equations it yields

$$\alpha\{DR_{ij}h_j^{(2)} + [DL_{ij} + U_{ijk}(\alpha u_k^{(2)} + \omega u_k^{(1)}]\varphi_j^{(2)}\} +$$
$$\omega\{DR_{ij}h_j^{(1)} + [DL_{ij} + U_{ijk}(\alpha u_k^{(2)} + \omega u_k^{(1)})]\varphi_j^{(1)}\} + \tag{6.9}$$
$$T_{ij}\left\{\frac{h_j^{(2)} - h_j^{(1)}}{\Delta t} + \alpha[\mu h_j^{(2)} + f_j^{(2)}] + \omega[\mu h_j^{(1)} + f_j^{(1)}]\right\} = 0 \quad i,j,k=1,2, \ 0 \le \alpha \le 1$$

or

$$\left[\alpha(DR_{ij} + \mu T_{ij}) + \frac{T_{ij}}{\Delta t}\right]h_j^{(2)} + \alpha\left[DL_{ij} + U_{ijk}\left(\alpha u_k^{(2)} + \omega u_k^{(1)}\right)\right]\varphi_j^{(2)} +$$
$$\left[\omega(DR_{ij} + \mu T_{ij}) - \frac{T_{ij}}{\Delta t}\right]h_j^{(1)} + \omega\left[DL_{ij} + U_{ijk}\left(\alpha u_k^{(2)} + \omega u_k^{(1)}\right)\right]\varphi_j^{(1)} + \tag{6.9}$$
$$T_{ij}[\alpha f_j^{(2)} + \omega f_j^{(1)}] = 0 \qquad i,j,k=1,2, \ 0 \le \alpha \le 1$$

By the second equivalent expression in eq. (6.9) we have grouped together coefficients of similar quantities. Also it is noted that the bracketed superscripts 2 and 1 denote the current and previous time levels, respectively, and $\Delta t = t_2 - t_1$ is the time step or the temporal element size.

When the velocity of the transporting fluid is uniform, that is $U(t) \equiv u(x,t)$ then it is no longer necessary to express the velocity in terms of a linear distribution of its nodal values. (We may need to point out that it is not practicable to have a transient uniform flow; only a steady uniform flow can be realized in practice.) In that case, the integral equation given by (6.5) becomes

$$D\left[-2\lambda h_i + G^*(x_2,x_i)h_2 - G^*(x_1,x_i)h_1 - G(x_2,x_i)\varphi_2 + G(x_1,x_i)\varphi_1\right] +$$
$$\int_{x_1}^{x_2} G(x,x_i)\Omega_j\left[U\varphi_j + \frac{dh_j}{dt} + \mu h_j + f_j\right]dx \tag{6.10}$$

The line integral is evaluated to obtain the coefficients of the unknown quantities, and the generalized two-level time scheme employed to simplify the temporal derivative. When these are done the outcome is a system of discrete element equations which are given as

$$
\begin{aligned}
&\alpha\left\{DR_{ij}h_j^{(2)} + [DL_{ij} + \left(\alpha U^{(2)} + \omega U^{(1)}\right)T_{ij}]\varphi_j^{(2)}\right\} + \\
&\omega\left\{DR_{ij}h_j^{(1)} + [DL_{ij} + \left(\alpha U^{(2)} + \omega U^{(1)}\right)T_{ij}]\varphi_j^{(1)}\right\} + \\
&T_{ij}\left\{\frac{h_j^{(2)} - h_j^{(1)}}{\Delta t} + \alpha[\mu h_j^{(2)} + f_j^{(2)}] + \omega[\mu h_j^{(1)} + f_j^{(1)}]\right\} = 0 \quad i,j,k=1,2,\ 0\leq\alpha\leq 1
\end{aligned}
\tag{6.11}
$$

or

$$
\boxed{
\begin{aligned}
&\left[\alpha(DR_{ij} + \mu T_{ij}) + \frac{T_{ij}}{\Delta t}\right]h_j^{(2)} + \alpha\left[DL_{ij} + \left(\alpha U^{(2)} + \omega U^{(1)}\right)T_{ij}\right]\varphi_j^{(2)} + \\
&\left[\omega(DR_{ij} + \mu T_{ij}) - \frac{T_{ij}}{\Delta t}\right]h_j^{(1)} + \omega\left[DL_{ij} + \left(\alpha U^{(2)} + \omega U^{(1)}\right)T_{ij}\right]\varphi_j^{(1)} + \\
&T_{ij}[\alpha f_j^{(2)} + \omega f_j^{(1)}] = 0 \qquad i,j,k=1,2,\ 0\leq\alpha\leq 1
\end{aligned}
}
\tag{6.11}
$$

We have adopted a consistent notation; for the sake of clarity $U^{(2)}$ represents the ambient fluid flow velocity at time t_2. Eq. (6.11) is the **QSGE** model of the transient 1-D. transport (diffusion-advection) equation under uniform flow..

6.3 Model 2 (TGE Model)

Our second **GE** formulation for the transport equation given by eq. (6.1) makes use of the free-space Green's function to the linear transient diffusion equation. This formulation is similar to the second Green element formulation developed for the transient diffusion equation in chapter 5. It uses the time-dependent complementary differential equation

$$
\frac{\partial^2 G}{\partial x^2} + \frac{1}{D}\frac{\partial G}{\partial t} = \delta(x - x_i)\delta(t_2 - t), \quad -\infty\leq x\leq\infty,\ t_2 > t \tag{6.12}
$$

whose solution is

$$
G(x,t;x_i,t_2) = \frac{H(t_2 - t)}{\sqrt{[4\pi D(t_2 - t)]}}\exp\left[-\frac{(x - x_i)^2}{4D(t_2 - t)}\right] \tag{6.13}
$$

which is essentially the same as eq. (5.25). As we did in chapter 5, the integral representation of (6.1) is obtained from a time integration of Green's identity applied to a typical spatial element $[x_1,x_2]$. That integral representation is

$$\int_{t_1}^{t_2}\int_{x_1}^{x_2}\left[h\left(\delta(x-x_i)\delta(t_2-t)-\frac{1}{D}\frac{\partial G}{\partial t}\right)-\frac{G}{D}\left(\frac{\partial h}{\partial t}+u(x,t)\frac{\partial h}{\partial x}+\mu h+f(x,t)\right)\right]dxdt = \int_{t_1}^{t_2}\left[h\frac{\partial G}{\partial x}-G\varphi\right]_{x_1}^{x_2}dt \tag{6.14}$$

The reader may wish to refer to eqs. (5.28a) and (5.28b) in simplifying eq. (6.14). With the use of those relationships eq. (6.14) now simplifies to

$$\begin{aligned}&-\lambda h_i^{(2)}+\int_{t_1}^{t_2}\left[h(x_2,t)G^*(x_2,t;x_i,t_2)-h(x_1,t)G^*(x_1,t;x_i,t_2)\right]dt-\\&\int_{t_1}^{t_2}\left[G(x_2,t;x_i,t_2)\varphi(x_2,t)-G(x_1,t;x_i,t_2)\varphi(x_1,t)\right]dt-\\&\frac{1}{D}\int_{x_1}^{x_2}G(x,t_1;x_i,t_2)h(x,t_1)dx+\\&\frac{1}{D}\int_{t_1}^{t_2}\int_{x_1}^{x_2}G(x,t;x_i,t_2)[u(x,t)\varphi(x,t)+\mu h(x,t)+f(x,t)]dxdt=0\end{aligned} \tag{6.15}$$

where

$$G^*(x,t;x_i,t_2) = \frac{\partial G}{\partial x} = -\frac{(x-x_i)}{2D(t_2-t)}G(x,t;x_i,t_2) \tag{6.16}$$

There is the need to prescribe a distribution for $h(x,t)$ and $\varphi(x,t)$ within the spatial element $[x_1, x_2]$ and the temporal element $[t_1, t_2]$. A linear interpolation function in both space and time is proposed for those quantities similar to the expression in eqs. (5.31). Because of the complexity of the free-space Green's function employed in this model, it will be difficult to evaluate the line integral of the advection term $u(x,t)\varphi(x,t)$ if both quantities are approximated by linear shape functions in space and time. To simplify this term we assume that the velocity $u(x,t)$ is uniform within a spatial element. That is

$$u(x,t) \approx \bar{u} = \frac{1}{4}[u_1^{(1)}+u_1^{(2)}+u_2^{(1)}+u_2^{(2)}] \tag{6.17}$$

Using the linear distribution of eq. (5.31), the expression in eq. (6.17), and the properties of the free-space Green's function, eq. (6.15) becomes

$$
\begin{aligned}
&-\lambda h_1^{(2)} + \Delta t \int_0^1 \Omega^{(m)}(\tau) h_2^{(m)} G^*(x_2, \Delta t(1-\tau); x_i, 0)\, d\tau \ - \\
&\Delta t \int_0^1 \Omega^{(m)}(\tau) h_1^{(m)} G^*(x_1, \Delta t(1-\tau); x_i, 0)\, d\tau \ - \\
&\Delta t \int_0^1 \Omega^{(m)}(\tau) \varphi_2^{(m)} G(x_2, \Delta t(1-\tau); x_i, 0)\, d\tau \ + \\
&\Delta t \int_0^1 \Omega^{(m)}(\tau) \varphi_1^{(m)} G(x_1, \Delta t(1-\tau); x_i, 0)\, d\tau \ - \\
&\frac{l}{D} \int_0^1 \Omega_j(\zeta) h_j^{(1)} G(l(\zeta-\zeta_i), \Delta t; 0,0)\, d\zeta \ + \\
&\frac{l\Delta t}{D} \int_0^1 \int_0^1 G(l(\zeta-\zeta_i), \Delta t(1-\tau); 0,0) \\
&\left\{\Omega^{(m)}(\tau) \Omega_j(\zeta) [\bar{u}\varphi_j^{(m)} + \mu h_j^{(m)} + f_j^{(m)}]\right\} d\zeta\, d\tau \ = \ 0; \quad i,j,m = 1,2
\end{aligned}
\tag{6.18}
$$

For the sake of clarity we may need to point out that two equations can be obtained from eq. (6.18); the first equation comes from the source point located at x_1

$$
\begin{aligned}
&-\frac{1}{2} h_1^{(2)} + \Delta t \int_0^1 \Omega^{(m)}(\tau) h_2^{(m)} G^*(l, \Delta t(1-\tau); 0,0)\, d\tau \ - \\
&\Delta t \int_0^1 \Omega^{(m)}(\tau) \varphi_2^{(m)} G(l, \Delta t(1-\tau); 0,0)\, d\tau \ + \\
&\Delta t \int_0^1 [\Omega^{(m)}(\tau) \varphi_1^{(m)} G(0, \Delta t(1-\tau); 0,0)\, d\tau \ - \\
&\frac{l}{D} \int_0^1 \Omega_j(\zeta) h_1^{(1)} G(l\zeta, \Delta t; 0,0)\, d\zeta \ + \\
&\frac{l\Delta t}{D} \int_0^1 \int_0^1 G(l\zeta, \Delta t(1-\tau); 0,0) \Omega^{(m)}(\tau) \Omega_j(\zeta) [\bar{u}\varphi_j^{(m)} + \mu h_j^{(m)} + f_j^{(m)}]\, d\zeta\, d\tau = 0
\end{aligned}
\tag{6.19a}
$$

and the second from the source point x_i located at x_2

$$
\begin{aligned}
&-\frac{1}{2}h_2^{(2)} - \Delta t\int_0^1 \Omega^{(m)}(\tau)h_1^{(m)}G^*(-l,\Delta t(1-\tau);0,0)d\tau \; - \\
&\Delta t\int_0^1 \Omega^{(m)}(\tau)\varphi_2^{(m)}G(0,\Delta t(1-\tau);0,0)d\tau \; + \\
&\Delta t\int_0^1 \Omega^{(m)}(\tau)\varphi_1^{(m)}G(-l,\Delta t(1-\tau);0,0)d\tau \; - \\
&\frac{l}{D}\int_0^1 \Omega_j(\zeta)h_j^{(1)}G(l(\zeta-1),\Delta t;0,0)d\zeta \; + \\
&\frac{l\Delta t}{D}\int_0^1\int_0^1 G(l(\zeta-1),\Delta t(1-\tau);0,0)\Omega^{(m)}(\tau)\Omega_j(\zeta)[\bar{u}\varphi_j^{(m)} + \mu h_j^{(m)} + f_j^{(m)}]d\zeta d\tau = 0
\end{aligned}
\tag{6.19b}
$$

where $h_j^{(2)}$, for instance, stands for the value of h at x_j and at time t_2. When eqs. (6.19a) and (6.19b) are combined, the system of discrete element equations is obtained for the model. It is given as

$$
[R_{ij}^{(m)} + \mu T_{ij}^{(m)}]h_j^{(m)} + [L_{ij}^{(m)} + \bar{u}T_{ij}^{(m)}]\varphi_j^{(m)} \; + \; T_{ij}^{(m)}f_j^{(m)} = 0, \;\; i,j,m=1,2 \tag{6.20}
$$

in which the expressions for all the element matrices in eq. (6.20) are obtained from eqs. (5.37a) to (5.37l). It should be noted that the definition for γ in those equations takes the form of eq. (5.38a) except that the parameter K should be replaced by the parameter D.

6.4 Model 3 (ADGE Model)

This third Green element formulation for the transport equation is herein derived for a part of the original differential equation (6.1) which is given by

$$
D\frac{\partial^2 h(x,t)}{\partial x^2} = U\frac{\partial h(x,t)}{\partial x} + \frac{\partial h(x,t)}{\partial t}, \quad on \; x_0 \le x \le x_L \tag{6.21}
$$

The form of eq. (6.21) suggests that the transport process occurs under uniform ambient flow velocity, and first-order chemical reaction is assumed absent. Also it assumes that there is no external distributed contaminant source, that is $f(x,t) = 0$. The complementary differential equation of the homogeneous advection-diffusion equation

(6.21) from which we derive the free-space Green's function is

$$\frac{\partial^2 G}{\partial x^2} + \frac{U}{D}\frac{\partial G}{\partial x} + \frac{1}{D}\frac{\partial G}{\partial t} = \delta(x-x_i)\delta(t_2-t), \quad on \ -\infty \le x \le \infty, \quad t_2 > t \tag{6.22}$$

The solution to eq. (6.22), which is the fundamental solution or the free-space Green's function or unit response function, is given by

$$G(x,t;x_i,t_2) = \frac{H(t_2-t)}{\sqrt{[4\pi D(t_2-t)]}} \exp\left[-\frac{(t_2-t)U^2}{4D} - \frac{(x-x_i)^2}{4D(t_2-t)} - \frac{U(x-x_i)}{2D} \right] \tag{6.23}$$

Comparing the expression for the free-space Green's function given by eq (6.23) to that of eq. (5.25) for linear transient diffusion, and then to eq. (2.30) for linear steady diffusion or Laplace differential operator, it is obvious that the current Green's function is the most complicated of the three. This complexity of G extends to the computational effort involved in forming the element equations which is the next stage in our formulation process. At this point this general observation can be made: the computational effort involved in Green element calculations is directly related to the complexity of the free-space Green's function employed in the analysis. When the expression for the free-space Green's function is simple, the element equations are not only easier to formulate but also more pliable to deal with, and the converse is the case when the free-space Green's function is complicated.

The element equations of the current formulation are obtained when a time-integrated form of the Green's identity given by eq. (2.25) is applied to eqs. (6.21) and (6.22) within the element $[x_1,x_2]$ and temporal element $[t_1,t_2]$.

$$\int_{t_1}^{t_2}\int_{x_1}^{x_2}\left[h\left(\delta(x-x_i)\delta(t_2-t) - \frac{1}{D}\frac{\partial G}{\partial t} - \frac{U}{D}\frac{\partial G}{\partial x} \right) - \frac{G}{D}\left(\frac{\partial h}{\partial t} + U\frac{\partial h}{\partial x} \right) \right] dx dt = \int_{t_1}^{t_2} \left[h\frac{\partial G}{\partial x} - G\varphi \right]_{x_1}^{x_2} dt \tag{6.24}$$

Using the property of the Dirac delta function discussed in chapter 2

$$\int_{t_1}^{t_2}\int_{x_1}^{x_2} h(x,t)\delta(x-x_i)\delta(t_2-t)dxdt = \lambda h(x_i,t_2) = \lambda h_i^{(2)} \tag{6.25a}$$

in which $\lambda = 1$ if x_i is within the interval $[x_1,x_2]$, and $\lambda = 0.5$ if x_i is at x_1 or x_2. Furthermore

$$\frac{1}{D}\int_{t_1}^{t_2}\int_{x_1}^{x_2}\left[-h\left(\frac{\partial G}{\partial t}+U\frac{\partial G}{\partial x}\right) - G\left(\frac{\partial h}{\partial t}+U\frac{\partial h}{\partial x}\right)\right]dxdt$$

$$= -\frac{1}{D}\int_{t_0}^{t}\int_{x_1}^{x_2}\left[\frac{\partial}{\partial t}(hG)+U\frac{\partial (hG)}{\partial x}\right]dxd\tau \tag{6.25b}$$

$$= \frac{1}{D}\int_{x_1}^{x_2} G(x,t_1;x_i,t_2)h(x,t_1)dx - \frac{U}{D}\int_{t_1}^{t_2}\left[hG\right]_{x_1}^{x_2}dt$$

In deriving eq. (6.25b) we have employed one of the characteristics of the Green's function, and that is $G(x,t_2;x_i,t_2) = 0$. We can now introduce the results of eqs. (6.25a) and (6.25b) into (6.24) to obtain the integral equation

$$\begin{aligned}
&-\lambda h_i^{(2)} + \int_{t_1}^{t_2}\left[h(x_2,t)G^*(x_2,t;x_i,t_2) - h(x_1,t)G^*(x_1,t;x_i,t_2)\right]dt \;+\\
&\int_{t_1}^{t_2}\left[G(x_1,t;x_i,t_2)\varphi(x_1,t) - G(x_2,t;x_i,t_2)\varphi(x_2,t)\right]dt \;-\\
&\frac{1}{D}\int_{x_1}^{x_2} G(x,t_1;x_i,t_2)h(x,t_1)dx \;+\\
&\frac{U}{D}\int_{t_1}^{t_2}\left[h(x_2,t)G(x_2,t;x_i,t_2) - h(x_1,t)G(x_1,t;x_i,t_2)\right]dt \;=\; 0
\end{aligned} \tag{6.26}$$

where

$$G^*(x,t;x_i,t_2) = \frac{\partial G}{\partial x} = \frac{-1}{2D}\left[\frac{x-x_i}{t_2-t} + U\right]G(x,t;x_i,t_2) \tag{6.27}$$

The quantities $h(x,t)$ and $\varphi(x,t)$ are approximated both in time and space by linear interpolation functions with expressions similar to that of eq. (5.31) using the local spatial coordinate ζ and local temporal coordinate τ. In terms of these local coordinates, the following quantities are simplified below

$$G(x,t;x_i,t_2) = G(l(\zeta-\zeta_i),\Delta t(1-\tau);0,0) \tag{6.28a}$$

$$G^*(x,t;x_i,t_2) = \frac{-1}{2D}\left[\frac{l(\zeta-\zeta_i)}{\Delta t(1-\tau)} + U\right]G(l(\zeta-\zeta_i),\Delta t(1-\tau);0,0) \tag{6.28b}$$

$$G(x,t_1;x_i,t_2) = G(l(\zeta-\zeta_i),\Delta t;0,0) \tag{6.28c}$$

When we make use of the expressions in eqs. (6.28a) to (6.28c) in our integral equation (6.26), we have two discrete equations to deal with; one equation arises when the source point is at x_1

$$\begin{aligned}
&-\frac{1}{2}h_1^{(2)} + \frac{\Delta t}{2D}\int_0^1\left[U - \frac{l}{\Delta t(1-\tau)}\right]\Omega^{(m)}(\tau)h_2^{(m)}G(l,\Delta t(1-\tau);0,0)d\tau \; - \\
&\frac{U\Delta t}{2D}\int_0^1\Omega^{(m)}(\tau)h_1^{(m)}G(0,\Delta t(1-\tau);0,0)d\tau \; + \\
&\Delta t\int_0^1\Omega^{(m)}(\tau)\varphi_2^{(m)}G(l,\Delta t(1-\tau);0,0)d\tau \; + \\
&\Delta t\int_0^1\Omega^{(m)}(\tau)\varphi_1^{(m)}G(0,\Delta t(1-\tau);0,0)d\tau \; - \\
&\frac{l}{D}\int_0^1\Omega_j(\zeta)h_j^{(1)}G(l\zeta,\Delta t;0,0)d\zeta = 0 \quad j,m=1,2
\end{aligned} \tag{6.29a}$$

and the second arises when the source point x_i is located at x_2

$$\begin{aligned}
&-\frac{1}{2}h_2^{(2)} - \frac{\Delta t}{2D}\int_0^1\left[U + \frac{l}{\Delta t(1-\tau)}\right]\Omega^{(m)}(\tau)h_1^{(m)}G(-l,\Delta t(1-\tau);0,0)d\tau \; + \\
&\frac{U\Delta t}{2D}\int_0^1\Omega^{(m)}(\tau)h_2^{(m)}G(0,\Delta t(1-\tau);0,0)d\tau \; + \\
&\Delta t\int_0^1\Omega^{(m)}(\tau)\varphi_1^{(m)}G(-l,\Delta t(1-\tau);0,0)d\tau \; - \\
&\Delta t\int_0^1\Omega^{(m)}(\tau)\varphi_2^{(m)}G(0,\Delta t(1-\tau);0,0)d\tau \; - \\
&\frac{l}{D}\int_0^1\Omega_j(\zeta)h_j^{(1)}G(l(\zeta-1),\Delta t;0,0)d\zeta = 0 \quad j,m=1,2
\end{aligned} \tag{6.29b}$$

Combining eqs. (6.29a) and (6.29b) gives the element matrix equation of the form

$$R_{ij}^{(m)} h_j^{(m)} + L_{ij}^{(m)} \varphi_j^{(m)} = 0, \quad i,j,m = 1,2 \tag{6.30}$$

Equation (6.30) is the third Green element model for the transient homogeneous transport equation which we shall denote as the **ADGE** model. It provides information on the solution within a spatial element $[x_1, x_2]$ and in an interval of time Δt. The expressions for its element matrices are obtained by standard integration procedures. They are:

$$\begin{aligned} R_{11}^{(2)} &= -\frac{1}{2} - \frac{U\Delta t}{2D}\int_0^1 [\Omega^{(2)}(\tau) G(0, \Delta t(1-\tau); 0,0) d\tau \\ &= \frac{1}{2}\left[-1 + \frac{\sqrt{\beta}}{\sqrt{\pi}}\int_0^1 \frac{\tau}{(1-\tau)^{1/2}} \exp[-\beta(1-\tau)] d\tau\right] \\ &= \frac{1}{2}\left[-1 + \frac{\exp(-\beta)}{\sqrt{\pi\beta}} + erf(\sqrt{\beta})(1 - \frac{1}{2\beta})\right] \end{aligned} \tag{6.31a}$$

$$\begin{aligned} R_{12}^{(2)} &= \frac{\Delta t}{2D}\int_0^1 [U - \frac{l}{\Delta t(1-\tau)}] \Omega^{(2)}(\tau) G(l, \Delta t(1-\tau); 0,0) d\tau \\ &= \frac{1}{2\sqrt{\pi}}\int_0^1 \left(\frac{\sqrt{\gamma}}{1-\tau} - \beta\right) \frac{\tau}{(1-\tau)^{1/2}} \exp\left[-\left(\beta(1-\tau) + \frac{\gamma}{1-\tau} + 2\sqrt{\gamma\beta}\right)\right] d\tau \\ &= \frac{1}{4}\left\{\left(\frac{\sqrt{\gamma}}{\sqrt{\beta}} + 2\right) erfc(\sqrt{\gamma} + \sqrt{\beta}) + \frac{\sqrt{\gamma}}{\sqrt{\beta}} \exp(-4\sqrt{\gamma\beta}). \right. \\ &\left. [erfc(\sqrt{\beta} - \sqrt{\gamma}) - 2] + 2\frac{\sqrt{\beta}}{\sqrt{\pi}} \exp(-2\sqrt{\gamma\beta}) \int_0^1 \sqrt{\tau} \exp\left[-\left(\beta\tau + \frac{\gamma}{\tau}\right)\right] d\tau \right\} \end{aligned} \tag{6.31b}$$

$$
\begin{aligned}
R_{21}^{(2)} &= \frac{\Delta t}{2D}\int_0^1 [U - \frac{l}{\Delta t(1-\tau)}]\Omega^{(2)}(\tau)G(-l,\Delta t(1-\tau);0,0)d\tau \\
&= \frac{1}{2\sqrt{\pi}}\int_0^1 \left(\frac{\sqrt{\gamma}}{1-\tau} - \beta\right)\frac{\tau}{(1-\tau)^{1/2}}\exp\left[-\left(\beta(1-\tau)+\frac{\gamma}{1-\tau}+2\sqrt{\gamma\beta}\right)\right]d\tau \\
&= \frac{1}{4}\left\{\frac{\sqrt{\gamma}}{\sqrt{\beta}}\exp(4\sqrt{\gamma\beta})erfc(\sqrt{\gamma}+\sqrt{\beta}) + \left(\frac{\sqrt{\gamma}}{\sqrt{\beta}}-2\right)[erfc(\sqrt{\beta}-\sqrt{\gamma})-2] - \right. \\
&\left. 2\frac{\sqrt{\beta}}{\sqrt{\pi}}\exp(2\sqrt{\gamma\beta})\int_0^1\sqrt{\tau}\exp\left[-\left(\beta\tau+\frac{\gamma}{\tau}\right)\right]d\tau\right\}
\end{aligned}
\tag{6.31c}
$$

$$
\begin{aligned}
R_{22}^{(2)} &= -\frac{1}{2} - \frac{U\Delta t}{2D}\int_0^1 [\Omega^{(2)}(\tau)G(0,\Delta t(1-\tau);0,0)d\tau \\
&= \frac{1}{2}\left[-1 + \frac{\sqrt{\beta}}{\sqrt{\pi}}\int_0^1 \frac{\tau}{(1-\tau)^{1/2}}\exp[-\beta(1-\tau)]d\tau\right] \\
&= -\frac{1}{2}\left[1 + \frac{\exp(-\beta)}{\sqrt{\pi\beta}} + erf(\sqrt{\beta})(1-\frac{1}{2\beta})\right]
\end{aligned}
\tag{6.31d}
$$

$$
\begin{aligned}
L_{11}^{(2)} &= \Delta t\int_0^1 [\Omega^{(2)}(\tau)G(0,\Delta t(1-\tau);0,0)d\tau \\
&= \frac{-l}{4\sqrt{\gamma\pi}}\int_0^1 \frac{\tau}{(1-\tau)^{1/2}}\exp[-\beta(1-\tau)]d\tau \\
&= \frac{-l}{4\sqrt{\pi\gamma}}\left[\frac{\exp(-\beta)}{\beta} + erf(\sqrt{\beta})\frac{\sqrt{\pi}}{\sqrt{\beta}}(1-\frac{1}{2\beta})\right]
\end{aligned}
\tag{6.31e}
$$

$$
\begin{aligned}
L_{12}^{(2)} &= -\Delta t\int_0^1 \Omega^{(2)}(\tau)G(l,\Delta t(1-\tau);0,0)d\tau \\
&= \frac{l}{4\sqrt{\gamma\pi}}\int_0^1 \frac{\tau}{(1-\tau)^{1/2}}\exp\left[-\left(\beta(1-\tau)+\frac{\gamma}{1-\tau}+2\sqrt{\gamma\beta}\right)\right]d\tau \\
&= \frac{-l\exp(-2\sqrt{\gamma\beta})}{4\sqrt{\pi\gamma}}\left\{\int_0^1\sqrt{\tau}\ \exp\left[-\left(\beta\tau+\frac{\gamma}{\tau}\right)\right]d\tau + \right. \\
&\left. \frac{\sqrt{\pi}\exp(2\sqrt{\gamma\beta})}{2\sqrt{\beta}}\left[erfc(\sqrt{\gamma}+\sqrt{\beta}) + \exp(-4\sqrt{\gamma\beta})[erfc(\sqrt{\beta}-\sqrt{\gamma})-2]\right]\right\}
\end{aligned}
\tag{6.31f}
$$

$$
\begin{aligned}
L_{21}^{(2)} &= \Delta t \int_0^1 \Omega^{(2)}(\tau) G(-l, \Delta t(1-\tau); 0,0)\, d\tau \\
&= -\frac{l}{4\sqrt{\gamma\pi}} \int_0^1 \frac{\tau}{(1-\tau)^{1/2}} \exp\left[-\left(\beta(1-\tau) + \frac{\gamma}{1-\tau} - 2\sqrt{\gamma\beta}\right)\right] d\tau \\
&= \frac{l \exp(2\sqrt{\gamma\beta})}{4\sqrt{\pi\gamma}} \left\{ \int_0^1 \sqrt{\tau}\, \exp\left[-\left(\beta\tau + \frac{\gamma}{\tau}\right)\right] d\tau + \right. \\
&\left. \frac{\sqrt{\pi}\exp(2\sqrt{\gamma\beta})}{2\sqrt{\beta}} \left[erfc(\sqrt{\gamma}+\sqrt{\beta}) + \exp(-4\sqrt{\gamma\beta})[erfc(\sqrt{\beta}-\sqrt{\gamma}) - 2]\right] \right\}
\end{aligned}
\tag{6.31g}
$$

$$
\begin{aligned}
L_{22}^{(2)} &= -\Delta t \int_0^1 [\Omega^{(2)}(\tau) G(0, \Delta t(1-\tau); 0,0)\, d\tau \\
&- \frac{l}{4\sqrt{\gamma\pi}} \int_0^1 \frac{\tau}{(1-\tau)^{1/2}} \exp[-\beta(1-\tau)]\, d\tau = -L_{11}^{(2)}
\end{aligned}
\tag{6.31h}
$$

$$
\begin{aligned}
R_{11}^{(1)} &= \frac{U\Delta t}{2D} \int_0^1 \Omega^{(1)}(\tau) G(0, \Delta t(1-\tau); 0,0)\, d\tau - \frac{l}{D} \int_0^1 \Omega_1(\zeta) G(l\zeta, \Delta t; 0,0)\, d\zeta \\
&= \frac{1}{2\sqrt{\pi}} \left\{ \sqrt{\beta} \int_0^1 (1-\tau)^{1/2} \exp[-\beta(1-\tau)]\, d\tau + \right. \\
&\left. 2\sqrt{\gamma} \int_0^1 (1-\zeta) \exp[-(\beta + \gamma\zeta^2 + 2\sqrt{\gamma\beta}\zeta)]\, d\zeta \right\} \\
&= \frac{1}{2\sqrt{\pi}} \left\{ \left[\frac{1}{2\beta} - \left(1 + \frac{\sqrt{\beta}}{\sqrt{\gamma}}\right)\right] \sqrt{\pi}\, erf(\sqrt{\beta}) - \left(\frac{1}{\sqrt{\beta}} + \frac{1}{\sqrt{\gamma}}\right) \exp(-\beta) + \right. \\
&\left. \frac{1}{\sqrt{\gamma}} \left[\exp[-(\sqrt{\gamma}+\sqrt{\beta})^2] + \sqrt{\pi}(\sqrt{\gamma}+\sqrt{\beta})\, erf(\sqrt{\gamma}+\sqrt{\beta})\right] \right\}
\end{aligned}
\tag{6.31i}
$$

$$R_{12}^{(1)} = \frac{\Delta t}{2D}\int_0^1 [U - \frac{l}{\Delta t(1-\tau)}]\Omega_1(\tau)G(l,\Delta t(1-\tau);0,0)d\tau - \frac{l}{D}\int_0^1 \Omega_2(\zeta)G(l\zeta,\Delta t;0,0)d\zeta$$

$$= \frac{1}{2\sqrt{\pi}}\left\{\int_0^1\left(\frac{\sqrt{\gamma}}{\sqrt{(1-\tau)}} - \sqrt{\beta(1-\tau)}\right)\exp\left[-\left(\beta(1-\tau)+\frac{\gamma}{1-\tau}+2\sqrt{\gamma\beta}\right)\right]d\tau + \right.$$

$$\left. +2\sqrt{\gamma}\int_0^1 \zeta\exp[-(\beta+\gamma\zeta^2+2\sqrt{\gamma\beta}\zeta)]d\zeta\right\} \qquad (6.31j)$$

$$= \frac{1}{2\sqrt{\pi}}\left\{\frac{\sqrt{\pi\gamma}}{2\sqrt{\beta}}[\exp(-4\sqrt{\gamma\beta})(2 - erfc(\sqrt{\beta}-\sqrt{\gamma}) - erfc(\sqrt{\gamma}+\sqrt{\beta})] + \right.$$

$$\frac{1}{\sqrt{\gamma}}\left[\exp(-\beta) - \exp[-(\sqrt{\gamma}+\sqrt{\beta})^2] + \sqrt{\pi\beta}[erf(\sqrt{\beta}) - erf(\sqrt{\gamma}+\sqrt{\beta})]\right] -$$

$$\left.\sqrt{\beta}\exp(-2\sqrt{\gamma\beta})\int_0^1\sqrt{\tau}\ \exp\left[-\left(\beta\tau+\frac{\gamma}{\tau}\right)\right]d\tau\right\}$$

$$R_{21}^{(1)} = \frac{-\Delta t}{2D}\int_0^1[U + \frac{l}{\Delta t(1-\tau)}]\Omega^{(1)}(\tau)G(-l,\Delta t(1-\tau);0,0)d\tau -$$

$$\frac{l}{D}\int_0^1\Omega_1(\zeta)G(l(1-\zeta),\Delta t;0,0)d\zeta$$

$$= \frac{1}{2\sqrt{\pi}}\left[\int_0^1\left(\frac{\sqrt{\gamma}}{\sqrt{(1-\tau)}} + \sqrt{\beta(1-\tau)}\right)\ \exp\left[-\left(\beta(1-\tau)+\frac{\gamma}{1-\tau}+2\sqrt{\gamma\beta}\right)\right]d\tau + \right.$$

$$\left. 2\sqrt{\gamma}\int_0^1(1-\zeta)\exp[-(\beta+\gamma(\zeta-1)^2+2\sqrt{\gamma\beta}(\zeta-1))]d\zeta\right] \qquad (6.31k)$$

$$= \frac{1}{2\sqrt{\pi}}\left\{\frac{\sqrt{\pi\gamma}}{2\sqrt{\beta}}\exp(4\sqrt{\gamma\beta})\left[\exp(-4\sqrt{\gamma\beta})[2 - erfc(\sqrt{\beta}-\sqrt{\gamma})] - erfc(\sqrt{\gamma}+\sqrt{\beta})\right] + \right.$$

$$\frac{1}{\sqrt{\gamma}}\left[\exp(-\beta) - \exp[-(\sqrt{\gamma}-\sqrt{\beta})^2] + \sqrt{\pi\beta}[erf(\sqrt{\beta}) - erf(\sqrt{\beta}-\sqrt{\gamma})]\right] +$$

$$\left.\sqrt{\beta}\exp(2\sqrt{\gamma\beta})\int_0^1\sqrt{\tau}\ \exp\left[-\left(\beta\tau+\frac{\gamma}{\tau}\right)\right]d\tau\right\}$$

$$
\begin{aligned}
R_{22}^{(1)} &= \frac{U\Delta t}{2D}\int_0^1 \Omega^{(1)}(\tau)G(0,\Delta t(1-\tau);0,0)d\tau - \frac{l}{D}\int_0^1 \Omega_2(\zeta)G(l(1-\zeta),\Delta t;0,0)d\zeta \\
&= \frac{1}{2\sqrt{\pi}}\left\{-\sqrt{\beta}\int_0^1 (1-\tau)^{1/2}\exp[-\beta(1-\tau)]d\tau \; + \right. \\
&\left. 2\sqrt{\gamma}\int_0^1 \zeta\exp\left[-\left(\beta+\gamma(\zeta-1)^2+2\sqrt{\gamma\beta}(\zeta-1)\right)\right]d\zeta\right] \\
&= \frac{1}{2\sqrt{\pi}}\left\{\left(1-\frac{1}{2\beta}-\frac{\sqrt{\beta}}{\sqrt{\gamma}}\right)\sqrt{\pi}erf(\sqrt{\beta}) + \left(\frac{1}{\sqrt{\beta}}-\frac{1}{\sqrt{\gamma}}\right)\exp(-\beta) \; + \right. \\
&\left. \frac{1}{\sqrt{\gamma}}\left[\exp[-(\sqrt{\beta}-\sqrt{\gamma\beta})^2]-\sqrt{\pi}(\sqrt{\gamma}-\sqrt{\beta})erf(\sqrt{\beta}-\sqrt{\gamma})\right]\right\}
\end{aligned}
\tag{6.31l}
$$

$$
\begin{aligned}
L_{11}^{(1)} &= \Delta t\int_0^1 [\Omega^{(1)}(\tau)G(0,\Delta t(1-\tau);0,0)d\tau \\
&= \frac{-l}{4\sqrt{\gamma\pi}}\int_0^1 (1-\tau)^{1/2}\exp[-\beta(1-\tau)]d\tau = \frac{l}{4\sqrt{\pi\gamma}}\left[\frac{\exp(-\beta)}{\beta} \quad erf(\sqrt{\beta})\frac{\sqrt{\pi}}{\sqrt{\beta^3}}\right]
\end{aligned}
\tag{6.31m}
$$

$$
\begin{aligned}
L_{12}^{(1)} &= -\Delta t\int_0^1 \Omega^{(1)}(\tau)G(l,\Delta t(1-\tau);0,0)d\tau \\
&= \frac{l}{4\sqrt{\gamma\pi}}\int_0^1 (1-\tau)^{1/2}\exp\left[-\left(\beta(1-\tau)+\frac{\gamma}{1-\tau}+2\sqrt{\gamma\beta}\right)\right]d\tau \\
&= \frac{l\exp(-2\sqrt{\gamma\beta})}{4\sqrt{\pi\gamma}}\int_0^1 \sqrt{\tau}\;\exp\left[-\left(\beta\tau+\frac{\gamma}{\tau}\right)\right]d\tau
\end{aligned}
\tag{6.31n}
$$

$$
\begin{aligned}
L_{21}^{(1)} &= \Delta t\int_0^1 \Omega^{(1)}(\tau)G(-l,\Delta t(1-\tau);0,0)d\tau \\
&= -\frac{l}{4\sqrt{\gamma\pi}}\int_0^1 (1-\tau)^{1/2}\exp\left[-\left(\beta(1-\tau)+\frac{\gamma}{1-\tau}-2\sqrt{\gamma\beta}\right)\right]d\tau \\
&= \frac{-l\exp(2\sqrt{\gamma\beta})}{4\sqrt{\pi\gamma}}\int_0^1 \sqrt{\tau}\;\exp\left[-\left(\beta\tau+\frac{\gamma}{\tau}\right)\right]d\tau
\end{aligned}
\tag{6.31o}
$$

$$
\begin{aligned}
L_{22}^{(1)} &= -\Delta t \int_0^1 [\Omega^1(\tau) G(0, \Delta t(1-\tau); 0,0) d\tau \\
&= \frac{l}{4\sqrt{\gamma\pi}} \int_0^1 (1-\tau)^{1/2} \exp[-\beta(1-\tau)] d\tau = -L_{11}^{(1)}
\end{aligned}
\tag{6.31p}
$$

where γ has earlier been defined by eq. (5.38a), *erf* and *erfc* are the error function and complementary error function which are given by eq. (5.38b) and the parameter β is defined as

$$
\beta = \frac{U^2 \Delta t}{4D} \tag{6.32}
$$

There are a number of line integrals which can not be evaluated exactly because, as in the case of the **TGE** model, the free-space Green's function is complicated. That was not the case with the first model in which the free-space Green's function was easier to handle.

6.5 Stability Characteristics of GEM for the Transport Equation

Here in this chapter we derived three time-marching formulations of the Green element method for the transient transport (diffusion-advection) equation. The numerical process uses the known data at the beginning of the time step to obtain their new values at the end of the time step on the basis of the numerical replication of the differential equation. For a well-posed initial-value problem, the solution at any time level is expected to be bounded, and that can only be achieved by a numerical scheme which does not allow the solution and accumulated numerical errors to grow out of bounds.

The stability characteristics of the three Green element models for the transport equation are assessed by the same approach adopted in section 5.6 for the transient linear diffusion equation (Gray and Pinder [1] and Gray and Lynch [2]). Using that approach, the general solution of the linear diffusion-advection equation is represented as an infinite series of Fourier wave components, and the stability characteristics of the numerical scheme are assessed by comparing the amount by which the numerical scheme modifies the amplitude and phase angle of each wave component in relation to their theoretical values. The ideal situation is for the numerical scheme to modify after each time step the amplitudes and phase angles of the Fourier wave components by the same amount as their corresponding analytic values. A scheme is said to be dissipative if it is unable to correctly modify the amplitude of each Fourier wave component after each time step. The numerical difficulty posed by problems with solution profiles that have large gradients or steep fronts, characterized predominantly by the short wavelengths, is that a dissipative scheme artificially introduces additional diffusion, thereby smearing these steep gradients. The inability of a numerical scheme to modify the phase angle of each wave component by the correct amount introduces dispersive effects. Dispersive effects cause the small wave lengths to be out phase with the longer wave lengths, thereby producing oscillatory features in the numerical solution. When the solution does not have steep gradients, then the long wavelengths predominantly characterize the solution, and the effects of numerical dispersion and dissipation on the solution become insignificant.

The stability analysis is carried out on eq. (6.1) under uniform ambient flow, and with no first-order decay nor external forcing function term. That is

$$D\frac{\partial^2 h}{\partial x^2} - U\frac{\partial h}{\partial x} - \frac{\partial h}{\partial t} = 0 \tag{6.34}$$

We propose a general solution similar to eq. (5.46) which is an infinite series of Fourier-type components of the form

$$h = \sum_{n=-\infty}^{\infty} \Phi_n \exp[\hat{i}(\sigma_n x + \beta_n t)] \tag{6.35}$$

where for each n-th Fourier wave component, $\beta_n = 2\pi/T_n$ is the temporal frequency, T_n is the period, $\sigma_n = 2\pi/L_n$ is the spatial frequency or wave number, and L_n is the wave length, and $\hat{i}$ is the imaginary representation of a complex variable. Because of the linear nature of eq. (6.34), one term of the solution is considered at a time. With that eq. (6.35) is then expressed as

$$h = h_i \sim \Phi \exp[\hat{i}(\sigma x + \beta t)] \tag{6.36}$$

We have dropped the subscript n from the parameters in eq. (6.36) because only one term of the series is considered. The relationship between the temporal frequency β and the wave number σ is obtained by introducing eq. (6.36) into the differential equation (6.34). When that is done the following relationship is obtained

$$\beta = \sigma(\hat{i}D\sigma - U) \tag{6.37}$$

Eq. (6.37) is introduced into the general solution given by eq. (6.36) to yield

$$h = h_t \sim \Phi \exp[\hat{i}\sigma(x - Ut)]\exp(-D\sigma^2 t) \tag{6.38}$$

The first exponential in eq. (6.38) indicates that each Fourier wave component is a translatory wave whose speed of propagation is U, while the second exponential describes the modification in amplitude of the wave with time. In other words after a simulation time Δt the Fourier wave would have been propagated a distance of $U\Delta t$, and its amplitude reduced by $\exp(-D\sigma^2\Delta t)$ of its original value. Eq. (6.38) gives the solution at the beginning of the time level t, and that at the end of the time step $t + \Delta t$ is

$$\begin{aligned} h_{t+\Delta t} &\sim \Phi \exp[\hat{i}\sigma(x - U(t+\Delta t))]\exp[-D\sigma^2(t+\Delta t)] \\ &= h_t \exp(\hat{i}\sigma U\Delta t)\exp(-D\sigma^2\Delta t) = h_t g_T \end{aligned} \tag{6.39}$$

where g_T is the theoretical amplification of each Fourier wave after one time step which, from eq. (6.39), has the expression

$$g_T = \exp(-\hat{i}\sigma U\Delta t)\exp(-D\sigma^2\Delta t) \tag{6.40}$$

In general g_T is a complex parameter that is expressed as

$$g_T = |g_T|\exp(\hat{i}\theta_T) \tag{6.41}$$

where θ_T is the theoretical phase change after one time step. Comparing eqs. (6.40) and (6.41), the magnitude or modulus of the theoretical amplification is

$$|g_T| = \exp(-D\sigma^2\Delta t) \tag{6.42}$$

and the theoretical phase change after one time step is

$$\theta_T = -\sigma U\Delta t \tag{6.43}$$

Our goal in this analysis is to compare the theoretical propagation characteristics of each Fourier wave component to those that arise from the three Green element replications of the differential equation. Let us denote the modulus and phase change of each Fourier wave component obtained from the numerical approximation of the differential equation as $|g_N|$ and θ_N, respectively, then the ratio of the modulus of the numerical amplification to that of theoretical amplification after one time step is

$$T = \left|\frac{g_N}{g_T}\right| = \frac{|g_N|}{\exp[-D\sigma^2\Delta t]} \tag{6.44}$$

where T is the amplification factor. And the phase lag at the end of one time step is

$$\Theta = \theta_N - \theta_T = \theta_N - \sigma U\Delta t \tag{6.45}$$

We may wish to evaluate the amplification factor and the phase lag at a time when the Fourier wave has propagated one wavelength at the theoretical speed, rather than at the end of a simulation time as in eqs. (6.44) and (6.45). To achieve this we need to know the number of time steps that would elapse for the Fourier wave to propagate one wavelength. It is given by

$$N_t = \frac{L}{U\Delta t} = \frac{L}{l}\frac{l}{U\Delta t} = \frac{L}{lCr} \tag{6.46}$$

where Cr is a dimensionless parameter, namely the Courant number. It is herein defined as

$$Cr = \frac{U\Delta t}{l} \tag{6.47}$$

With eq. (6.46) the amplification factor at the end of one wavelength is

$$T = \left|\frac{g_N}{g_T}\right|^{N_t} = \left[\frac{|g_N|}{\exp(-D\sigma^2\Delta t)}\right]^{N_t} = \left[\frac{|g_N|}{\exp[-4\pi^2 E l^2/L^2]}\right]^{N_t} \tag{6.48}$$

in which E is another dimensionless parameter that is referred to as the *element dimensionless diffusion coefficient*. It has the expression

$$E = \frac{D\Delta t}{l^2} \tag{6.49}$$

Both dimensionless parameters Cr and E can be combined to give the *element Peclet number* Pe which is defined as

$$Pe = \frac{Ul}{D} = \frac{Cr}{E} \tag{6.50}$$

The element Peclet number is indicative of the relative magnitude of advection transport to diffusion. The corresponding expression for the phase lag at the end of one wavelength propagation of the Fourier wave component is

$$\Theta = 2\pi\left[\frac{\theta_N}{\theta_T}\right] - 2\pi = 2\pi\left[\frac{\theta_N}{\sigma l Cr} - 1\right] \tag{6.51}$$

To evaluate T and Θ that are given by the expressions (6.48) and (6.51) for the three Green element models proposed for the diffusion-advection equation, we first assume that the computational region is discretized uniformly by linear elements each of which has a length l (see Figure 5.3). All three Green element replications of the differential equation, which for the **QSGE** model is eq. (6.11), the **TGE** model is eq. (6.20), and for the **ADGE** model is eq. (6.30), can be expressed as

$$P_{ij}^{(m)} h_j^{(m)} + N_{ij}^{(m)} \varphi_j^{(m)} = 0, \quad i,j,m = 1,2, \tag{6.52}$$

where for the **QSGE** model

$$P_{ij}^{(2)} = \alpha R_{ij} + \frac{T_{ij}}{D\Delta t}, \quad P_{ij}^{(1)} = \omega R_{ij} - \frac{T_{ij}}{D\Delta t},$$
$$N_{ij}^{(2)} = \alpha[L_{ij} + UT_{ij}], \quad N_{ij}^{(1)} = \omega[L_{ij} + UT_{ij}] \tag{6.53}$$

and for both the **TGE** and **ADGE** models these equivalent relations apply

$$P_{ij}^{(m)} \equiv R_{ij}^{(m)}, \quad N_{ij}^{(m)} \equiv L_{ij}^{(m)}, \quad i,j,m=1,2 \tag{6.54}$$

We recognize that two quantities h and φ are computed at each node by the numerical scheme, requiring two discrete equations to be generated at the node x_i; the first discrete equation is obtained by considering the segment $[x_{i-1},x_i]$ with the source node at x_i

$$P_{22}^{(m)}h_i^{(m)} + P_{21}^{(m)}h_{i-1}^{(m)} + N_{22}^{(m)}\varphi_i^{(m)} + N_{21}^{(m)}\varphi_{i-1}^{(m)} = 0 \tag{6.55}$$

and the second discrete equation comes from evaluating eq. (6.52) in the element $[x_i,x_{i+1}]$ with the source node at x_i

$$P_{11}^{(m)}h_i^{(m)} + P_{12}^{(m)}h_{i+1}^{(m)} + N_{11}^{(m)}\varphi_i^{(m)} + N_{12}^{(m)}\varphi_{i+1}^{(m)} = 0 \tag{6.56}$$

It should be noted that φ also satisfies the differential equation (6.34) so that a general solution like that of eq (6.36) can be written for φ, that is

$$\varphi = \varphi_t \sim X\exp(\hat{i}\sigma x)\exp(-D\sigma^2 t) \tag{6.57}$$

From the prescribed general solutions of eqs. (6.36) and (6.57), the following relations are deduced

$$h_{i+1}^{(2)} = h_i^{(2)}\exp(\hat{i}\sigma l) = g_N h_i^{(1)}\exp(\hat{i}\sigma l) \tag{6.58a}$$

$$h_{i-1}^{(2)} = h_i^{(2)}\exp(-\hat{i}\sigma l) = g_N h_i^{(1)}\exp(-\hat{i}\sigma l) \tag{6.58b}$$

$$\varphi_{i+1}^{(2)} = \varphi_i^{(2)}\exp(\hat{i}\sigma l) = g_N \varphi_i^{(1)}\exp(\hat{i}\sigma l) \tag{6.58c}$$

$$\varphi_{i-1}^{(2)} = \varphi_i^{(2)}\exp(-\hat{i}\sigma l) = g_N \varphi_i^{(1)}\exp(-\hat{i}\sigma l) \tag{6.58d}$$

Introducing the above relations into the system of discrete element equations of (6.55) and (6.56), we obtain the numerical stability matrix equation which is given by

$$\begin{bmatrix} a_{11}g_N+b_{11} & a_{12}g_N+b_{12} \\ a_{21}g_N+b_{21} & a_{22}g_N+b_{22} \end{bmatrix} \begin{pmatrix} h_i^{(1)} \\ \varphi_i^{(1)} \end{pmatrix} = 0 \tag{6.59}$$

and where the elements of the matrix are

$$a_{11} = a_{11}^r + \hat{i}a_{11}^i = P_{22}^{(2)} + P_{21}^{(2)}\cos(\sigma l) - \hat{i}P_{21}^{(2)}\sin(\sigma l) \tag{6.60a}$$

$$a_{12} = a_{12}^r + \hat{i}a_{12}^i = N_{22}^{(2)} + N_{21}^{(2)}\cos(\sigma l) - \hat{i}N_{21}^{(2)}\sin(\sigma l) \tag{6.60b}$$

$$a_{21} = a_{21}^r + \hat{i}a_{21}^i = P_{11}^{(2)} + P_{12}^{(2)}\cos(\sigma l) + \hat{i}P_{12}^{(2)}\sin(\sigma l) \tag{6.60c}$$

$$a_{22} = a_{22}^r + \hat{i}a_{22}^i = N_{11}^{(2)} + N_{12}^{(2)}\cos(\sigma l) + \hat{i}N_{12}^{(2)}\sin(\sigma l) \tag{6.60d}$$

$$b_{11} = b_{11}^r + \hat{i}b_{11}^i = P_{22}^{(1)} + P_{21}^{(1)}\cos(\sigma l) - \hat{i}P_{21}^{(1)}\sin(\sigma l) \tag{6.60e}$$

$$b_{12} = b_{12}^r + \hat{i}b_{12}^i = N_{22}^{(1)} + N_{21}^{(1)}\cos(\sigma l) - \hat{i}N_{21}^{(1)}\sin(\sigma l) \tag{6.60f}$$

$$b_{21} = b_{21}^r + \hat{i}b_{21}^i = P_{11}^{(1)} + P_{12}^{(1)}\cos(\sigma l) + \hat{i}P_{12}^{(1)}\sin(\sigma l) \tag{6.60g}$$

$$b_{22} = b_{22}^r + \hat{i}b_{22}^i = N_{11}^{(1)} + N_{12}^{(1)}\cos(\sigma l) + \hat{i}N_{12}^{(1)}\sin(\sigma l) \tag{6.60h}$$

The superscripts r and i on the elements of the matrix denote the real and imaginary components of those elements. We eliminate $h_i^{(1)}$ and $\varphi_i^{(1)}$ from eq. (6.59) to obtain an expression for the numerical amplification g_N. It is given by the quadratic equation

$$Ag_N^2 + Bg_N + C = 0 \tag{6.61}$$

where

$$A = a_{22}a_{11} - a_{12}a_{21} \tag{6.62a}$$

$$B = a_{22}b_{11} + a_{11}b_{22} - a_{12}b_{21} - a_{21}b_{12} \tag{6.62b}$$

$$C = b_{11}b_{22} - b_{12}b_{21} \tag{6.62c}$$

The expression for g_N is readily obtained from

$$g_N = \frac{-B \pm \sqrt{B^2 - 4AC}}{2A} \tag{6.63}$$

Two values of g_N are obtained from eq. (6.63). One of them is inadmissible; that inadmissible root of g_N has the character of being either unacceptably too large or too small or uniform in value for most values of the Courant number. A numerical scheme is said to be stable if

$$|g_N| < 1 \tag{6.64}$$

The numerical phase change after a time step Δt is given as

$$\exp(\hat{i}\theta_N) = \frac{g_N}{|g_N|} \tag{6.65}$$

The expressions for g_N from eq. (6.63) and θ_N from eq. (6.65) are used to compute the amplification factor T and phase lag Θ that are given by eqs. (6.48) and (6.51), respectively. If the numerical scheme correctly reproduces the propagation of each Fourier wave component in amplitude and speed, then $T = 1$ and $\Theta = 0$. If $T > 1$, then the numerical scheme amplifies the Fourier wave amplitude more than its theoretical rate, and conversely if $T < 1$. Whenever the value of T deviates from the ideal value, the numerical scheme is said to be dissipative, and when Θ deviates from the ideal value, the scheme is said to be dispersive.

Although we have only given the general expressions for g_N and θ_N by eqs. (6.63) and (6.65) for the three Green element models of the transport equation, it may be necessary to indicate the parameters on which those general expressions depend. That information comes from examining the contents of the element matrices $P_{ij}^{(m)}$ and $N_{ij}^{(m)}$ for the three models. The **QSGE** model depends on the temporal weighting factor α, the Courant number Cr, and the element Peclet number Pe, whereas the **TGE** and **ADGE** models depend only on the two latter parameters.

Our next step in this analysis involves using selected values of the Courant number Cr and element Peclet number Pe to evaluate, assess, and compare the propagation characteristics of the three Green element models. Two modes of the transport process are examined: the first has a Peclet value of 2 which represents a marginal advection-dominant transport process, and the second represents a strong advection-dominant transport with a Peclet value of 50. The influence of temporal discretization on the numerical solutions, represented by the value of the Courant number, is assessed for these two transport modes. The numerical calculations use Courant number values of 0.2 and 1.0. A small value of Courant number is indicative of a numerical computation with small temporal element size.

Our first set of results is presented in Figures 6.1a, 6.1b, 6.2a, and 6.2b, and they show the variations of the modulus of the numerical amplification g_N with respect to the ratio of the wavelength to the spatial element size L/l. Because the performance of the **QSGE** model depends on the temporal weighting factor α, four selected values of α

are incorporated into the model and examined. These are: $\alpha = 0.5, 0.67, 0.75$, and 1.0. The results in those figures show that all three models (**QSGE**, **TGE**, and **ADGE**) are stable for all values of L/l. Our assessment here of stability of these schemes is based on the condition expressed by eq. (6.64). A broader perspective on stability of these schemes is provided by the propagation characteristics of the Fourier wave components *vis-a-vis* their theoretical propagation values, and these can be found in the next set of plots in Figures 6.3 through 6.6. These figures show the variation of the amplification factor T and phase lag Θ at the end of one wavelength of propagation with respect to L/l. The results in Figures 6.3a and 6.3b for the marginal advection-dominant transport ($Pe = 2$) with a Courant number of 0.2 show that the Crank-Nicholson scheme ($\alpha = 0.5$) and **ADGE** model exhibit the least phase lag, but the latter is more dissipative. At this value of Courant number harmonics smaller than 3.5 times the spatial element size are poorly propagated in speed by all the numerical schemes which, apart from the fully implicit scheme, damp out the amplitudes of these harmonics. The **TGE** and **ADGE** give the most dissipative schemes, while the Crank-Nicholson scheme is least dissipative. When the Courant number is increased to unity (indicative of numerical calculations with larger time step) for the same transport mode ($Pe = 2$), the propagation characteristics of the Fourier harmonics by the numerical schemes become slightly different (see Figures 6.4a and 6.4b). Only harmonics with wavelengths larger than 4.5 times the spatial element size l are correctly propagated in speed by the Crank-Nicholson scheme, the **TGE** and **ADGE** models. Unlike the case when $Cr = 0.2$, the amplitudes of these short wavelengths are amplified. The surprising result is that the **TGE** model better propagates the Fourier harmonics in speed and amplitude at this larger Courant value of unity than at 0.2, while the **ADGE** shows this same trend with the propagation of the amplitudes of the harmonics. One is yet to find a theoretical justification for this feature. It is also observed that at unit value of Courant number all schemes of the **QSGE** model do poorly in propagating the amplitudes of the harmonics, whereas the **TGE** and **ADGE** models provide better results.

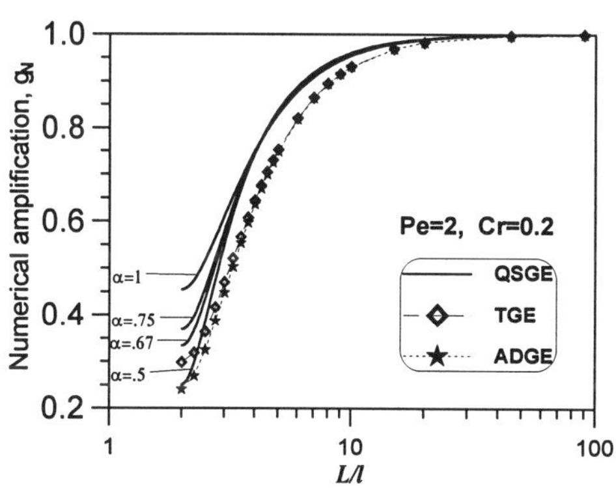

Figure 6.1a Modulus of numerical amplitude modification for $Pe = 2$ and $Cr = 0.2$.

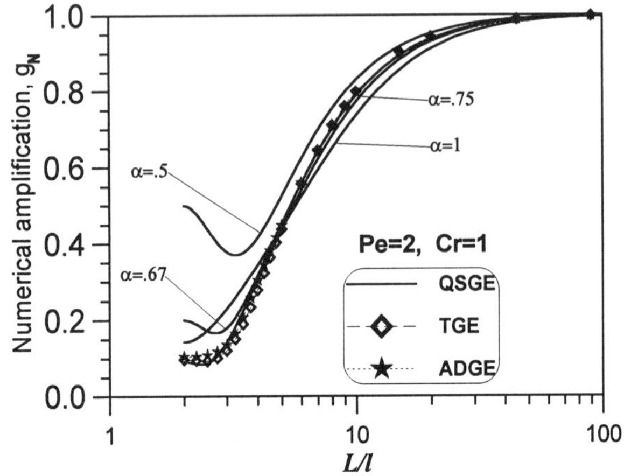

Figure 6.1b Modulus of numerical amplitude modification for $Pe = 2$ and $Cr = 1$.

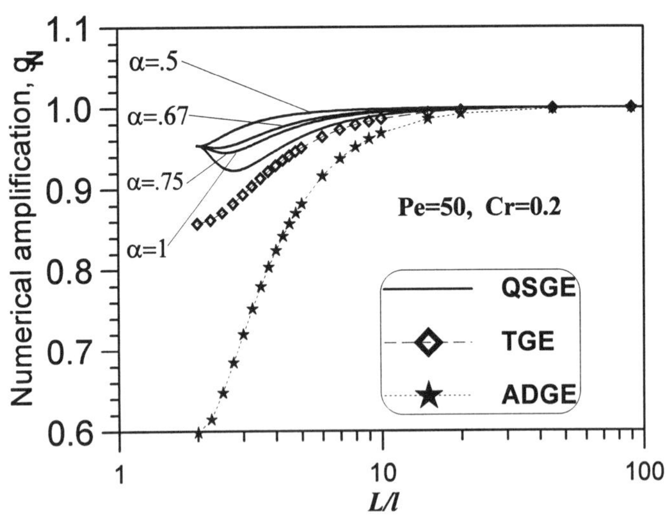

Figure 6.2a Modulus of numerical amplitude modification for $Pe = 50$ and $Cr = 0.2$.

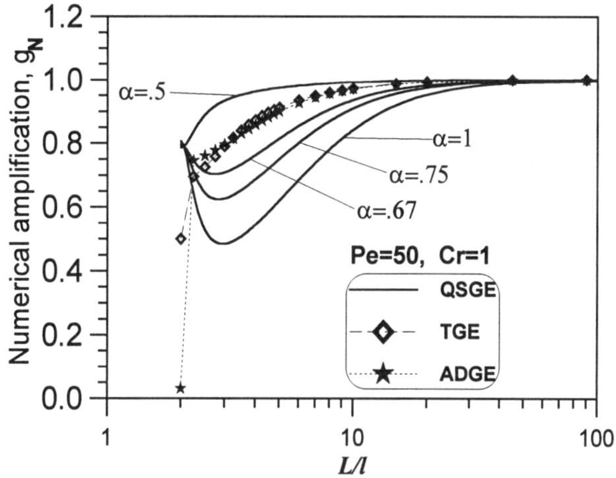

Figure 6.2b Modulus of numerical amplitude modification for $Pe = 50$ and $Cr = 1$.

For the strong advection-dominant transport mode ($Pe = 50$), the numerical schemes with Courant number value of 0.2 exhibit a trend similar to that of the marginal advection-dominant transport, except that of the **ADGE** model whose phase lag is now larger than that of the Crank-Nicholson scheme (see Figures 6.5a and 6.5b). All four **QSGE** schemes exhibit virtually identical phase angles. The Crank-Nicholson scheme offers optimal propagation of the Fourier harmonics. That scheme correctly propagates in speed and amplitude all harmonics whose wavelengths are larger than 4 times the spatial element size l. For the same transport mode ($Pe = 50$) with larger value of Courant number ($Cr = 1$), all **QSGE** schemes are more dissipative and dispersive than at smaller Courant number (Figures 6.6a and 6.6b). The Crank-Nicholson scheme and the **ADGE** model give the least phase lag, but the former amplifies the amplitudes of the small harmonics. The **TGE** and **ADGE** models again propagate the harmonics at unit value of Cr better than at 0.2. The above results are summarized below:

1. The Crank-Nicholson scheme gives optimal stability characteristics at small values of Courant number for all transport modes.
2. The stability characteristics of the **TGE** and **ADGE** are improved when a unit value Courant number is used.
3. For strong advection-dominant transport, none of the numerical schemes correctly propagates the small Fourier wavelengths when a larger time step is used.

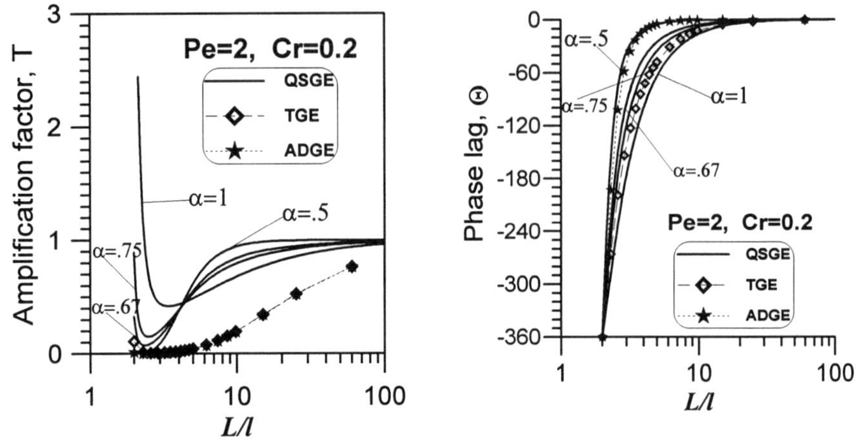

(a) (b)

Figure 6.3 **(a)** Amplification factor, **(b)** Phase lag for $Pe = 2$ and $Cr = 0.2$.

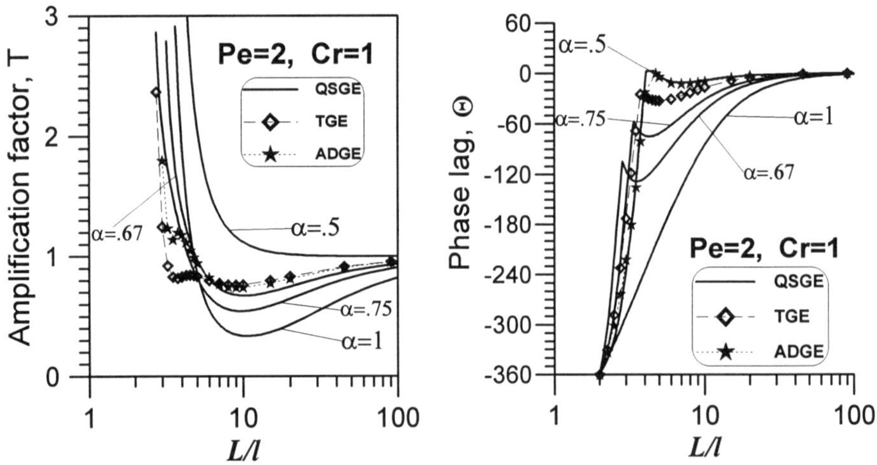

(a) (b)

Figure 6.4 **(a)** Amplification factor, **(b)** Phase lag for $Pe = 2$ and $Cr = 1$.

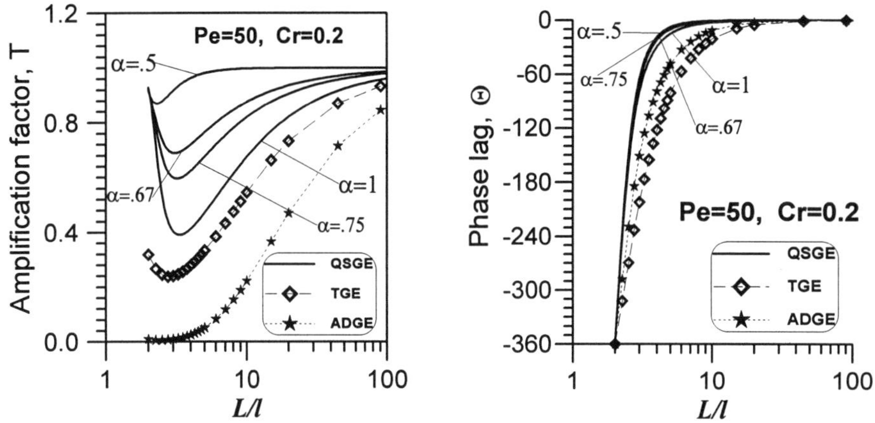

(a) **(b)**

Figure 6.5 **(a)** Amplification factor, **(b)** Phase lag for $Pe = 50$ and $Cr = 0.2$.

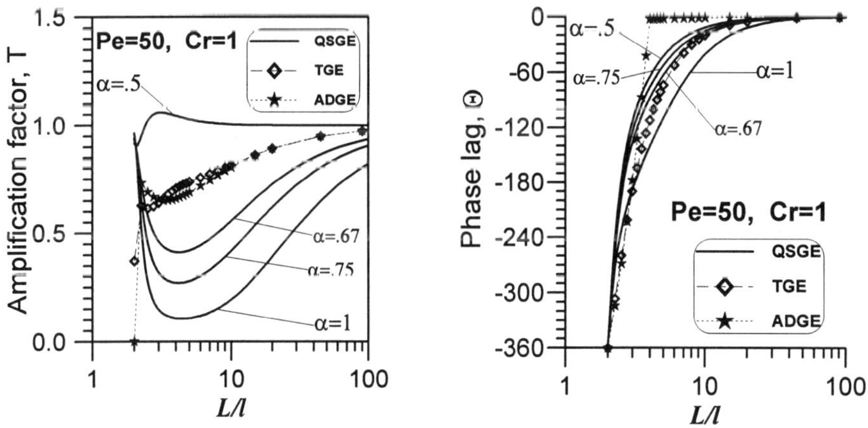

(a) **(b)**

Figure 6.6: **(a)** Amplification factor, **(b)** Phase lag for $Pe = 50$ and $Cr = 1$.

6.6 Numerical Examples

Three numerical examples of transient 1-D. transport (diffusion-advection) problems are used to demonstrate the capabilities and characteristics of the three Green element models. Since one of the unique features of the solution to the transport problem is that the initial concentration profile is maintained with time when advection is dominant (large value of Peclet number), the examples have been so designed that we can evaluate how well the models behave under such situations, especially for initial concentration profiles with steep gradients.

Example 1

This first example is the classical 1-D. transport problem which is used quite often to assess the performance of many numerical schemes. The problem is governed by eq. (6.1) with $\mu = 0$ and $f(x,t) = 0$, and uniform ambient velocity $u(x,t) \equiv U$. The boundary and initial conditions are

$$h(0,t) = 1, \quad \frac{\partial h(x=\infty,t)}{\partial x} = 0, \text{ and } \quad h(x,t=0) = 0 \tag{6.66}$$

The solution to this transport problem with the above conditions is well known and restated below (Ogata and Banks [3])

$$h(x,t) = \frac{1}{2}\left\{ erfc\left(\frac{x - Ut}{2\sqrt{Dt}} \right) + \exp\left(\frac{Ux}{D} \right) erfc\left(\frac{x + Ut}{2\sqrt{Dt}} \right) \right\} \tag{6.67}$$

To assess the performance of the three earlier derived Green element models in simulating this example, two cases or modes of transport, the same as those used in the stability analysis of section 6.5, are examined. One case is marginal advection-dominant transport of element Peclet number value of 2, and the second case represents advection-dominant transport of element Peclet number of 50. The numerical computations are carried out with uniform ambient velocity $U = 1$ and uniform spatial element size $l = 0.025$. The solution of the concentration front is presented at $t = 0.5$. With these simulation parameters the size of the temporal element is measured by the value of the Courant number. We examine two values of Courant number: $Cr = 0.2$ and $Cr = 1$. Whereas Pe and Cr are adequate to evaluate the performance of the **TGE** and **ADGE** models, additional information on the value of α is required to evaluate that of the **QSGE** model. To this end four values of α ($\alpha = 0.5$, 0.67, 0.75, and 1) are incorporated into the model. To evaluate the quality of the numerical solutions, three error estimates are employed. They are the L_2 norm of the deviations between the exact and numerical solutions, denoted as e_l, the error in peak concentration e_p (indicative of the overshoot), and maximum negative concentration value e_n. (The last two error estimates are, for this example, indicative of spurious oscillations or wiggles.) The expressions for these three error estimates are

$$e_l = \left\{ \sum_i [h_i^N(t) - h_i^E]^2 \right\}^{1/2} \tag{6.68a}$$

$$e_p = |\max h_i^N(t) - \max h_i^E(t)| \tag{6.68b}$$

$$e_n = \max|h_i^N(t)<0| \tag{6.68c}$$

where the superscripts N and E represent the numerical and exact solutions, and the subscript i indicates the spatial nodal locations where the concentration values are computed. These error estimates are presented in Table 6.1 for the marginal advection-dominant transport ($Pe = 2$) when $Cr = 0.2$ and $Cr = 1$ are incorporated in the numerical calculations. All the numerical schemes exhibit negligible spurious oscillations or wiggles. The breakthrough concentration curves of the exact and numerical solutions at $t = 0.5$ are presented in Figures 6.7a and 6.7b.

The same surprising feature of the **TGE** and **ADGE** models, that was observed with the stability analysis, is again noticed with these solutions; their solutions are better with $Cr = 1$ than with $Cr = 0.2$. The solutions of the four schemes of the **QSGE** model are virtually the same, and they correctly reproduce the concentration front.

For the strong advection-dominant transport mode ($Pe = 50$), the numerical solutions are again presented in two forms: in terms of the error estimates (Table 6.2), and as breakthrough concentration profiles (Figures 6.8a and 6.8b). While the Crank-Nicholson scheme ($\alpha = 0.5$) gives optimal results when $Cr = 0.2$, the **ADGE** model offers comparable results that are free of oscillations or wiggles when $Cr = 1$. These results are consistent with those obtained from the stability analysis. When $Cr = 0.2$ the **ADGE** model gives the most dissipative solution (smeared concentration front), and when $Cr = 1$ the fully implicit scheme of the **QSGE** model gives the most dissipative solution. The oscillations of the Crank-Nicholson solution are amplified with increased value of the Courant number.

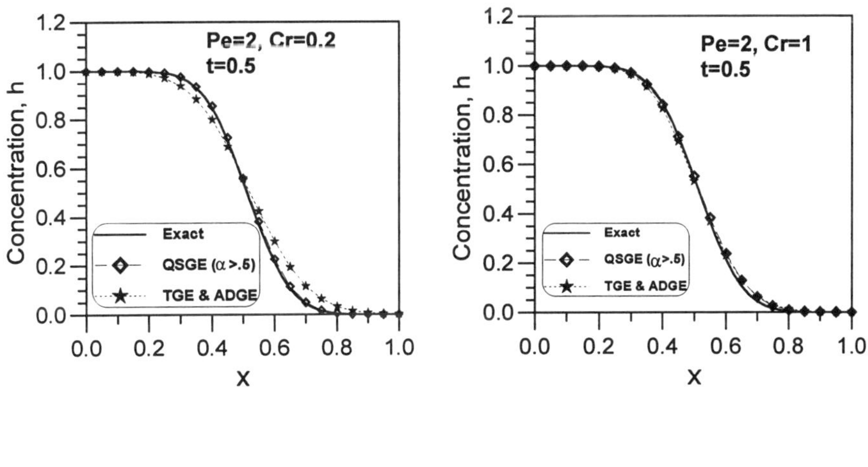

(a) **(b)**

Figure 6.7 GEM and exact solutions of example 1 — **(a)** $Pe = 2$ and $Cr = 0.2$, **(b)** $Pe = 2$ and $Cr = 1.0$.

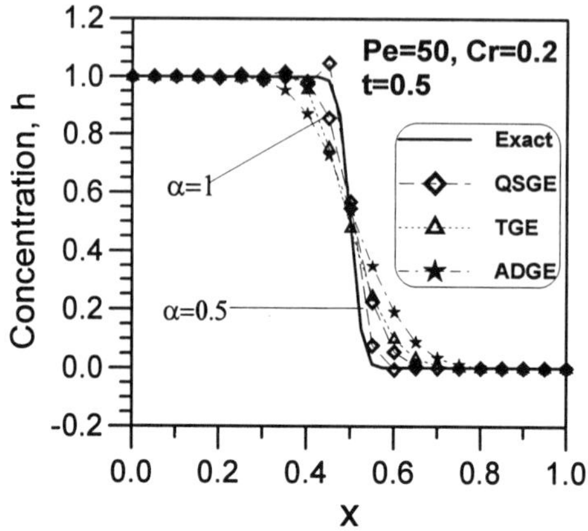

(a)

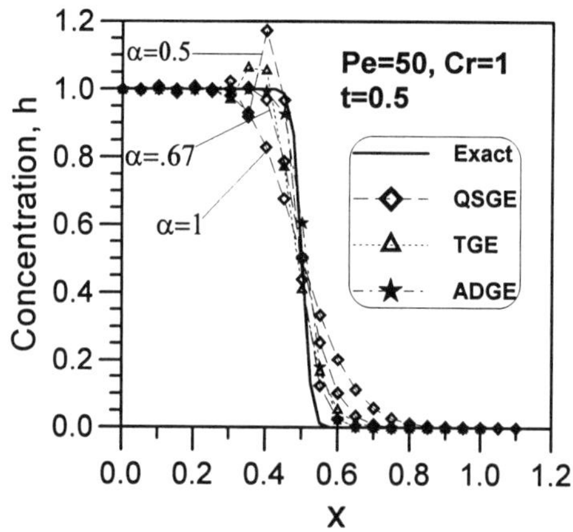

(b)

Figure 6.8 GEM and exact solutions of example 1 — **(a)** $Pe = 50$ and $Cr = 0.2$, **(b)** $Pe = 50$ and $Cr = 1.0$.

These results indicate that for advection dominant transport, small values of Courant number or small time steps have to be incorporated into the **QSGE** calculations to achieve reasonably good results, while a Courant number of unity is required for the **TGE** and **ADGE** models to offer optimal solutions.

Table 6.1 Error estimates of Green element solutions of example 1 ($Pe = 2$)

Numerical scheme		$Cr = 0.2$			Numerical scheme		$Cr = 1$		
		e_l x10^{-2}	e_p x10^{-5}	e_n			e_l x10^{-2}	e_p x10^{-5}	e_n
QSGE: α	=0.5	4.29	0.0	0.0	**QSGE**: α	=0.5	3.53	0.20	0.0
	=0.67	4.95	9.54	0.0		=0.67	12.30	0.0	0.0
	=0.75	5.70	0.0	0.0		=0.75	17.34	9.54	0.0
	=1.0	8.57	0.0	0.0		=1.0	30.36	0.0	0.0
TGE		26.18	0.0	0.0	**TGE**		7.70	0.0	0.0
ADGE		26.56	0.0	0.0	**ADGE**		8.14	0.0	0.0

Table 6.2 Error estimates of Green element solutions of example 1 ($Pe = 50$)

Numerical scheme		$Cr = 0.2$			Numerical scheme		$Cr = 1$		
		e_l x10^{-1}	e_p x10^{-3}	e_n x10^{-3}			e_l x10^{-1}	c_p x10^{-3}	c_n
QSGE: α	=0.5	1.81	47.55	5.44	**QSGE**: α	=0.5	3.70	173.2	0.0
	=0.67	2.57	5.76	0.58		=0.67	5.05	5.04	0.0
	=0.75	3.01	1.35	0.16		=0.75	6.00	0.15	0.0
	=1.0	4.07	0.52	0.0		=1.0	7.91	0.015	0.0
TGE		5.2	16.78	0.0	**TGE**		4.52	92.15	0.0
ADGE		7.62	0.0	0.0	**ADGE**		3.28	0.0	0.0

Example 2

Our second example is one in which an initial Gauss-hill concentration profile is subjected to hydrodynamic dispersion and advection. The governing differential equation is

$$D\frac{\partial^2 h(x,t)}{\partial x^2} = U\frac{\partial h(x,t)}{\partial x} + \frac{\partial h(x,t)}{\partial t} \qquad -\infty \leq x \leq \infty \tag{6.69}$$

The boundary conditions and initial Gauss-hill profile are given by

$$\varphi(x=-\infty,t) = 0, \ \varphi(\infty,t) = 0, \quad t>0, \ h(x,0) = \exp\left[-\frac{(x-x_0)^2}{2\sigma_0^2}\right] \tag{6.70}$$

These results indicate that for advection dominant transport, small values of Courant number or small time steps have to be incorporated into the **QSGE** calculations to achieve reasonably good results, while a Courant number of unity is required for the **TGE** and **ADGE** models to offer optimal solutions.

The exact solution is

$$h(x,t) = \frac{\sigma_0}{\sigma}\exp\left[-\frac{(x-\bar{x})^2}{2\sigma^2}\right] \tag{6.71}$$

in which

$$\sigma^2 = \sigma_0^2 + 2Dt \quad \text{and} \quad \bar{x} = x_0 + \int_0^t u(s)ds \tag{6.72}$$

We have used the following parameters in our simulations $x_0 = 0$ and $\sigma_0 = 0.1$. The numerical solutions are obtained at time $t = 0.5$. As in the first example, we examine two transport modes: one is marginal advection-dominant transport with $Pe = 2$, and the other is strong advection-dominant transport with $Pe = 50$. Both cases are simulated with values of Courant number of 0.2 and 1 incorporated into the Green element models. The quality of the numerical solutions is again assessed by the error estimates e_l, e_p and e_n. The interpretation of e_p and e_n for this example are, respectively, error in peak concentration and oscillations or wiggles upstream and downstream of the concentration hill. With the two values of Cr incorporated in the numerical models, the error estimates of their solutions are presented in Table 6.3 for the marginal advection-dominant transport, while the concentration breakthrough curves at $t = 0.5$ can be found in Figures 6.9a and 6.9b. For both $Cr = 0.2$ and $Cr = 1$, the Crank-Nicholson scheme reproduces the exact solution, and the fully implicit scheme gives the most diffused solution of all the **QSGE** schemes. The **TGE** and **ADGE** solutions are identical, being more diffused at $Cr = 0.2$ than at $Cr = 1$ — a trend which has now become associated with these two models. In fact at $Cr = 1$ both models closely reproduce the exact solution, but their solutions are inferior to that of the Crank-Nicholson scheme. The results for e_n in Table 6.3 indicate that none of the Green element solutions exhibit any significant oscillations or wiggles upstream or downstream of the concentration hill. With increased dominance of advection on the transport process ($Pe = 50$), the Crank-Nicholson scheme ($\alpha = 0.5$) still shows superior results at both $Cr = 0.2$ and $Cr = 1$ (see Figures 6.10a and 6.10b), while the fully implicit scheme of the **QSGE** model exhibits the most diffused solution of all the **QSGE** schemes. While both the **TGE** and **ADGE** models perform poorly at $Cr = 0.2$, they give significantly better solutions at $Cr = 1$.

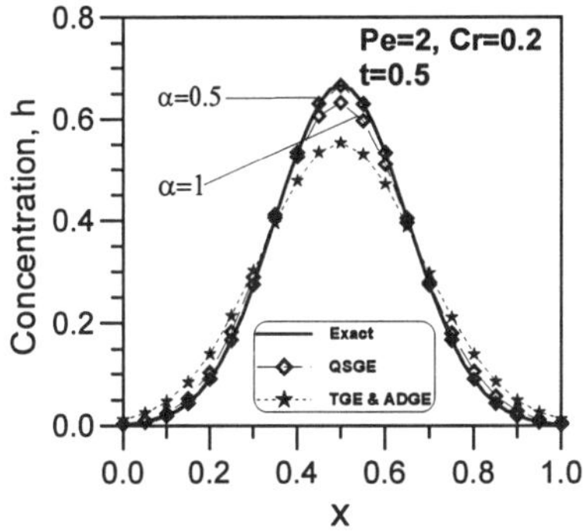

(a)

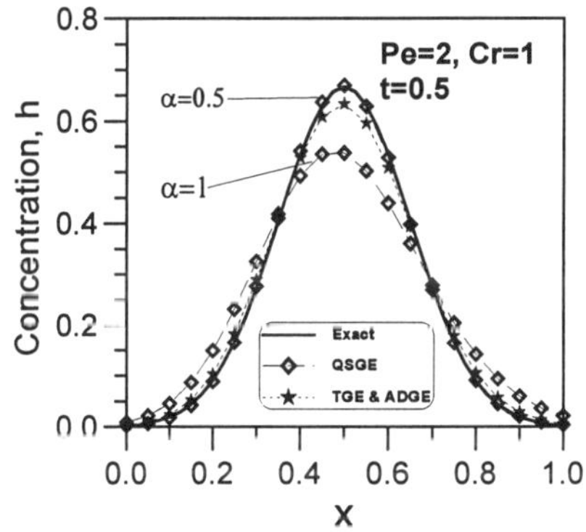

(b)

Figure 6.9 GEM and exact solutions of example 2 with uniform advection; — **(a)** $Pe = 2$ and $Cr = 0.2$, **(b)** $Pe = 2$ and $Cr = 1.0$.

Table 6.3 Error estimates of Green element solutions of example 2 with uniform advection ($Pe = 50$)

Numerical scheme		$Cr = 0.2$			Numerical scheme		$Cr = 1$		
		e_l x10^{-3}	e_p x10^{-3}	e_n			e_l 10^{-2}	e_p x10^{-3}	e_n
QSGE: α	=0.5	3.30	1.13	0.0	**QSGE**: α	=0.5	2.25	2.27	0.0
	=0.67	37.28	12.80	0.0		=0.67	15.91	51.82	0.0
	=0.75	53.90	18.42	0.0		=0.75	22.62	74.42	0.0
	=1.0	101.6	34.47	0.0		=1.0	39.58	126.20	0.0
TGE		334.7	113.60	0.0	**TGE**		10.17	33.46	0.0
ADGE		335.0	113.70	0.0	**ADGE**		10.81	36.69	0.0

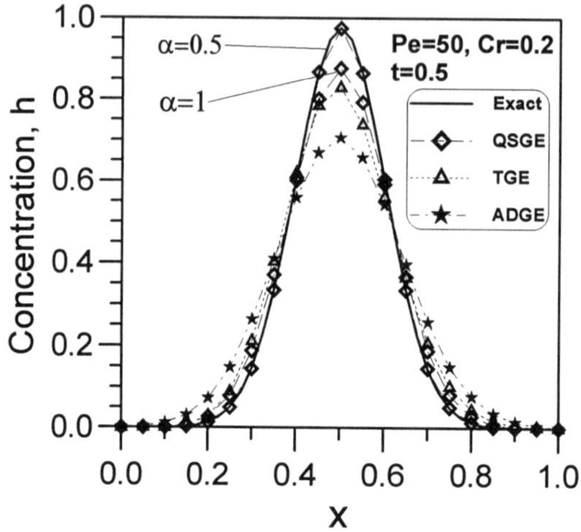

(a)

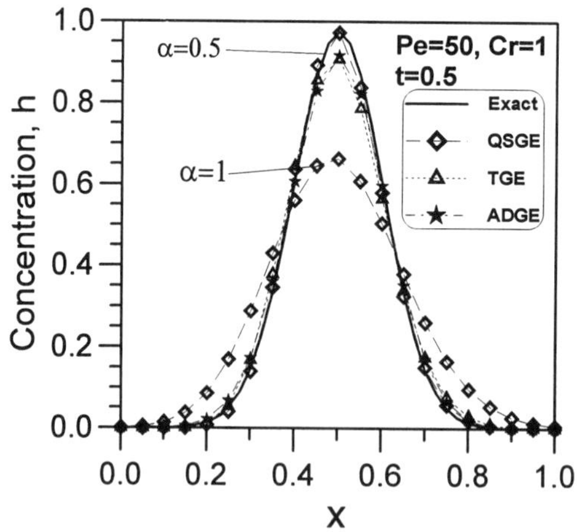

(b)

Figure 6.10 GEM and exact solutions of example 2 with uniform advection — **(a)** $Pe = 50$ and $Cr = 0.2$, **(b)** $Pe = 50$ and $Cr = 1.0$.

Table 6.4: Error estimates of Green element solutions of example 2 with uniform advection ($Pe = 50$)

Numerical scheme		$Cr = 0.2$			Numerical scheme		$Cr = 1$		
		e_l x10^{-3}	e_p x10^{-3}	e_n x 10^{-7}			e_l x10^{-2}	e_p x10^{-3}	e_n x10^{-7}
QSGE: α	=0.5	4.83	0.36	7.98	**QSGE**: α	=0.5	8.579	3.32	26.19
	=0.67	86.77	36.76	5.30		=0.67	36.48	148.30	0.18
	=0.75	126.7	53.46	6.61		=0.75	49.68	200.70	2.14
	=1.0	236.8	98.79	5.72		=1.0	79.41	312.40	1.52
TGE		365.3	147.50	2.28	**TGE**		193.5	67.08	20.94
ADGE		674.3	270.00	0.00	**ADGE**		142.7	59.71	0.0

Using this same second example with $D = 5\text{x}10^{-4}$, the ability of the numerical models to handle a situation where the velocity is time dependent is examined. The imposed ambient velocity is given by

$$U(t) = 4\sin(4\pi t) \tag{6.73}$$

Two sets of simulations are carried out; one uses a uniform time step of 0.005, and the other a time step of 0.025. The numerical and exact solutions at $t = 0.5$ are presented in Figures 6.11a and 6.11b, while the error parameters of the numerical solutions are found in Table 6.5. The **ADGE** model is simply not suited for this transport problem, and as such no solution of the model is presented. With both time steps the Crank-Nicholson scheme gives optimal results, while the fully implicit scheme gives highly dissipative solutions. The **TGE** solution is dissipative when the time step is 0.005, but overshoots the peak concentration at the larger time step of 0.025.

Table 6.5: Error estimates of Green element solutions of example 2 with advection velocity $U = 4\sin(4\pi t)$ ($D = 0.0005$)

Numerical scheme		$\Delta t = 0.005$			Numerical scheme		$\Delta t = 0.025$		
		e_l x10^{-2}	e_p x10^{-3}	e_n x10^{-6}			e_l	e_p x10^{-3}	e_n x10^{-6}
QSGE: α	= 0.5	4.54	0.019	0.001	**QSGE**: α	= 0.5	0.23	4.67	20.400
	= 0.67	51.88	210.50	0.021		= 0.67	1.23	471.70	0.213
	= 0.75	68.78	275.10	0.015		= 0.75	1.45	549.30	0.026
	= 1.0	102.9	400.50	0.610		= 1.0	1.77	660.90	0.000
TGE		34.39	141.90	0.000	**TGE**		1.04	445.50	0.000

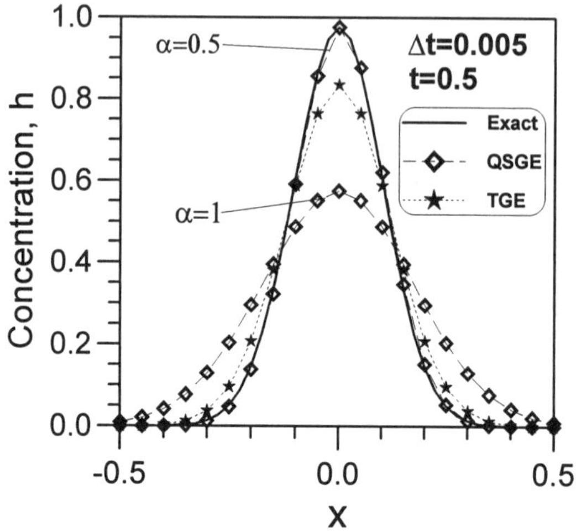

(a)

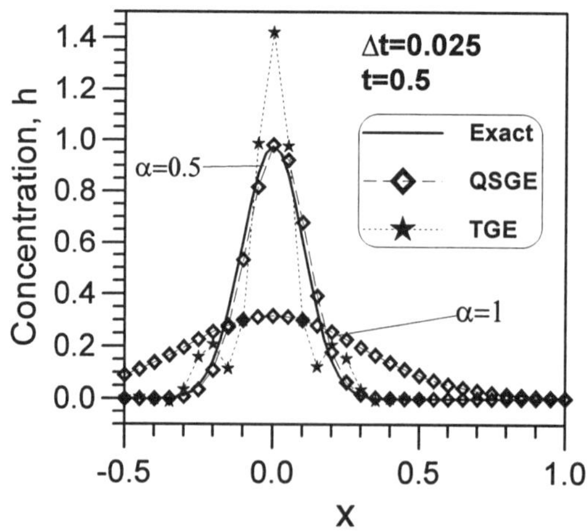

(b)

Figure 6.11 GEM and exact solutions of example 2 with time dependent advection — **(a)** $\Delta t = 0.005$, **(b)** $\Delta t = 0.025$.

Example 3

The third example of transient 1-D. transport employed to validate our models is one in which first-order decay of the advecting contaminant is allowed to take place. The physical interpretation of this transport example, which brings out its engineering significance, is that of a contaminated stream being cleaned or decontaminated using mechanical means of advection from a freshwater source and an induced biological/chemical first-order decay of the pollutant.

$$D\frac{\partial^2 h(x,t)}{\partial x^2} = U\frac{\partial h(x,t)}{\partial x} + \frac{\partial h(x,t)}{\partial t} + \mu h(x,t) \qquad 0 \leq x \leq \infty \tag{6.74}$$

The initial and boundary conditions are:

$$h(x,t=0) = 1, \; h(x=0,t>0) = 0, \text{ and } \varphi(x=\infty,t>0) = 0 \tag{6.75}$$

The exact solution is given by van Genuchten and Alves [4]

$$h(x,t) = \exp(-\mu t)\left\{1 - \frac{1}{2}erfc\left[\frac{x-Ut}{2\sqrt{Dt}}\right] - \frac{1}{2}\exp(\frac{Ux}{D})erfc\left[\frac{x+Ut}{2\sqrt{Dt}}\right]\right\} \tag{6.76}$$

Two cases of this example with the following parameters are examined: $U = 1$, $D = 0.025$ with corresponding element Peclet number value of 2, $\mu = 0.2$, and $U = 1$, $D = 0.0005$ with corresponding element Peclet number value of 50, and $\mu = 0.2$. For this example the **ADGE** model is not applied because the first-order decay term was not considered in it. The numerical calculations with the **QSGE** model incorporates the Crank-Nicholson scheme, and a uniform spatial element size of 0.025. The numerical and exact solutions, obtained at times of 0.5 and 1.0, are presented in Figure 6.12a for the first case, while those of the second case are presented in Figure 6.12b. For both cases the Crank-Nicholson scheme reproduced the exact solution more closely than the **TGE** model. The numerical solutions have trailing oscillations for the strong advection-dominant transport but their strength appears to decrease as simulation time increases.

6.7 Remarks

The three Green element models presented in this chapter for the transport equation have arisen from the different portions of the equation from which the free-space Green's function was derived. The **QSGE** model derived its free-space Green's function from the diffusive term, while the **TGE** model derived its from the diffusive and temporal terms, and the **ADGE** from the diffusion-advection and temporal terms. While the derivation of their Green element equations followed a uniform approach, the main difference between the models is the nature of the expressions of their free-space Green's functions; that of the **QSGE** model being the simplest, and the **ADGE** model the most complicated. The nature of these expressions is reflected in the computational effort involved in evaluating the global coefficient matrix.

The results of the stability analysis and those from simulating three numerical examples indicate that the Crank-Nicholson scheme ($\alpha = 0.5$) of the **QSGE** model provides optimal solutions of the transport equation. However, acceptable solutions from that scheme for the strong advection-dominant transport can only be guaranteed if the Courant number value incorporated into the scheme is much less than unity. Optimal solutions from the **TGE** and **ADGE** models are achieved when the Courant number is unity, but their solutions are in most cases inferior to those of the **QSGE** model with $\alpha = 0.5$.

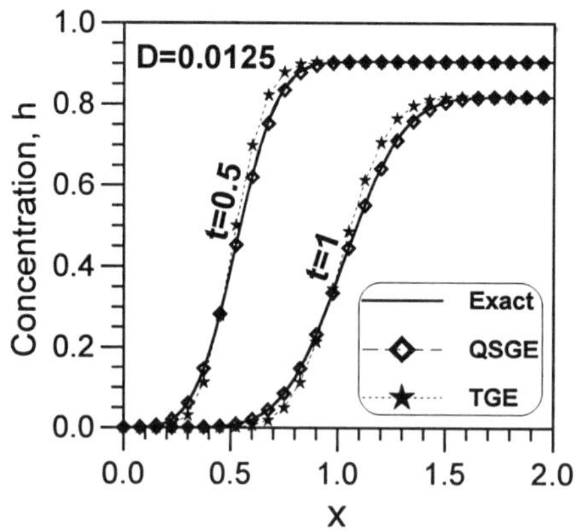

(a)

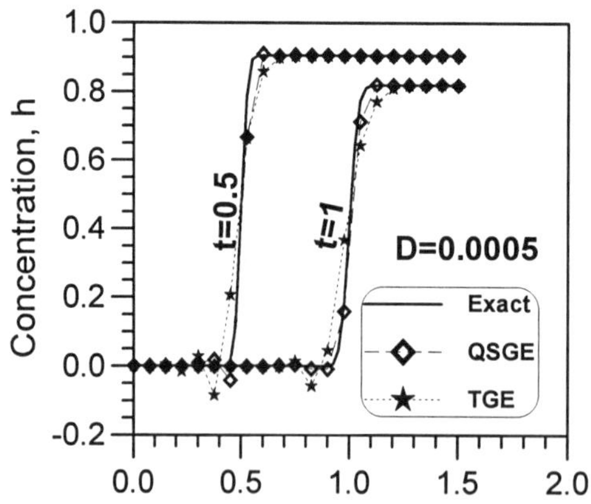

(b)

Figure 6.12 GEM and exact solutions of example 3 — **(a)** $\mu = 0.2$ and $D = 0.0125$, **(b)** $\mu = 0.2$ and $D = 0.0005$.

6.8 Exercises

6.1 Derive the Green element equations for the 1-D. transient transport equation so that there is only one degree of freedom at each node. Use these new models to solve the examples presented in this chapter, and compare with our results.

6.2 Examine the stability characteristics of the Green element models derived in Exercise 6.1. Comment on your results.

6.3 Examine the performance of the **QSGE** model for the transport equation when the temporal derivative is approximated by the relationship

$$\left.\frac{dh_j}{dt}\right]_{t=t_2} = \frac{\alpha}{\Delta t}\left(h_j^{(2)} - h_j^{(1)}\right) + (1-\alpha)\left.\frac{dh_j}{dt}\right]_{t=t_1} \quad ; \quad 1 \le \alpha \le 2$$

Compare its performance to that of the generalized 2-level time discretization scheme presented in this chapter.

6.4 Employ any Green element model to solve the transport problem

$$D\frac{\partial^2 h}{\partial x^2} = \frac{\partial h}{\partial t} + U\frac{\partial h}{\partial x}; \qquad 0 \le x \le \infty$$

with the initial and boundary conditions

$$h(x,0) = \begin{cases} 2 & 0 \le x < x_1 \\ 1 & x \ge x_1 \end{cases}$$

$h(0,t) = 2$ and $\varphi(\infty,t) = 0; \quad t>0$
Use the parameters $U = 1$, $D = 10^{-3}$, and $x_1 = 1$ in your computations. Compare your solutions with the exact solution

$$h(x,t) = 1 + \frac{1}{2}\left[erfc\left(\frac{x - x_1 - Ut}{2\sqrt{Dt}}\right) + \exp\left(\frac{Ux}{D}\right) erfc\left(\frac{x + x_1 + Ut}{2\sqrt{Dt}}\right)\right].$$

6.9 References

1. Gray, W.G. and G.F. Pinder, "An Analysis of the Numerical Solution of the Transport Equation," *Water Resources research*, **12**, pp. 547-555, 1976.

2. Gray, W.G. and D.R. Lynch, "Time-Stepping Schemes for Finite Element Tidal Model Computations," in *Surface Flow*, ed. W.G. Gray, pp.1-14, CML Pub. Ltd., Southampton UK, 1984.

3. Ogata, A. and R.B. Banks, "A solution of the differential equation of longitudinal dispersion in porous media", *U.S. Geol. Surv. Prof. Pap. 411-A.*, 7pp, 1976.

4. van Genuchten, M. Th. and W.J. Alves, "Analytical solutions of the One-dimensional Convective-Dispersive solute transport equation", *Technical Bulletin No. 1661,* 151p., U.S. Dept. of Agric., 1982.

7 BURGERS EQUATION

7.1 Introduction

In the previous chapter (chapter 6) we derived the Green element equations for the mathematical statement that describes the storage and movement of species or contaminants in a fluid. The transported specie can be considered to be the momentum of the flow, and in that case the relevant differential equations become the *Navier-Stokes equations* [1]. Simplifications of the Navier-Stokes equations can be carried out to achieve easier versions of the equations. One of such second-order simplifications yields the transient one-dimensional form of the equations which provides a useful model for many physical fluid flow phenomena — nonlinear propagating shock wave with viscous dissipation, turbulence, propagating shock waves in gases, propagating flame in a combustion chamber, etc. As a result of the extensive research works carried out by Burgers in modeling of turbulence, the simplified transient nonlinear momentum transport equation in one spatial dimension is popularly referred to as *Burgers equation* [2,3]. The nonlinear nature of Burgers equation has been exploited as a useful prototype differential equation for modeling many divers and rather unrelated phenomena such as shock flows, wave propagation in combustion chambers, vehicular traffic movement, acoustic transmission, etc. In fact, Burgers equation can be considered applicable to any flow phenomenon in which there exist the balancing effects of viscous and inertia or convective forces. It is probably one of the simplest nonlinear transient partial differential equations which exhibits some very unique features. When inertia or convective forces are dominant, its solution resembles that of the kinematic wave equation which displays a propagating wave front and boundary layers. In that case Burgers equation essentially behaves as a *hyperbolic* partial differential equation. In contrast, when viscous forces are dominant, it behaves as a parabolic equation, and any propagating wave front is smeared and diffused due to viscous action. Because of these different forms that Burgers equation can assume, coupled with its nonlinear characteristics, it has become a model equation for assessing and evaluating the performance of many computational techniques. It is for the same reason that we have devoted this chapter to the development of a number of schemes of the Green element method for the solution of the Burgers equation.

In our derivations of the various Green element models for Burgers equation, we shall proceed along familiar lines which had been set forth in earlier chapters. These involve using a suitable form of the differential operator to obtain the free-space Green's function, invoking Green's identity to derive the integral representation of the differential equation within typical spatial and temporal elements, and assembling the integral equations for all the elements in the computational domain to produce the global matrix equation that is solved for the nodal unknowns [4] In the case of Burgers equation, there are quite a number of exact solutions for various combinations of initial and boundary conditions with which to validate our Green element solutions. An excellent summary of these exact solutions has been carried out by Benton and Platzman [5]. We shall take advantage of the availability of these solutions when testing the Green element models.

Our first Green element model employs the free-space Green's function of the 1-D. Laplace differential operator to derive its element equations so that the model closely resembles the **QSGE** model of chapters 5 and 6, while the second model uses the free-space Green's function of the 1-D. linear transient diffusion equation so that this model is essentially the **TGE** model of chapters 5 and 6 [6]. The third model employs the free-space Green's function of the linear diffusion-advection differential operator, and that makes it mandatory to linearize the nonlinear convective term of the equation.

The mathematical statement that describes one-dimensional nonlinear fluid flow phenomenon of shock or wave propagation with viscous dissipation, widely known as the Burgers equation, is given by Lighthill [7]

$$\nu\frac{\partial^2 h(x,t)}{\partial x^2} - h(x,t)\frac{\partial h(x,t)}{\partial x} - \frac{\partial h(x,t)}{\partial t} = 0, \quad on \quad x_0 \le x \le x_L \tag{7.1}$$

in which $\nu = 1/R_e$ is the reciprocal of the flow Reynolds number which is a dimensionless parameter that reflects the relative magnitude of inertia forces to viscous forces, and it is defined by appropriate length and velocity scales of the flow. $h(x,t)$ is the dependent variable which denotes the velocity field, and $L = x_L - x_0$ is considered the length of the flow domain. Fletcher [8] discussed the important features of equation (7.1) as a prototype of the force balance between the nonlinear acceleration or inertia or convective term $h\partial h/\partial x$ or $h\varphi$ and the diffusive term $\nu\partial^2 h/\partial x^2$. The suitability of the Green element models depends on their ability to effectively handle this balance between these two terms of the equation. To obtain unique solutions of eq. (7.1) appropriate boundary and initial data similar to those described in section 5.3 have to be prescribed.

7.2 Model 1 for the Burgers Equation

Our first Green element model of the Burgers equation is based on the Fredholm singular integral theory which employs the free-space Green's function of the Laplace differential operator [6]. Because we have encountered similar forms of this model right from chapter 2 we can afford to skip a number of steps in our derivation. As in previous chapters, the procedures for deriving the Green element equations are:

i. Obtain an appropriate complimentary differential operator of the differential equation of interest so that its solution (fundamental solution) is available.
ii. Identify a suitable element to discretize the computational domain.
iii. Invoke Green's identity to transform the differential equation into an integral one within the element (element equations).
iv. Represent any known and unknown functional quantities in terms of their nodal values by appropriate interpolating functions.
v. Assemble the element equations for all elements that are used to discretize the computational domain to obtain the global matrix equation.
vi. Where the element equations are nonlinear, employ either the Picard or Newton-Raphson algorithm to linearize them before embarking on the previous step.

For our current model, the complementary differential operator employed is eq. (2.20) and its fundamental solution is eq. (2.30). With the use of Green's identity on eq. (7.1), it is transformed into an integral equation which applies to a typical linear element $[x_1, x_2]$

$$\begin{aligned} &\nu\Big[-\lambda h(x_i,t) + G^*(x_2,x_i)h(x_2,t) - G^*(x_1,x_i)h(x_1,t) \\ &\quad - G(x_2,x_i)\varphi(x_2,t) + G(x_1,x_i)\varphi(x_1,t)\Big] + \\ &\int_{x_1}^{x_2} G(x,x_i)\left[h(x,t)\varphi(x,t) + \frac{\partial h(x,t)}{\partial t}\right]dx = 0; \qquad i = 1,2 \end{aligned} \tag{7.2}$$

As in earlier chapters $\varphi(x,t) \equiv \partial h(x,t)/\partial x$, $G(x,x_i)$ denotes the fundamental solution which is given by eq. (2.30), and $G^*(x,x_i) = dG(x,x_i)/dx$. To simplify the line integral of eq. (7.2) some functional distribution has to be prescribed for $h(x,t)$ and $\varphi(x,t)$ over each element. We elect to use linear element shape functions to approximate these quantities

$$h(x,t) \approx \Omega_j(\zeta)h_j(t), \qquad \varphi(x,t) \approx \Omega_j(\zeta)\varphi_j(t), \quad j = 1,2 \tag{7.3}$$

We introduce eq. (7.3) into eq. (7.2) to give

$$\nu\left[-\lambda h_i + G^*(x_2,x_i)h_2 - G^*(x_1,x_i)h_1 - G(x_2,x_i)\varphi_2 + G(x_1,x_i)\varphi_1\right] + \int_{x_1}^{x_2} G(x,x_i)\left[\Omega_j h_j \Omega_n \varphi_n + \Omega_j \frac{dh_j}{dt}\right] dx = 0; \quad i,j,n = 1,2 \tag{7.4}$$

Recalling the expressions for $G(x,x_i)$ and $G^*(x,x_i)$ from eqs. (2.30) and (2.31) in chapter 2, the following relations exist:

$$G(x_1,x_1) = G(x_2,x_2) = \frac{\tilde{l}}{2}, \quad G(x_2,x_1) = G(x_1,x_2) = \frac{\tilde{l}+l}{2}$$

$$G^*(x_1,x_1) = G^*(x_2,x_2) = 0, \quad G^*(x_2,x_1) = -G^*(x_1,x_2) = \frac{1}{2}$$

where $\tilde{l}$ is the length of the longest spatial element. Introducing the above relations into eq. (7.4) gives for $i=1$

$$\nu\left[-h_1 + h_2 + \tilde{l}\varphi_1 - (\tilde{l}+l)\varphi_2\right] + l\int_0^1 \Omega_j\left(\Omega_n h_n \varphi_j + \frac{dh_j}{dt}\right)(\tilde{l}+l\zeta)d\zeta = 0, \quad j,n=1,2 \tag{7.5a}$$

and for $i=2$

$$\nu\left[h_1 - h_2 + (\tilde{l}+l)\varphi_1 - \tilde{l}\varphi_2\right] + l\int_0^1 \Omega_j\left(\Omega_n h_n \varphi_j + \frac{dh_j}{dt}\right)[\tilde{l}+l(1-\zeta)]d\zeta = 0, \quad j,n=1,2 \tag{7.5b}$$

in which $l = x_2 - x_1$. Eqs. (7.5a) and (7.5b) are combined to form the element matrix equations of the first Green element model of Burgers equation.

$$\nu R_{ij}h_j + (\nu L_{ij} + U_{inj}h_n)\varphi_j + T_{ij}\frac{dh_j}{dt} = 0, \quad i,j,n=1,2 \tag{7.6}$$

This Green element model is also referred to as the quasi-steady formulation and denoted, as that in chapters 5 and 6, **QSGE**. The element matrices in eq. (7.6) have been encountered in earlier chapters and are summarized in eqs. (6.8a) through (6.8d).

The element equations of (7.6) are not only a system of first-order ordinary differential equations in time but also nonlinear. The time derivative is simplified by the generalized 2-level time discretization scheme which was first proposed for the transient diffusion equation in chapter 5. The scheme approximates the temporal derivative at $t=t_{1+\alpha}=t_1+\alpha\Delta t$ (see Figure 5.2) by a difference expression that is given by eq. (5.20). That approximation is herein incorporated into the system of discrete equations to yield

$$\left[\alpha\nu R_{ij}+\frac{T_{ij}}{\Delta t}\right]h_j^{(2)}+\alpha\left[\nu L_{ij}+U_{ijn}h_n^{(2)}\right]\varphi_j^{(2)}+ \\ \left[\omega\nu R_{ij}-\frac{T_{ij}}{\Delta t}\right]h_j^{(1)}+\omega\left[\nu L_{ij}+U_{ijn}h_n^{(1)}\right]\varphi_j^{(1)}\equiv g_i=0 \quad i,j,n=1,2,\ 0\leq\alpha\leq 1 \tag{7.7}$$

in which α, as in chapters 5 and 6, represents a time weighting factor. Also it is noted that the bracketed superscripts 2 and 1 denote the current and previous time levels, respectively, and $\Delta t=t_2-t_1$ is the time step. Eq. (7.7) is still a system of discretized nonlinear equations. We present both a Picard solution and a Newton-Raphson solution of the system of nonlinear equations. In applying the Picard solution algorithm we need only direct our attention at the nonlinear acceleration term which is simplified by evaluating h at the previous iteration k, and φ at the current iteration $k+1$. When that is done, eq. (7.7) becomes

$$\left[\alpha\nu R_{ij}+\frac{T_{ij}}{\Delta t}\right]h_j^{(2,k+1)}+\alpha\left[\nu L_{ij}+U_{ijn}h_n^{(2,k)}\right]\varphi_j^{(2,k+1)}= \\ -\left[\omega\nu R_{ij}-\frac{T_{ij}}{\Delta t}\right]h_j^{(1)}-\omega\left[\nu L_{ij}+U_{ijn}h_n^{(1)}\right]\varphi_j^{(1)} \quad i,j,n=1,2,\ 0\leq\alpha\leq 1 \tag{7.8}$$

Eq. (7.8) is now linear in h and φ, and can be assembled for all elements which are used to discretize the computational domain to give the global matrix equation from which the values of $h_j^{(2,k+1)}$ and $\varphi_j^{(2,k+1)}$ are evaluated at the nodes. Those computed nodal values are compared with their corresponding values at the previous iteration, and convergence is said to be attained when two successive iterates are "close" to each other.

For the N-R algorithm, we proceed along the usual approach of implementing the algorithm, which is to first propose a solution $\{h_j^{(2,k)},\varphi_j^{(2,k)}\}^T$ that is then updated by

$$\begin{Bmatrix} h_j^{(2,k+1)} \\ \varphi_j^{(2,k+1)} \end{Bmatrix}=\begin{Bmatrix} h_j^{(2,k)} \\ \varphi_j^{(2,k)} \end{Bmatrix}+\begin{Bmatrix} \Delta h_j^{(2,k+1)} \\ \Delta\varphi_j^{(2,k+1)} \end{Bmatrix} \tag{7.9}$$

The increment, given by the second term on the right hand side of eq. (7.9), is obtained from the matrix equation:

$$J_{ij}^{(2,k)} \begin{Bmatrix} \Delta h_j^{(2,k+1)} \\ \Delta \varphi_j^{(2,k+1)} \end{Bmatrix} = -g_i^{(2,k)} \tag{7.10}$$

in which the Jacobian $J_{ij}^{(2,k)}$ takes the expression

$$J_{ij}^{(2,k)} = \begin{cases} \left.\dfrac{\partial g_i}{\partial h_j}\right|_{h_j = h_j^{(2,k)}} = \alpha(\nu R_{ij} + U_{ijn}\varphi_n^{(2,k)}) + \dfrac{T_{ij}}{\Delta t} \\ \\ \left.\dfrac{\partial g_i}{\partial \varphi_j}\right|_{\varphi_j = \varphi_j^{(2,k)}} = \alpha(\nu L_{ij} + U_{inj}h_n^{(2,k)}) \end{cases} \qquad 0 \le \alpha \le 1 \tag{7.11}$$

Eq. (7.10) is solved at each time step for as many number of times until convergence is achieved.

7.3 Model 2 for the Burgers Equation

Our second **GE** formulation for Burgers equation given by eq. (7.1) makes use of the free-space Green's function to the linear transient diffusion equation. This formulation, which is similar to the second Green element formulation developed for the transient diffusion equation in chapter 5, uses the time-dependent complementary differential eq. (5.24) whose solution is eq. (5.25). When Green's identity is applied in a typical spatial element $[x_1,x_2]$ and integrated in time between t_1 and t_2, it yields

$$\begin{aligned} &-\lambda h_i^{(2)} + \int_{t_1}^{t_2} \left[h(x_2,t)G^*(x_2,t;x_i,t_2) - h(x_1,t)G^*(x_1,t;x_i,t_2) \right] dt - \\ &\int_{t_1}^{t_2} \left[G(x_2,t;x_i,t_2)\varphi(x_2,t) - G(x_1,t;x_i,t_2)\varphi(x_1,t) \right] dt - \\ &\frac{1}{\nu} \left\{ \int_{x_1}^{x_2} G(x,\Delta t;x_i,0)h(x,t_1)dx + \int_{t_1}^{t_2}\int_{x_1}^{x_2} G(x,t;x_i,t_2)h(x,t)\varphi(x,t)dxdt \right\} = 0 \end{aligned} \tag{7.12}$$

A linear distribution is prescribed for $h(x,t)$ and $\varphi(x,t)$ within the spatial element $[x_1,x_2]$ and the temporal element $[t_1,t_2]$ along the line of eq. (5.31). Because of the complexity of the free-space Green's function employed in this model, it will be computationally difficult to evaluate the line integral of the nonlinear acceleration term $h(x,t)\,\varphi(x,t)$ if both quantities are approximated by linear shape functions in space and time. To simplify this term we assume that $h(x,t)$ is uniform within the spatial element so that it simplifies to

$$h(x,t)\,\varphi(x,t) \approx \bar{h}\varphi(x,t) = \frac{1}{4}[h_1^{(1)} + h_1^{(2)} + h_2^{(1)} + h_2^{(2)}]\varphi(x,t) \tag{7.13}$$

With a linear distribution for the functional quantities, eq. (7.12) simplifies to

$$\begin{aligned}
&-\lambda\varphi_i^{(2)} + \Delta t\int_0^1 \Omega^{(m)}(\tau)h_2^{(m)}G^*(x_2,\Delta t(1-\tau);x_i,0)d\tau - \\
&\Delta t\int_0^1 \Omega^{(m)}(\tau)h_1^{(m)}G^*(x_1,\Delta t(1-\tau);x_i,0)]d\tau - \\
&\Delta t\int_0^1 \Omega^{(m)}(\tau)\varphi_2^{(m)}G(x_2,\Delta t(1-\tau);x_i,0)d\tau + \\
&\Delta t\int_0^1 [\Omega^{(m)}(\tau)\varphi_1^{(m)}G(x_1,\Delta t(1-\tau);x_i,0)d\tau - \\
&\frac{l}{\nu}\int_0^1 \Omega_j(\zeta)\varphi_j^{(1)}G(l(\zeta-\zeta_i),\Delta t;0,0)d\zeta + \frac{l\Delta t}{\nu}\int_0^1\int_0^1 G(l(\zeta-\zeta_i),\Delta t(1-\tau);0,0) \\
&\qquad \left\{\bar{h}\Omega^{(m)}(\tau)\Omega_j(\zeta)\varphi_j^{(m)}\right\}d\zeta d\tau = 0, \quad i,j,m = 1,2
\end{aligned} \tag{7.14}$$

When the line integrals in eq. (7.14) are implemented using the expressions for the interpolating functions at each source node, the outcome are the discretized element equations which in matrix form is

$$R_{ij}^{(m)}h_j^{(m)} + [L_{ij}^{(m)} + \bar{h}T_{ij}^{(m)}]\varphi_j^{(m)} = 0, \quad i,j=1,2 \tag{7.15}$$

The element matrices in eq. (7.15) for this model are the same as those for the **TGE** model of chapters 5 and 6, and as such their expressions are obtained from eqs. (5.37a) to (5.37l). Eq. (7.15) is nonlinear, and can only be solved by applying a nonlinear solution scheme. The Picard scheme is employed, while application of the Newton-Raphson algorithm is left as an exercise to the reader. Applying the Picard algorithm to eq. (7.15), the system of element equations for the Burgers equation based on our second Green element model is

$$
\begin{aligned}
&R_{ij}^{(2)} h_j^{(2,k+1)} + [L_{ij}^{(2)} + \bar{h}^{(k)} T_{ij}^{(2)}] \varphi_j^{(2,k+1)} = \\
&- R_{ij}^{(1)} h_j^{(1)} - [L_{ij}^{(1)} + \bar{h}^{(k)} T_{ij}^{(1)}] \varphi_j^{(1)}, \quad i,j = 1,2
\end{aligned}
\tag{7.16}
$$

in which, as in other similar circumstances, k stands for the iteration number.

7.4 Model 3 for the Burgers Equation

As in the earlier model, the derivation of this third Green element formulation for Burgers equation follows that of the **ADGE** model of chapter 6. The free-space Green's function comes from the linear diffusion-advection differential operator whose expression is given by eq. (6.23). The integral equation of the current model is obtained by invoking Green's identity within a spatial element $[x_1, x_2]$ and carrying out a time integration in the interval $[t_1, t_2]$. It is given by

$$
\begin{aligned}
&-\lambda h_i^{(2)} + \int_{t_1}^{t_2} \left[h(x_2,t) G^*(x_2,t;x_i,t_2) - h(x_1,t) G^*(x_1,t;x_i,t_2) \right] dt - \\
&\int_{t_1}^{t_2} \left[G(x_2,t;x_i,t_2) \varphi(x_2,t) - G(x_1,t;x_i,t_2) \varphi(x_1,t) \right] dt - \\
&\frac{1}{\nu} \int_{x_1}^{x_2} G(x,t_1;x_i,t_2) h(x,t_1) dx + \frac{\bar{h}}{\nu} \int_{t_1}^{t_2} \left[h(x_2,t) G(x_2,t;x_i,t_2) - \right. \\
&\left. h(x_1,t) G(x_1,t;x_i,t_2) \right] dt = 0, \quad i=1,2
\end{aligned}
\tag{7.17}
$$

In arriving at eq. (7.17), we approximated the convective term by assuming that h takes an average value h within the spatial and temporal elements. We could proceed, as in chapter 6, along the line of approximating $h(x,t)$ and $\varphi(x,t)$ by linear interpolation functions in time and space. However, we found that this approximation exhibits numerical instability for certain values of ν. A stable scheme is obtained by approximating the distribution of $h(x,t)$ by linear shape functions in space and time, and that of $\varphi(x,t)$ by its value at the current time. These approximations make the current model similar to that of Kakuda and Tasaka [9]. Using the relations given by eqs. (6.28a) through (6.28c), the integral eq. (7.17), with the source point at x_1, simplifies to

$$
\begin{aligned}
&-\frac{1}{2}h_1^{(2)} + \frac{\Delta t}{2\nu}\int_0^1 [\bar{h} - \frac{l}{\Delta t(1-\tau)}]\Omega^{(m)}(\tau)h_2^{(m)}(\tau)G(l,\Delta t(1-\tau);0,0)d\tau - \\
&\frac{\bar{h}\Delta t}{2\nu}\int_0^1 \Omega^{(m)}(\tau)h_1^{(m)}G(0,\Delta t(1-\tau);0,0)d\tau - \\
&\Delta t\int_0^1 \varphi_2^{(2)}G(l,\Delta t(1-\tau);0,0)d\tau + \Delta t\int_0^1 \varphi_1^{(2)}G(0,\Delta t(1-\tau);0,0)d\tau - \\
&\frac{l}{\nu}\int_0^1 \Omega_j(\zeta)h_j^{(1)}G(l\zeta,\Delta t;0,0)d\zeta = 0
\end{aligned}
\tag{7.18a}
$$

and with the source point x_i at x_2

$$
\begin{aligned}
&-\frac{1}{2}h_2^{(2)} - \frac{\Delta t}{2\nu}\int_0^1 [\bar{h} + \frac{l}{\Delta t}]\Omega^{(m)}(\tau)h_1^{(m)}G(-l,\Delta t(1-\tau);0,0)d\tau + \\
&\frac{\bar{h}\Delta t}{2\nu}\int_0^1 \Omega^{(m)}(\tau)h_2^{(m)}G(0,\Delta t(1-\tau);0,0)d\tau + \\
&\Delta t\int_0^1 \varphi_1^{(2)}G(-l,\Delta t(1-\tau);0,0)d\tau - \Delta t\int_0^1 \varphi_2^{(2)}G(0,\Delta t(1-\tau);0,0)d\tau - \\
&\frac{l}{\nu}\int_0^1 \Omega_j(\zeta)h_j^{(m)}G(l(\zeta-1),\Delta t;0,0)d\zeta = 0
\end{aligned}
\tag{7.18b}
$$

Combining eqs. (7.18a) and (7.18b), and implementing those line integrations gives the element matrix equation which takes the form

$$
\boxed{R_{ij}^{(m)}\varphi_j^{(m)} + L_{ij}^{(2)}\varphi_j^{(2)} = 0, \quad i,j,m = 1,2}
\tag{7.19}
$$

Eq. (7.19) constitutes the system of discrete element equations of the third Green element model for Burgers equation. The element matrices denoted by $R_{ij}^{(m)}$ have the same expressions as those given in eqs. (6.31a) through (6.31l), except for the fact that the advecting velocity U is replaced by the average quantity $\bar{h}$. The expressions for $L_{ij}^{(2)}$ are given below:

$$
\begin{aligned}
L_{11}^{(2)} &= \Delta t \int_0^1 G(0,\Delta t(1-\tau);0,0)\,d\tau = \frac{-l}{4\sqrt{\gamma\pi}} \int_0^1 \frac{\exp[-\beta(1-\tau)]}{(1-\tau)^{1/2}}\,d\tau \\
&= \frac{-l}{4\sqrt{\gamma\pi}} erf(\sqrt{\beta}) = -L_{22}^{(2)}
\end{aligned}
\tag{7.20a}
$$

$$
\begin{aligned}
L_{12}^{(2)} &= -\Delta t \int_0^1 G(l,\Delta t(1-\tau);0,0)\,d\tau \\
&= \frac{l}{4\sqrt{\gamma\pi}} \int_0^1 \frac{1}{(1-\tau)^{1/2}} \exp\left[-\left(\beta(1-\tau) + \frac{\gamma}{1-\tau} + 2\sqrt{\gamma\beta}\right)\right] d\tau \\
&= \frac{-l}{8\sqrt{\gamma\pi}} \left\{ erfc(\sqrt{\gamma}+\sqrt{\beta}) + \exp(-4\sqrt{\gamma\pi})\left[erfc(\sqrt{\beta}-\sqrt{\gamma}) - 2\right] \right\}
\end{aligned}
\tag{7.20b}
$$

$$
\begin{aligned}
L_{21}^{(2)} &= \Delta t \int_0^1 G(-l,\Delta t(1-\tau);0,0)\,d\tau \\
&= \frac{-l}{4\sqrt{\gamma\pi}} \int_0^1 \frac{1}{(1-\tau)^{1/2}} \exp\left[-\left(\beta(1-\tau) + \frac{\gamma}{1-\tau} - 2\sqrt{\gamma\beta}\right)\right] d\tau \\
&= \frac{l}{8\sqrt{\gamma\pi}} \left\{ \exp(4\sqrt{\gamma\pi})\left[erfc(\sqrt{\gamma}+\sqrt{\beta}) - 2\right] + erfc(\sqrt{\beta}-\sqrt{\gamma}) \right\}
\end{aligned}
\tag{7.20c}
$$

Since the value of $\bar{h}$ depends on the solution h, which is being sought, it means that the system of equations provided by (7.19) is nonlinear. While we elect to use the Picard algorithm to solve the system of nonlinear equations, we may also point out that the Newton-Raphson technique can be applied. (We leave the application of the N-R algorithm as an exercise to the reader.) Applying the Picard algorithm to eq. (7.19) yields

$$
\boxed{R_{ij}^{(2,k)} h_j^{(2,k+1)} + L_{ij}^{(2,k)} \varphi_j^{(2,k+1)} = -R_{ij}^{(1)} h_j^{(1)} = 0, \quad i,j = 1,2}
\tag{7.21}
$$

where as before k stands for the iteration number, and $L_{ij}^{(2,k)}$ means the value of L_{ij} at the current time but evaluated at the previous iterate.

7.5. Numerical Examples

Three numerical examples are used to demonstrate the capabilities of these current numerical models. The first two examples have analytic solutions which serve as bases for comparison with our numerical results, while the third example, which does not have an exact solution, had been simulated by the finite element method.

Example 1

The first example is an initial sinusoidal wave that is allowed to propagate and diffuse within a confined flow domain in the x-dimension. The problem has the following boundary conditions:

$$h(0,t) = 0, \quad h(1,t) = 0, \quad h(x,0) = \sin\pi x \tag{7.22}$$

The exact solution to this problem, in the form of an infinite series, has been provided by Cole [10].

Example 2

The second example is an initial discontinuous wave form which is allowed to diffuse into a continuous wave form, while at the same time being propagated in time along the x-direction. The initial and boundary conditions are given by

$$h(x,0) = \begin{cases} 1, & x \le 0 \\ 0, & x > 0 \end{cases} \tag{7.23}$$

$$h(\ \infty,t) = 1, \quad h(\infty,t) - 0$$

The exact solution, obtained by Cole-Hoph transform, is given by Lighthill [7].

$$h(x,t) = \frac{1}{1 + \exp\left[\dfrac{x - t/2}{2\nu}\right]\dfrac{erfc\left[\dfrac{-x}{2\sqrt{\nu t}}\right]}{erfc\left[\dfrac{x-t}{2\sqrt{\nu t}}\right]}} \tag{7.24}$$

Example 3

The third example used in demonstrating the capabilities of the current numerical models, which had also been employed by other investigators (Varoglu and Finn [11], and Kakuda and Tasaka [9]), has the following initial and boundary conditions

$$h(x,0) = \begin{cases} 1, & x < 5 \\ 6 - x, & 5 \le x < 6 \\ 0, & x > 6 \end{cases} \tag{7.25}$$

$$h(-100,t) = 1, \quad h(20,t) = 0$$

7.6 Numerical Experiments and Results

The time, spatial discretization and other simulation data for the three numerical examples are summarized in Table 7.1. The results from the numerical simulations of example 1 by the three Green element models are presented in tabular form in Tables 7.2a through 7.2c for values of the flow parameter $\nu = 1,\ 10^{-1},\ 10^{-2}$, and in Figures 7.1a and 7.1b for $\nu = 10^{-2}$ and 10^{-3}, respectively.

With the exact solution tabulated at a number of times, the accuracy of the three models is assessed by the mean absolute error and presented in those tables. In all the numerical calculations with the first model, the Crank-Nicholson scheme is employed for the time discretization. The results show that the first model gives the most accurate solution, and then followed by the second model. By incorporating a timer into the computer programs which were run on a personal computer, we observe that the computing times of the first, second, and third models are in the approximate ratio of 2:1:10. The speed of the second model is enhanced by the ease with which its global coefficient matrix is assembled. The **GEM** solutions of the velocity profiles for $\nu = 10^{-2}$ and $\nu = 10^{-3}$ for example 1, presented in Figures 7.1a and 7.1b, reflect the general behavior expected of the propagation of the velocity which exhibits a sharp front in the vicinity of $x = 1$ at early times, and, as result of pronounced viscous action in this region, a smeared front at later times. The solution of the first model for $\nu = 10^{-3}$ in the neighborhood of $x = 1$ exhibits oscillations due to its inability to accurately represent the large spatial gradients and incorporate them into the numerical scheme in a stable fashion. However, in Figure 7.2 we show that a more stable solution of this first model for $\nu = 10^{-3}$ can be obtained by generating the same number of elements with the *Chebyshev polynomials* so that the nodes cluster in the vicinity of $x = 1$.

Table 7.1: Data for **GEM** simulations

Example	End point coordinates x_0	x_l	ν	Δt	No. of elements N	Spatial discretization	Max. no. of iterations	Figure or Table no.
1	0	1	1	0.01	20	Uniform	4	Table 7.2a
	0	1	10^{-1}	0.025	40	Uniform	4	Table 7.2b
	0	1	10^{-2}	0.01	100	Uniform	4	Table 7.2c; Figure 7.1a
	0	1	10^{-2}	0.01	100	Uniform	5	Figure 7.1b
2	-1.5	2.5	10^{-1}	0.025	100	Uniform	5	Figure 7.3a
	-0.7	1.3	10^{-2}	0.01	100	Uniform	5	Figure 7.3b
3	-100	20	1	0.2	49	Nonuniform	4	Figure 7.5a
	-100	20	10^{-1}	0.1	87	Nonuniform	5	Figure 7.5b
	-100	20	10^{-2}	0.05	149	Nonuniform	5	Figure 7.5c

Table 7.2a: **GEM** and Exact solutions for example 1 (ν = 1)

Position x	$h(x,0.02)$				$h(x,0.04)$			
	Model 1	Model 2	Model 3	Exact	Model 1	Model 2	Model 3	Exact
0.0	0.00000	0.00000	0.00000	0.0000	0.00000	0.00000	0.00000	0.00000
0.1	0.24335	0.24383	0.24616	0.24369	0.19664	0.19703	0.19997	0.19700
0.2	0.46558	0.46636	0.47052	0.46618	0.37698	0.37760	0.38296	0.37763
0.3	0.64673	0.64748	0.65260	0.64742	0.52532	0.52591	0.53284	0.52613
0.4	0.76926	0.76967	0.77477	0.76988	0.62743	0.62773	0.63518	0.62828
0.5	0.81966	0.81952	0.82377	0.82010	0.67173	0.67153	0.67851	0.67248
0.6	0.79017	0.78947	0.79242	0.79039	0.65078	0.65008	0.65585	0.65137
0.7	0.68058	0.67954	0.68118	0.68064	0.56315	0.56212	0.56632	0.56354
0.8	0.49947	0.49845	0.49912	0.49943	0.41487	0.41386	0.41647	0.41510
0.9	0.26431	0.26368	0.26385	0.26427	0.22010	0.21947	0.22068	0.22019
1.0	0.00000	0.00000	0.00000	0.00000	0.00000	0.00000	0.00000	0.00000
Mean abs. error x 10^{-3}	0.27727	0.43273	2.1682		0.42909	0.58635	0.33691	

Table 7.2a Contd. **GEM** and Exact solutions for example 1 (ν = 1)

Position x	$h(x,0.10)$				$h(x,0.22)$			
	Model 1	Model 2	Model 3	Exact	Model 1	Model 2	Model 3	Exact
0.0	0.00000	0.00000	.00000	0.00000	0.00000	0.00000	0.00000	0.00000
0.1	0.10924	0.10925	.11214	0.10954	0.03433	0.03426	0.03607	0.03454
0.2	0.20922	0.20921	.21472	0.20979	0.06547	0.06534	0.06894	0.06586
0.3	0.29110	0.29102	.29864	0.29190	0.09046	0.09027	0.09530	0.09101
0.4	0.34696	0.34677	.35577	0.34792	0.10686	0.10664	0.11264	0.10751
0.5	0.37054	0.37021	.37974	0.37158	0.11296	0.11273	0.11915	0.11367
0.6	0.35802	0.35760	.36672	0.35905	0.10802	0.10780	0.11399	0.10870
0.7	0.30900	0.30855	.31636	0.30991	0.09234	0.09215	0.09750	0.09294
0.8	0.22714	0.22676	.23248	0.22782	0.06736	0.06721	0.07115	0.06779
0.9	0.12033	0.12010	.12314	0.12069	0.03550	0.03543	0.03748	0.03573
1.0	0.00000	0.00000	.00000	0.00000	0.00000	0.00000	0.00000	0.00000
Mean abs. error x 10^{-3}	0.60455	0.79364	4.6827		0.40455	0.53818	3.1336	

Table 7.2b GEM and Exact solutions for example 1 ($\nu = 0.1$)

Position x	$h(x,0.05)$				$h(x,0.25)$			
	Model 1	Model 2	Model 3	Exact	Model 1	Model 2	Model 3	Exact
0.0	.00000	0.00000	0.00000	0.00000	.00000	0.00000	0.00000	0.00000
0.1	.25847	0.25889	0.26117	0.25869	.16011	0.16039	0.16327	0.16026
0.2	.49976	0.50032	0.50403	0.50005	.31597	0.31640	0.32173	0.31623
0.3	.70630	0.70655	0.71031	0.70646	.46271	0.46303	0.46995	0.46300
0.4	.85996	0.85955	0.86200	0.85990	.59388	0.59372	0.60101	0.59409
0.5	.94259	0.94143	0.94187	0.94237	.69997	0.69885	0.70500	0.70001
0.6	.93763	0.93601	0.93473	0.93743	.76551	0.76293	0.76637	0.76530
0.7	.83349	0.83188	0.83003	0.83343	.76412	0.75975	0.75937	0.76372
0.8	.62891	0.62775	0.62643	0.62896	.65411	0.64848	0.64470	0.65373
0.9	.33932	0.33875	0.33825	0.33936	.39274	0.38823	0.38422	0.39259
1.0	.00000	0.00000	0.00000	0.00000	.00000	0.00000	0.00000	0.00000
Mean abs. error x 10^{-4}	1.2000	5.9273	20.5000		1.9001	15.6090	45.6272	

Table 7.2b Contd GEM and Exact solutions for example 1 ($\nu = 0.1$)

Position x	$h(x,0.75)$				$h(x,1.5)$			
	Model 1	Model 2	Model 3	Exact	Model 1	Model 2	Model 3	Exact
0.0	0.00000	0.00000	0.00000	0.00000	0.00000	0.00000	0.00000	0.00000
0.1	0.08340	0.08330	0.08673	0.08343	0.04374	0.04342	0.04413	0.04375
0.2	0.16547	0.16522	0.16799	0.16552	0.08575	0.08509	0.08650	0.08575
0.3	0.24442	0.24393	0.24788	0.24448	0.12393	0.12290	0.12498	0.12394
0.4	0.31734	0.31644	0.32129	0.31738	0.15555	0.15412	0.15681	0.15556
0.5	0.37891	0.37739	0.38277	0.37892	0.17687	0.17505	0.17824	0.17691
0.6	0.41924	0.41687	0.42230	0.41922	0.18319	0.18108	0.18455	0.18327
0.7	0.42093	0.41765	0.42256	0.42093	0.16951	0.16733	0.17071	0.1962
0.8	0.35888	0.35521	0.35901	0.35896	0.13242	0.13054	0.13332	0.13253
0.9	0.21298	0.21037	0.21247	0.21312	0.07316	0.07205	0.07365	0.07323
1.0	0.00000	0.00000	0.00000	0.00000	0.00000	0.00000	0.00000	0.00000
Mean abs. error x 10^{-4}	0.39089	10.8182	18.4909		0.4182	10.1637	7.5909	

Table 7.2c: GEM and Exact solutions for example 1 ($\nu = 0.01$)

Position x	$h(x,0.40)$				$h(x,0.80)$			
	Model 1	Model 2	Model 3	Exact	Model 1	Model 2	Model 3	Exact
0.0	0.00000	0.00000	0.00000	0.00000	0.00000	0.00000	0.00000	0.00000
0.2	0.27449	0.27485	0.27628	0.27452	0.17734	0.17761	0.17854	0.17736
0.4	0.53788	0.53834	0.54061	0.53792	0.35279	0.35327	0.35501	0.35282
0.6	0.77347	0.77334	0.77510	0.77345	0.52398	0.52452	0.52679	0.52401
0.8	0.94117	0.93882	0.93769	0.94100	0.68708	0.68737	0.68963	0.68708
0.9	0.95277	0.94699	0.94407	0.94891	0.76311	0.76285	0.76451	0.76295
0.92	0.93734	0.92964	0.92674	0.93655	0.77645	0.77540	0.77635	0.77586
0.94	0.90330	0.89154	0.88891	0.89998	0.78319	0.77936	0.77824	0.78119
0.96	0.81619	0.79573	0.79380	0.81227	0.75388	0.74166	0.73569	0.74905
0.98	0.56080	0.53349	0.53342	0.56583	0.55927	0.53647	0.52671	0.55546
1.0	0.00000	0.00000	0.00000	0.00000	0.00000	0.00000	0.00000	0.00000
Mean abs. error x 10^{-4}	15.6181	58.6363	78.1909		10.4273	27.6363	50.7363	

Table 7.2c Contd. GEM and Exact solutions for example 1 ($\nu = 0.01$)

Position x	$h(x,1.20)$				$h(x,3.0)$			
	Model 1	Model 2	Model 3	Exact	Model 1	Model 2	Model 3	Exact
0.0	0.00000	0.00000	0.00000	0.00000	0.00000	0.00000	0.00000	0.00000
0.2	0.13091	0.13111	0.06036	0.13092	0.06009	0.06036	0.13173	0.06009
0.4	0.26126	0.26164	0.12068	0.26128	0.12016	0.12068	0.26285	0.12016
0.6	0.39041	0.39092	0.18095	0.39044	0.18017	0.18095	0.39264	0.18018
0.8	0.51750	0.51806	0.23906	0.51753	0.23867	0.23906	0.52014	0.23863
0.9	0.57812	0.57779	0.23766	0.57781	0.24183	0.23766	0.57953	0.24159
0.92	0.58555	0.58360	0.22050	0.58472	0.22636	0.22050	0.58469	0.22612
0.94	0.57956	0.57368	0.18998	0.57779	0.19708	0.18998	0.57340	0.19690
0.96	0.52772	0.51482	0.14229	0.52524	0.14919	0.14229	0.51257	0.14911
0.98	0.35167	0.33577	0.07699	0.35061	0.08139	0.07699	0.33298	0.08139
1.0	0.00000	0.00000	0.00000	0.00000	0.00000	0.00000	0.00000	0.00000
Mean abs. error x10^{-4}	5.9454	29.1728	39.6636		0.71820	27.0730	26.9820	

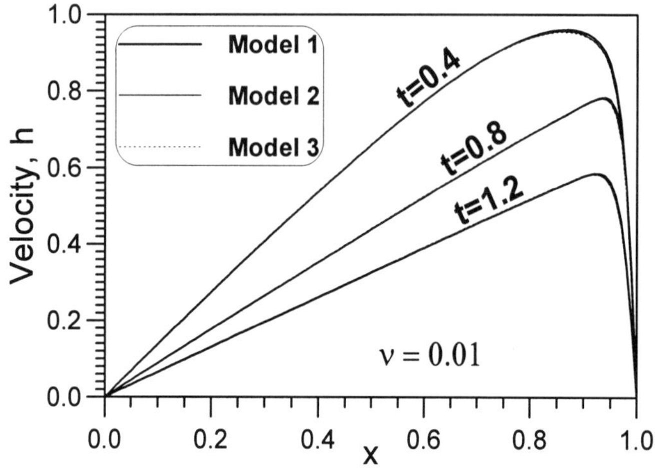

Figure 7.1a GEM solutions of example 1 with $\nu = 10^{-2}$.

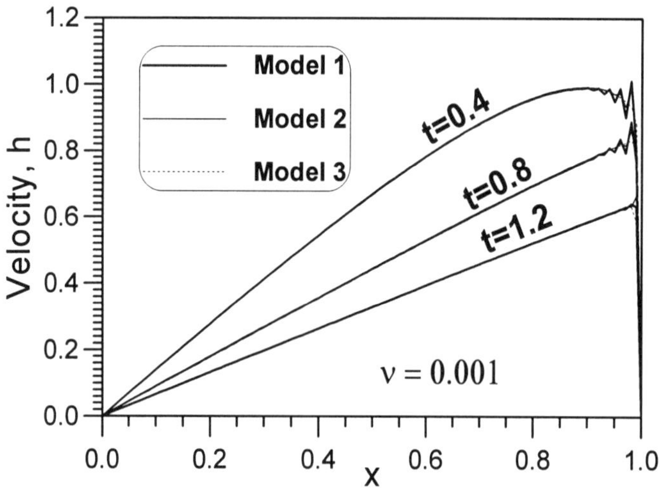

Figure 7.1b GEM solutions of example 1 with $\nu = 10^{-3}$.

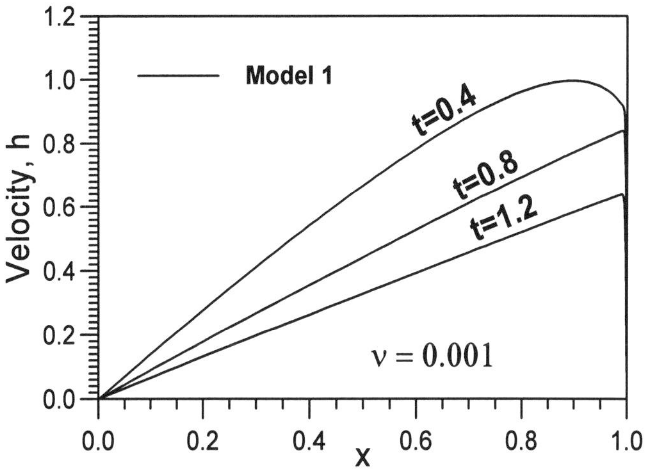

Figure 7.2 Solution of example 1 for $\nu = 10^{-3}$ by model 1 of **GEM** using Chebyshev polynomials for spatial discretization.

Figures 7.3a and 7.3b indicate the **GEM** and exact solutions of example 2 at different times for cases $\nu = 10^{-1}$ and $\nu = 10^{-2}$. Although the flow domain for this example is infinitely extensive, the numerical simulations are carried out using finite domains: $-1.5 \le x \le 2.5$ for the case $\nu = 10^{-1}$ and $-0.7 \le x \le 1.3$ for the case $\nu = 10^{-2}$, so that the exact solutions at the end nodes satisfy the specified boundary conditions at $x=-\infty$ and $x=\infty$ throughout the simulation times. All the **GEM** solutions are identical, and that is the reason why only the solutions from one model are presented along side the exact solutions. The L_2 norm estimates of the error between the numerical and exact solutions are the same for the three models when $\nu = 10^{-1}$ and differ marginally when $\nu = 10^{-2}$, as indicated in Figure 7.4. That figure shows a trend in accuracy which is similar to that obtained earlier for example 1 — the first model being the most superior in accuracy, and then followed by the second model.

The solutions of the three **GE** models for example 3 are compared to the finite element solutions of Varoglu and Finn [11] for values of $\nu = 1$, 10^{-1}, and 10^{-2} in Figures 7.5a, 7.5b, and 7.5c, respectively. We observe good agreement between the finite element solutions and those from the first **GE** model for all the three values of ν. However, the second and third **GE** models produce, for the case $\nu = 10^{-2}$, a velocity front that advances faster than that of the **FEM** and the first **GE** model.

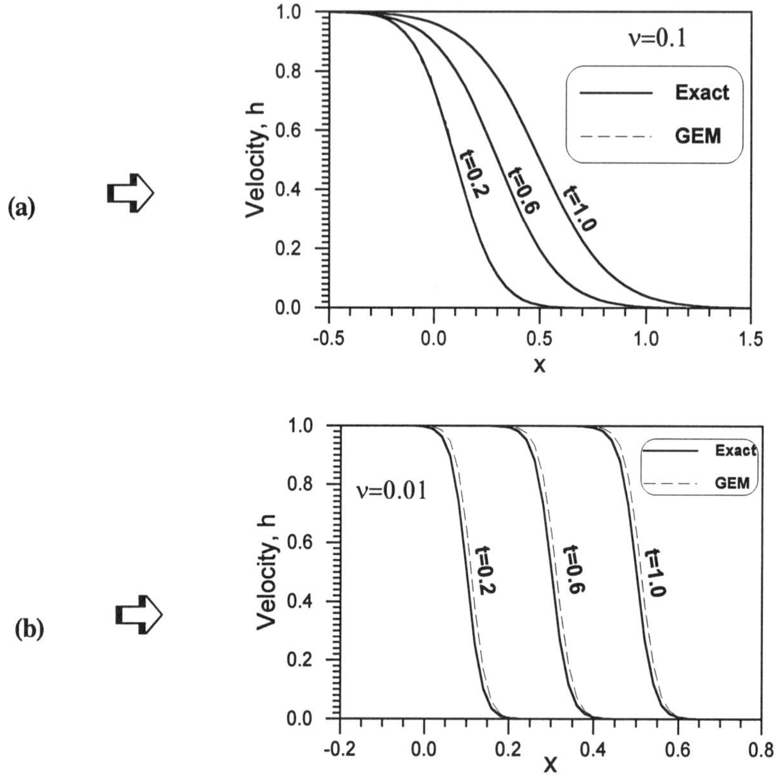

Figure 7.3 **GEM** solutions of example 2 — **(a)** $\nu = 10^{-1}$ and **(b)** $\nu = 10^{-2}$.

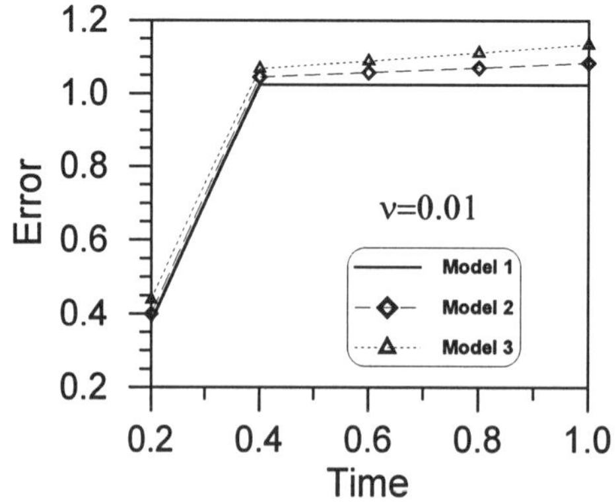

Figure 7.4 Error plots of the three **GEM** solutions of example 2 for $\nu = 10^{-2}$.

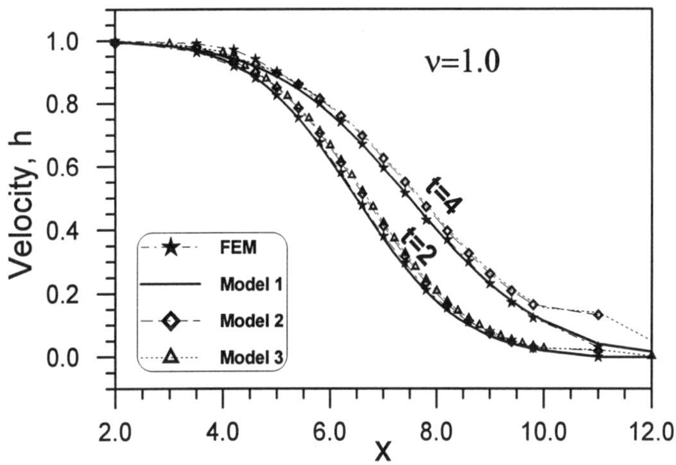

Figurc 7.5a FEM and GEM solutions of example 3 for $\nu = 1$.

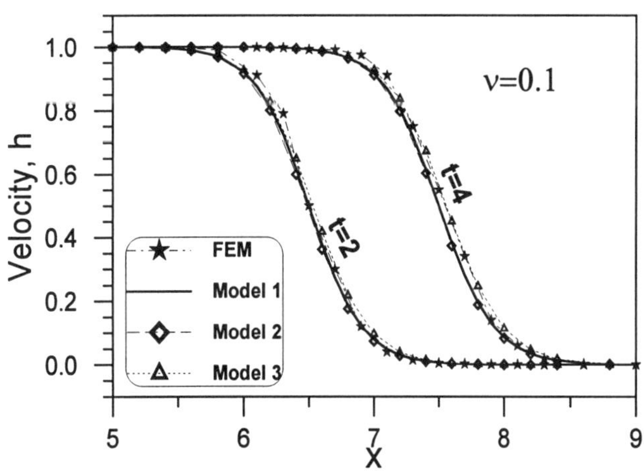

Figure 7.5b FEM and GEM solutions of example 3 for $\nu = 10^{-1}$.

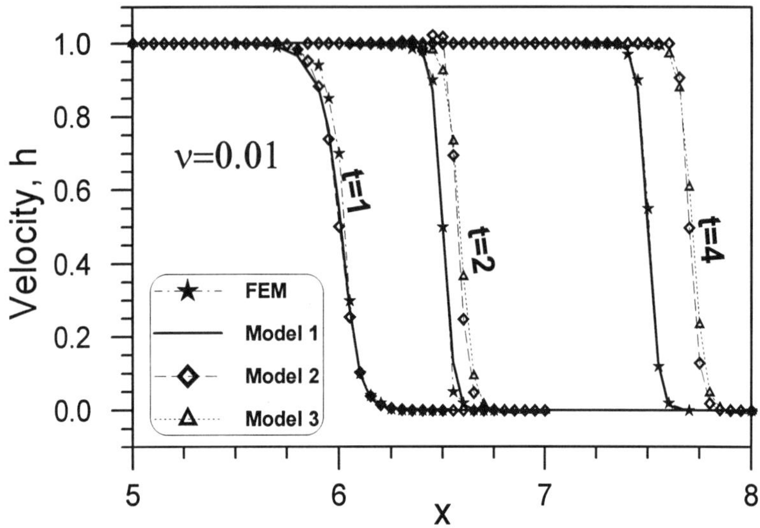

Figure 7.5c FEM and GEM solutions of example 3 for $\nu = 10^{-2}$.

7.7 Remarks

As in chapter 6, we have derived three Green element models for Burgers equation which have arisen from three different differential operators. These three differential operators provide the Green's functions that are incorporated in the integral equations. The first model uses the simple Green's function of the Laplace operator, evaluates the terms of the integral equation exactly, and linearizes the equation with a nonlinear solution algorithm only after the global matrix is formed. In contrast, the other 2 models linearize the equations within each element. These approximations are reflected in the accuracy of the numerical solutions. The first model has an advantage over the other two models in terms of accuracy, but uses about twice the computing time of the second model which gives the fastest computing speed. The first model has the additional advantage that it uses the same Green's function for both the steady and transient Burgers equation, so that one computer code of this model serves for both states of flow.

7.8 Exercises

7.1 Derive the Newton-Raphson algorithm for the second and third models of **GEM** presented in this chapter for the solution of Burgers equation. How well does this algorithm perform in comparison to the Picard algorithm?

7.2 Examine the performance of the first **GE** model for the Burgers equation when the temporal derivative is approximated by the relationship

$$\left.\frac{dh_j}{dt}\right]_{t=t_2} = \frac{\alpha}{\Delta t}\left(h_j^{(2)} - h_j^{(1)}\right) + (1-\alpha)\left.\frac{dh_j}{dt}\right]_{t=t_1} \quad ; \quad 1 \le \alpha \le 2$$

Compare its performance to that of the generalized 2-level time discretization scheme presented in this chapter.

7.3 Examine the performance of the second **GE** model for Burgers equation when both $h(x,t)$ and $\varphi(x,t)$ are approximated by linear shape functions in space and time. Compare the performance of this approximation to that presented in this chapter in which $h(x,t)$ is replaced by a constant value within an element.

7.9 References

1. Schlichting, H., *Boundary-Layer Theory*, McGraw-Hill Book Comp., 7th Ed., 1979.

2. Burgers, J.M., "A mathematical Model illustrating the theory of Turbulence", *Advances in Applied Mech.*, (**1**), Ed. by R. von Mises and T. von Karman, Academy Press, NY., pp. 171-199, 1948.

3. Hopf, E., "The partial differential equation $u_t + u u_x = \mu u_{xx}$", *Comm. Pure Appl. Math.*, 3, pp. 201-230, 1950.

4. Taigbenu, A.E., "The Green Element Method," *Int. J. for Numerical Methods in Engineering*, Vol. 38, pp 2241-2263, 1995.

5. Benton, E.R. and G.W. Platzman, "A table of solutions of the one-dimensional Burgers equation", *Quarterly of Applied Mathematics*, pp. 195-212, July, 1972.

6. Taigbenu,, A.E. and O.O. Onyejekwe, "A Mixed Green element formulation for transient Burgers' equation", *Int. J. Num. Methods in Fluids*, **24**, 563-578, 1997.

7. Lighthill, M.J., "Viscosity effects in sound waves of finite amplitude", in *Surveys in Mechanics*, Ed.: G.K. Batchelor and R.M. Davis, cambridge Univ. Press, 1956.

8. Fletcher, C.A.J., "Burger's Equation: A model for all reasons", *Numerical Sol. of Partial Differential Equations*, Ed. by J. Noye, North-Holland Pub. Comp. pp.139-225, 1982.

9. Kakuda, K. and N. Tosaka, "The generalized boundary element approach to Burgers equation", *Int.*

J. Num. Meth. in Engrg., **29**, pp. 245-261, 1990.

10. Cole J.D., "On a quasi-linear parabolic equation occurring in aerodynamics", *Quarterly of Applied mathematics,* **23**, pp. 225-236 1951.

11. Varoglu, E. and W.D.L. Finn, Space-Time finite elements incorporating characteristics for the Burgers' equations, *Int. J. Num. Meth. in Engrg.,* **16**, 171-184, 1980.

8 UNSATURATED FLOW (RICHARDS EQUATION)

8.1 Introduction

We pay attention in this chapter to the solution to the problem of flow in variably saturated soils because of its importance in many fields of engineering such as drainage, irrigation, environmental, soil and petroleum engineering. The flow in unsaturated soils is essentially a two-phase flow of two immiscible fluids — air and water. The physical processes that give rise to this kind of flow are infiltration of surface water through the upper layers of soil which enriches the soil moisture, and subsurface flow through soils which are partially filled with air. The interaction of roots of plants with this flow, and the advection and dispersion of fertilizers and pesticides within the unsaturated zone make this flow phenomenon of considerable interest to soil scientists, agronomists, and irrigation engineers. In petroleum explorations of underground reservoirs where immiscible fluid flows are also encountered, the fluids involved are water, oil, and gas. Although the partial differential equations that govern the two-phase immiscible flow in unsaturated formations are similar to those that govern the three-phase immiscible flow in underground reservoirs, it is usual in the case of the former to assume that the dynamics of flow within the air phase play an insignificant role in determining water movement and storage in the unsaturated zone. In the absence of this assumption, the solution would have required a simultaneous solution of two partial differential equations, each of which describes the flow in each fluid phase.

The mathematical statement that describes the flow of water in a variably saturated soil is obtained by combining the statement of moisture conservation and the simplified momentum equation which is given by Darcy law. Depending on whether the hydraulic pressure or the moisture content is the dependent variable, three versions of the flow equation exist and are commonly referred to in the groundwater literature as the *Richards equation.* The Richards equation is highly nonlinear because the soil parameters are highly dependent on the flow variables, and that makes it difficult to obtain analytic solutions of the differential equation. The few analytic solutions of the Richards equation have been obtained after imposing severe restrictions on the soil constitutive relations. These analytic solutions have not only performed the traditional role of validating the performance of numerical schemes, but also in aiding our understanding of the intricate processes of flow in unsaturated soils, and the interdependencies of the soil parameters and the flow variables.

Numerical solution techniques appear to offer the most effective way of solving this highly nonlinear flow problem. However, the application of the boundary element theory to unsaturated flow has been far from successful compared to its application to saturated flow. The reasons for this situation have been due to earlier perception that the theory is suited only for linear differential operators, and the way the theory has been implemented whereby nodal data are coupled in a global sense, thereby making it impossible to accommodate spatial variability of soil parameters. Liggett and Liu [1] recognized this limitation of conventional **BEM**, and adopted a combined **BEM/FEM** model for saturated and unsaturated flow in which the former solves the linear saturated flow portion, and the latter the nonlinear unsaturated problem. Essentially the above approach completely avoided tackling the nonlinear unsaturated flow problem. The effort of Raats [2] in applying **BEM** to the steady unsaturated flow problem assumed the soil parameters were of uniform value throughout the flow domain so that the flow differential equation is elliptic in nature, while Pullan [3] made a similar assumption which transformed the transient unsaturated flow equation to a diffusion-type one.

The earlier applications of **BEM** to saturated flow in heterogeneous media by Butterfield and Tomlin [4], Lafe *et al.* [5], Clements [6], Cheng [7] and Lafe and Cheng [8] could have served as bases for extending the boundary element theory to the nonlinear unsaturated flow problem. While Butterfield and Tomlin [4] and Lafe *et al.* [5] employed the zoning technique, Clements [6] and Cheng [7] used a direct Green function formulation which incorporates polynomial distributions of the medium property into the free-space Green's function to obtain solutions to steady flows in saturated heterogeneous media. Recognizing the restrictive nature of the direct formulation that incorporates the variation of the hydraulic conductivity into the Green's function, Lafe and Cheng [8] proposed a perturbation-based formulation for steady groundwater flow in heterogeneous aquifers. That formulation, which decomposes the steady heterogeneous flow problem into a sequence of Laplace and Poisson equations, has the demerit of treating a linear problem as a nonlinear one in the sense that the solution is achieved by iteration. Furthermore, the authors admitted the possibility of failure of the perturbation formulation when the hydraulic conductivity varies rapidly within the domain, and that rules out its extension to the unsaturated flow problem where the hydraulic conductivity not only varies rapidly but has strong dependence on the flow variables. Another boundary element formulation which could have been pursued for the unsaturated flow problem was that used by Taigbenu and Liggett [9] for the nonlinear Boussinesq equation, but because its solution procedure couples information from every node in the computational domain it will be inappropriate for a problem with rapidly varying medium properties.

Here in this chapter we demonstrate how **GEM** is applied to unsaturated flow using the boundary element theory, without adopting any of those severe simplifications earlier used in **BEM** applications (see also Taigbenu and Onyejekwe [10]). One model of **GEM** is proposed for the problem, and it is derived using the free-space Green's function of the 1-D. Laplace operator. Before embarking on the derivation of this model, we shall examine how the Richards equation is obtained from the physical processes of infiltration flow in unsaturated soils.

8.2 Derivation of the Unsaturated Flow Equation

The flow in an unsaturated soil formation is illustrated in Figure 8.1. The movement of water occurs within the connected pores of the soil medium which are also partially filled with air. It is assumed that the flow occurs in one predominant direction in the downward direction so that flow in other directions can be said to be negligibly small. This assumption is valid for many infiltration processes in homogeneous soils. Also it is assumed that the dynamics of the flow in the air phase has negligible influence on the flow in the water phase. Our control volume has an infinitesimal length in the vertical direction but is infinitely extensive in other orthogonal directions. The statement of mass conservation is that the net efflux of mass across the control surface equals the temporal rate of change of mass within the control volume. Since flow is assumed to take place only in the z-direction, the upper and lower planes (AB and CD) that are normal to the z-direction constitute the control surface across which the flow occurs, and the net efflux of mass across a unit area of the control surface is given as

$$\rho_w q - \rho_w \left(q + \frac{\partial q}{\partial z} dz \right) = -\rho_w \frac{\partial q}{\partial z} dz \tag{8.1}$$

in which q, referred to as the *Darcy flux*, is the volumetric flux per unit area of soil. The time rate of change of mass of water within the control volume per unit area of soil is given as

$$\frac{d}{dt}\left(\rho_w h dz\right) = \rho_w \frac{\partial h}{\partial t} dz \tag{8.2}$$

where h is the *volumetric moisture content* which is defined as the ratio of the volume of water in the pores to the total volume of soil. By the statement of mass conservation, the quantities in eqs. (8.1) and (8.2) are equivalent. That is

$$\frac{\partial h}{\partial t} + \frac{\partial q}{\partial z} = 0 \tag{8.3}$$

The simplified form of the momentum equation for this flow comes from Darcy law which takes cognisance of the fact that the flow is slow, and as such the convective or acceleration force can be neglected. The Darcy law relates the flux to the gradient of the hydraulic head. That is

$$q = -K \frac{\partial \theta}{\partial z} \tag{8.4}$$

in which θ denotes the *hydraulic head* and K is the hydraulic conductivity. The hydraulic head is a sum of the suction head ψ (the head that accounts for the surface adhesion between the soil grains and water molecules) and the gravity or potential or elevation head. It is expressed as

$$\theta = \psi + H - z \tag{8.5}$$

in which H is the depth of an assumed datum from ground level. Introducing eq. (8.5) into eq. (8.4) yields

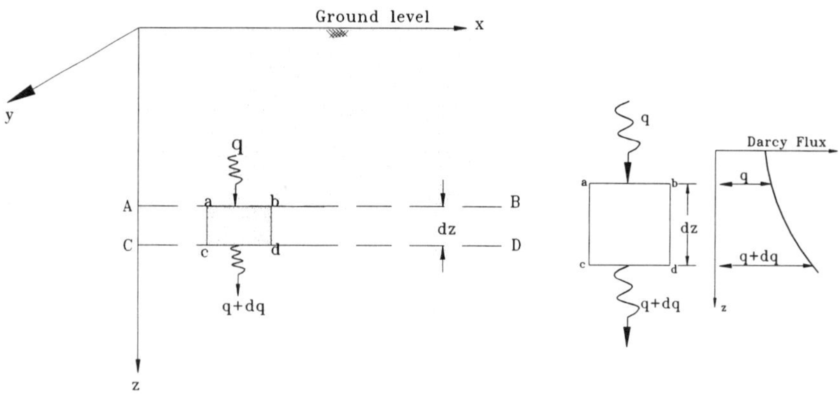

Figure 8.1 Definition sketch of unsaturated flow medium.

$$q = -K\frac{\partial(\psi - z)}{\partial z} = -K\frac{\partial \psi}{\partial z} + K$$

or

$$q = -K\frac{d\psi}{dh}\frac{\partial h}{\partial z} + K = -D\frac{\partial h}{\partial z} + K \tag{8.6}$$

where $D = Kd\psi/dh$ is the *soil moisture diffusivity*, and $dh/d\psi = C$ is referred to as the *specific moisture capacity*. The expression for q in eq. (8.6) is introduced in eq. (8.3) to have

$$\frac{\partial}{\partial z}\left(D\frac{\partial h}{\partial z} - K\right) = \frac{\partial h}{\partial t} \tag{8.7}$$

which is the 1-D. form of the Richards equation or the partial differential equation for transient flow in a variably saturated soil which was first presented by Richards [11]. It should be pointed out that, for a typical soil, the suction ψ and the hydraulic conductivity K vary appreciably within a range of several orders of magnitude for small changes in the soil moisture content, as illustrated in Figure 8.2. In general

$$\psi = \psi(h) \tag{8.8a}$$

and

$$K = K(h) \tag{8.8b}$$

The works of Brooks and Corey [12], Mualem [13], Haverkamp *et al.* [14], van Genuchten [15] and others provide some closed-form analytic expressions for the soil constitutive relations of some soils. A viable numerical scheme for the unsaturated flow problem should be capable of accommodating these numerous analytic expressions in the literature. That has been the handicap of earlier boundary element applications to the unsaturated flow problem.

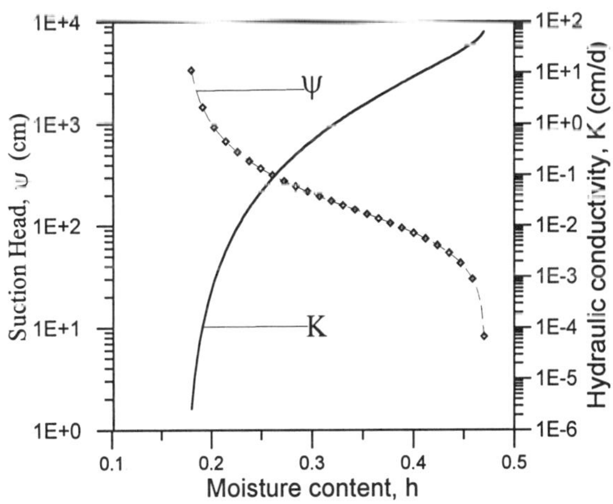

Figure 8.2 Typical soil constitutive relations.

The conditions that can be specified for the soil moisture content h at the ends of the flow length and within the flow length at initial time are similar to those of eqs. (5.8) and (5.11). However, with respect to the flux, the boundary condition has to be expressed in a manner consistent with the expression in eq. (8.6). In other words, if the coordinates of the boundaries are z_0 and z_L, the Neumann or flux conditions are

$$\left[-D\frac{\partial h(z,t)}{\partial n}+K\frac{\partial z}{\partial n}\right]_{z=z_0}=q_0(t) \text{ and } \left[-D\frac{\partial h(z,t)}{\partial n}+K\frac{\partial z}{\partial n}\right]_{z=z_L}=q_L(t) \tag{8.9}$$

These conditions have to be specified along side the governing partial differential equation (8.7) in order that unique solutions can be obtained.

8.3 Green Element Formulation

The transient Richards equation given by eq. (8.7) can be re-expressed as

$$\frac{\partial^2 h}{\partial z^2} = -\frac{\partial\Theta}{\partial z}\frac{\partial h}{\partial z} + \Psi\left(\frac{\partial K(h)}{\partial z} + \frac{\partial h}{\partial t}\right) \tag{8.10}$$

in which $\Theta(h) = \ln D(h)$ and $\Psi(h) = 1/D(h)$. Based on the boundary element theory which uses the complementary differential equation to eq. (8.10), that is $d^2G/dz^2 = \delta(z-z_i)$, and whose solution in the infinite z-space is given by eq. (2.30) or $G(z,z_i) = (|z-z_i| + \tilde{l})/2$, the integral representation of the flow differential equation within a typical linear element $[z_1,z_2]$ is given by

$$\begin{aligned}&-\lambda h(z_i,t) + G^*(z_2,z_i)h(z_2,t) - G^*(z_1,z_i)h(z_1,t)\\&- G(z_2,z_i)\varphi(z_2,t) + G(z_1,z_i)\varphi(z_1,t) +\\&\int_{z_1}^{z_2} G(z,z_i)\left[\Psi\left(\frac{\partial h(z,t)}{\partial t} + \frac{\partial K}{\partial z}\right) - \frac{\partial\Theta}{\partial z}\frac{\partial h}{\partial z}\right]dz = 0\end{aligned} \tag{8.11}$$

As in earlier chapters $\varphi(z,t) \equiv \partial h(z,t)/\partial z$ and $G^*(z,z_i) = dG(z,z_i)/dz$. The line integral over the element in eq. (8.11) has to be evaluated, and to achieve this there is the need to prescribe a distribution for $h(x,t)$, $\varphi(x,t)$, K, Θ, and Ψ over each element. As we have done in previous chapters linear interpolating functions are used to approximate these quantities. That is,

$$\begin{aligned}&h(z,t) \approx \Omega_j(\zeta)h_j(t), \qquad \varphi \approx \Omega_j(\zeta)\varphi_j(t)\\&K(h) \approx \Omega_j(\zeta)K_j(t); \quad \Theta \approx \Omega_j(\zeta)\Theta_j(t), \textit{ and } \Psi \approx \Omega_j(\zeta)\Psi_j(t); \quad j=1,2\end{aligned} \tag{8.12}$$

in which the element interpolating functions $\Omega_1(\zeta)$ and $\Omega_2(\zeta)$ are given by eq. (3.27). We substitute the expressions in eq. (8.12) into eq. (8.11) to give

$$-\lambda h_i + G^*(z_2,z_i)h_2 - G^*(z_1,z_i)h_1 - G(z_2,z_i)\varphi_2 + G(z_1,z_i)\varphi_1 +$$

$$\int_{z_1}^{z_2} G(z,z_i)\left\{\Omega_j\Psi_j[\Omega_n\frac{dh_n}{dt} + \frac{d\Omega_n}{l\,d\zeta}K_n] - \frac{d\Omega_j}{l\,d\zeta}\Theta_j\Omega_n\varphi_n\right\}dz = 0; \quad i,j,n=1,2 \tag{8.13}$$

Recalling the expressions for $G(z,z_i)$ and $G^*(z,z_i)$, their values at the end points of each element are

$$G(z_1,z_1) = G(z_2,z_2) = \frac{\tilde{l}}{2}, \quad G(z_2,z_1) = G(z_1,z_2) = \frac{\tilde{l}+l}{2}$$

$$G^*(z_1,z_1) = G^*(z_2,z_2) = 0, \quad G^*(z_2,z_1) = -G^*(z_1,z_2) = \frac{1}{2}$$

where $l = z_2 - z_1$. Using the above relations simplifies eq. (8.13) to

$$-h_1 + h_2 + \tilde{l}\varphi_1 - (\tilde{l}+l)\varphi_2 +$$

$$l\int_0^1\left[\Omega_j\Psi_j\left(\Omega_n\frac{dh_n}{dt} + \frac{d\Omega_n}{l\,d\zeta}K_n\right) - \Omega_j\varphi_j\frac{d\Omega_n}{l\,d\zeta}\Theta_n\right](\tilde{l}+l\zeta)d\zeta = 0, \quad j,n=1,2 \tag{8.14a}$$

when the source node z_i is at z_1, and to

$$h_1 - h_2 + (\tilde{l}+l)\varphi_1 - \tilde{l}\varphi_2 +$$

$$l\int_0^1\left[\Omega_j\Psi_j\left(\Omega_n\frac{dh_n}{dt} + \frac{d\Omega_n}{l\,d\zeta}K_n\right) - \Omega_j\varphi_j\frac{d\Omega_n}{l\,d\zeta}\Theta_n\right][\tilde{l}+l(1-\zeta)]d\zeta = 0, \quad j,n=1,2 \tag{8.14b}$$

when the source node z_i is at z_2. In matrix form, eqs. (8.14a) and (8.14b) become

$$\boxed{R_{ij}h_j + (L_{ij} - V_{inj}\Theta_n)\varphi_j + V_{inj}\Psi_j K_n + U_{inj}\Psi_n\frac{dh_j}{dt} = 0; \quad i,n,j=1,2} \tag{8.15}$$

Eq. (8.15) is the system of element equations for the Richards equation. The Green element model based on eq. (8.15) can also be referred to as the quasi-steady formulation since it uses the fundamental solution of the steady 1-D. flow equation of chapter 2. The element matrices in eq. (8.15), except for V_{inj}, are summarized in eqs. (6.8a) through (6.8d). The expression for V_{inj} is found in eq. (3.42c).

It is noted that eq. (8.15) is a system of discrete first-order (in time) nonlinear ordinary differential equations which can be solved for the nodal unknowns by simplifying the temporal derivative, and by adopting a nonlinear solution algorithm.

As we did in section 5.4, the 2-level generalized time discretization scheme is used to simplify the time derivative. That scheme uses the difference expression of eq. (5.20) to approximate the temporal derivative at $t = t_{1+\alpha} = t_1 + \alpha\Delta t$. The outcome of this simplification is

$$\left[\alpha R_{ij} + \frac{U_{inj}[\alpha\Psi_n^{(2)} + \omega\Psi_n^{(1)}]}{\Delta t}\right]h_j^{(2)} + \left[\omega R_{ij} - \frac{U_{inj}[\alpha\Psi_n^{(2)} + \omega\Psi_n^{(1)}]}{\Delta t}\right]h_j^{(1)} +$$
$$\alpha\left[L_{ij} - V_{inj}[\alpha\Theta_n^{(2)} + \omega\Theta_n^{(1)}]\right]\varphi_j^{(2)} + \omega\left[L_{ij} - V_{inj}[\alpha\Theta_n^{(2)} + \omega\Theta_n^{(1)}]\right]\varphi_j^{(1)} + \tag{8.16}$$
$$V_{inj}[\alpha\Psi_j^{(2)} + \omega\Psi_j^{(1)}][\alpha K_n^{(2)} + \omega K_n^{(1)}] = 0 \quad i,n,j=1,2,\ 0 \le \alpha \le 1$$

Eq. (8.16) is still a system of nonlinear equations. We have the option of either applying the Newton-Raphson or the Picard algorithm to solve the system of equations. We shall adopt the Picard algorithm here and, as an exercise, leave the reader to apply the Newton-Raphson algorithm. In applying the Picard algorithm, we shall group terms of eq. (8.16) in such a way that the unknown quantities $h_j^{(2)}$ and $\varphi_j^{(2)}$ can be evaluated from information at the previous iteration and at the previous time level. That is

$$\left[\alpha R_{ij} + \frac{U_{inj}[\alpha\Psi_n^{(2,k)} + \omega\Psi_n^{(1)}]}{\Delta t}\right]h_j^{(2,k+1)} +$$
$$\alpha\left[L_{ij} - V_{inj}[\alpha\Theta_n^{(2,k)} + \omega\Theta_n^{(1)}]\right]\varphi_j^{(2,k+1)} =$$
$$-\left[\omega R_{ij} - \frac{U_{inj}[\alpha\Psi_n^{(2,k)} + \omega\Psi_n^{(1)}]}{\Delta t}\right]h_j^{(1)} - \omega\left[L_{ij} - V_{inj}[\alpha\Theta_n^{(2,k)} + \omega\Theta_n^{(1)}]\right]\varphi_j^{(1)} - \tag{8.17}$$
$$V_{inj}[\alpha\Psi_j^{(2,k)} + \omega\Psi_j^{(1)}][\alpha K_n^{(2,k)} + \omega K_n^{(1)}] \quad i,n,j=1,2,\ 0 \le \alpha \le 1$$

The superscripts k and $k+1$ denote the previous and current iteration levels, respectively. Eq. (8.17) can now be assembled for all the linear elements used to discretize the computational domain to give the global coefficient matrix equation which takes the form

$$A_{ij}^{(2,k)}\begin{Bmatrix} h_j^{(2,k+1)} \\ \varphi_j^{(2,k+1)} \end{Bmatrix} = S_i^{(2,k)} \tag{8.18}$$

The above equation is solved at each time level t_2 for a number of times until convergence of the solution is achieved.

8.4 Numerical Examples

A computer program is developed, based on the forgoing theoretical formulation, to simulate one-dimensional flows in unsaturated soil media for which the analytical expressions of the soil constitutive relations are specified. To demonstrate the capability of the current numerical model we have chosen three numerical examples of unsaturated flow whose constitutive relationships cover a wide range of the analytical expressions found in the literature. The first example is a 1-D. horizontal infiltration flow into a soil column for which a semi-analytic solution has been provided by Philip [16]. Our solution for this example is compared to the analytic solution and a finite element solution provided by Huyakorn *et al.* [17]. The second and third examples are 1-D. vertical infiltration problems for which finite element solutions are available.

Example 1

In this example, flow takes place between two ends of a horizontal soil column of 20cm length that are maintained at different values of moisture levels: $h = 0.45$ and $h = 0.2$. Initially the moisture content is 0.2 along the entire length of the soil column. The soil constitutive relations for this example are:

$$
\begin{aligned}
K(h) &= \frac{10h}{3} - 0.5 \ cm/day \\
\psi &= \frac{10^3 h}{3} - 150 \ cm \\
C(h) &= \frac{dh}{d\psi} = 3x10^{-3} \ cm^{-1}
\end{aligned}
\tag{8.19}
$$

Because the flow in this example takes place in the horizontal direction, the term $\partial K(h)/\partial z$ of eq. (8.10) is zero. This example is simulated by the one-dimensional Green element model on a 20cm soil column that is discretized into 20 equal linear elements each of which has a length of 1cm. We have used the same time step of 0.01day that was employed by Huyakorn *et al.* [17] in their finite element model SATURN, while a convergence tolerance of 10^{-4}, based on the mean absolute error in two consecutive iterates, is imposed at each time step. It was observed that convergence was always achieved within 5 iterations. Since the 2-level generalized time discretization scheme gives us the liberty to choose the time level within a time step at which to approximate the temporal derivative, we have infinite number of schemes to incorporate in our model. We have adopted the Crank-Nicholson scheme with $\alpha = 0.5$ in our model. Figure 8.3 shows a comparison of the solutions for h obtained from **GEM**, the semi-analytic results of Philip [16], and **FEM**. The rate of influx of moisture, computed by the finite element model of Huyakorn *et al.* [17], is compared to that obtained from **GEM**, and presented in Table 8.1 and Figure 8.4. There are discrepancies in both numerical solutions during the first five time steps of the simulation, but thereafter both solutions become almost the same. It is reasonable to expect the **GEM** prediction of the flux to be one order accuracy higher than that of the finite element method because the flux is computed along with the primary variable in the **GEM** model whereas in the **FEM** the flux is computed by numerically differentiating the primary variable.

Table 8.1: **GEM** and **FEM** solutions of inflow flux for example 1.

	Inflow rate (cm^3/day/cm)	
Time (days)	**GEM**	**FEM** (SATURN)
0.01	29.15	38.41
0.02	18.10	18.93
0.03	13.90	14.14
0.04	11.67	11.78
0.05	10.25	10.31
0.06	9.247	9.282
0.07	8.489	8.511
0.08	7.891	7.905
0.09	7.404	7.412
0.10	6.997	7.002
0.11	6.650	6.652
0.12	6.350	6.351
0.13	6.087	6.086
0.14	5.855	5.852
0.15	5.647	5.642

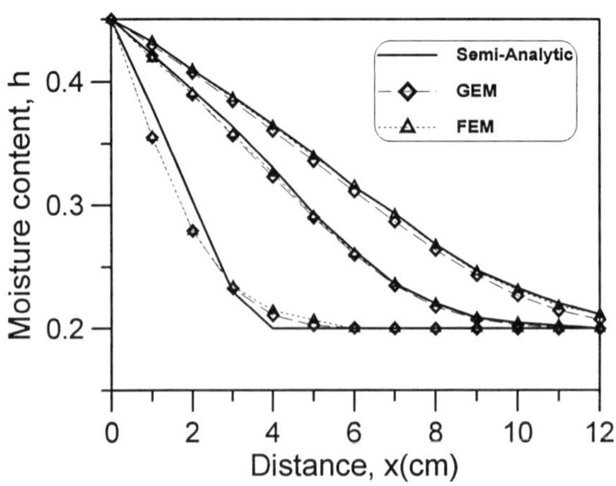

Figure 8.3 GEM, FEM and semi-analytic solutions of example 1.

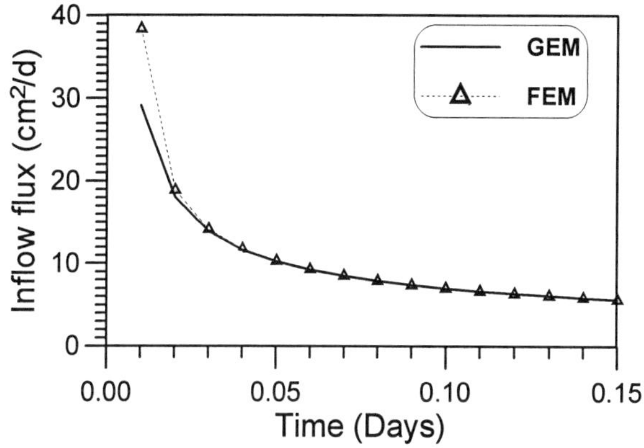

Figure 8.4 GEM and FEM solutions for the flux of example 1.

Example 2

In this example, infiltration into a dry vertical soil column is examined. Initially the soil is dry throughout the 80cm length of the soil column, and infiltration of moisture is generated by a saturation condition that is maintained from the top of the soil column. The soil constitutive relations, which are highly nonlinear, were reported and presented by van Genuchten [18] to approximate the Glendale clay loam soil, and are herein written in the form consistent with the moisture-based form of the differential equation

$$\begin{aligned} K(h) &= 75\left(\frac{h}{0.52}\right)^{13} \; cm/day \\ \psi &= 5.4\left(\frac{0.52}{h}\right)^{5} \; cm \\ C(h) &= \frac{dh}{d\psi} = 1.926x10^{-2}\left(\frac{h}{0.52}\right)^{6} \; cm^{-1} \end{aligned} \tag{8.20}$$

The boundary and initial data for this example are given as:

$$h(z{=}0,t) = 0.52, \quad h(z{=}80cm,t) = 0.275, \quad h(z,t_0) = 0.275 \tag{8.21}$$

The 80cm length of flow domain is discretized into 40 uniform linear elements each of which has a size of 2cm in the **GEM** simulations. The time history of the solution is discretized the same way as in the finite element solution to this problem: a time step of 7.143x10^{-4}days is used for the time interval, $0<t\leq 0.55$ days, and another time step of 10^{-3} days is employed within the interval $0.55\leq t\leq 0.127$ days. The **GEM** solutions

obtained at 0.055 and 0.127days are presented in Figure 8.5 along side a finite element solution that uses Hermitian basis functions for spatial interpolation of the primary variable. A difference weighting factor of 0.5 and a convergence tolerance of 10^{-4} based on the mean absolute deviation between the current and previous iterates are incorporated in the **GEM** simulations. Convergence was achieved with no more than 6 iterations at any time step. The **GEM** solution indicates less absorption of moisture by the soil than the **FEM** solution, although the latter exhibits some oscillations at the tail of the advancing moisture front.

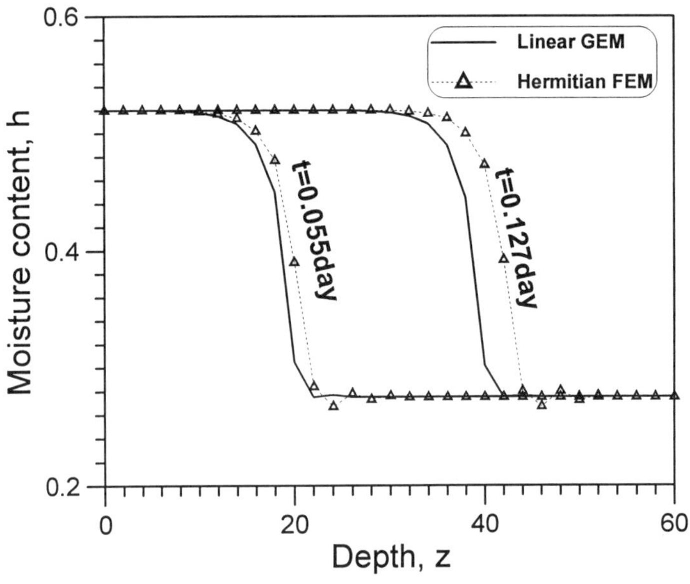

Figure 8.5 GEM and FEM solutions of example 2.

Example 3

The third example is again a vertical infiltration problem similar to the second example, but we have elected this example to demonstrate that the performance of the **GEM** model developed here is not dependent on the adopted form of soil constitutive relations. The example is a vertical infiltration flow into a dry soil column whose soil constitutive data were first reported by Haverkamp *et al.* [14] and was used by Celia *et al.* [19] in their numerical simulations. The soil constitutive data are:

$$
\begin{aligned}
K(h) &= K_s \frac{A}{A + \theta^{\frac{\gamma}{\beta}}} \; cm/s \\
\psi(h) &= \theta^{\frac{1}{\beta}} \; cm \\
C(h) &= \frac{dh}{d\psi} = \frac{\beta}{\omega} \left[\frac{(h - h_r)^2}{(h_s - h_r)\theta^{\frac{1-\beta}{\beta}}} \right] cm^{-1}
\end{aligned}
\tag{8.22}
$$

where $\theta = \omega(h_s - h)/(h_s - h_r)$, $\omega = 1.611x10^6$, $h_s = 0.287$, $h_r = 0.075$, $\beta = 3.96$, $K_s = 9.44x10^3 cm/s$, $A = 1.175x10^6$, and $\gamma = 4.74$. The corresponding boundary and initial data are:

$$
h(z=0,t) = 0.268, \quad h(z=40cm,t) = 0.1, \quad h(z,t=0) = 0.1 \tag{8.23}
$$

The Green element model simulates this example using 40 linear elements each of which has a size of 1cm, a time step of 30s, and a difference weighting factor of 0.5. A convergence tolerance of 10^{-4} based on the mean absolute deviation is specified at each time step. The accuracy of our solution is compared to that of the Galerkin solution which was presented by Celia *et al.* [19]. This comparison is done in Figure 8.6. The result of another simulation with a finer spatial discretization of the computational domain serves as bench mark result. We observe that for the same space and time discretization, the **GEM** solution reproduces the bench mark solution better than the **FEM** solution. This result is not surprising because the second-order accuracy of the singular theory of the boundary element theory is widely acknowledged.

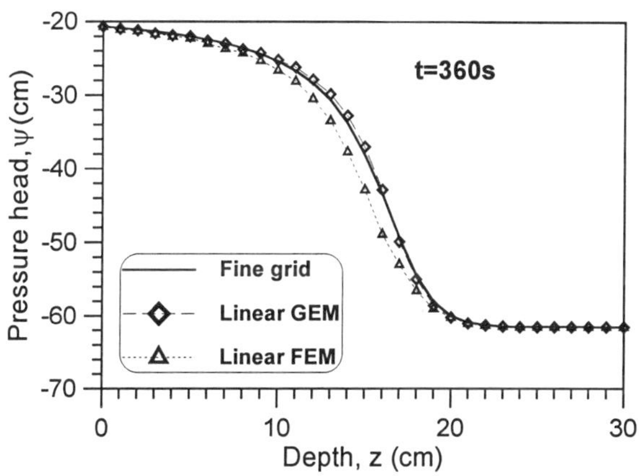

Figure 8.6 GEM and FEM solutions of example 3.

8.5 References

1. Liggett, J.A. and P.L-F. Liu, *The Boundary Integral Equation Method for Porous Media Flow*, George Allen & Unwin, London, 1983.

2. Raats, P.A.C., Steady Infiltration from line sources and furrows, *Soil Sci. Soc. Am. J.*, **34**, pp. 709-714, 1970.

3. Pullan, A.J., "Boundary Element Solutions of Quasilinearized time-dependent Infiltration", *Appl. Math. Modelling*, **12**, pp. 9-17, 1988.

4. Butterfield, R., and G.R. Tomlin, "Integral techniques for solving zoned anisotropic continuum problems", in *Proc. of the Int. Conference on Variational Methods in Engrg.*, **2**, Southampton University, Highfield, 1972.

5. Lafe, E.O., J.A. Liggett and P.L-F. Liu, "BIEM solutions of leaky, layered, confined, unconfined, nonisotropic aquifers", *Water Resources Res.*, **17**(5), pp. 1431-1444, 1981.

6. Clements, D.L., "A Boundary Integral Equation Method for the Numerical Solution of a Second-order Elliptic Equation with Variable Coefficients", *J. Austrl. Math. Soc.*, **22B**, pp. 218-228, 1980.

7. Cheng, A. H-D., "Darcy's flow with variable permeability - A Boundary Integral Solution", *Water Resources Res.*, **20**(7), pp. 980-984, 1984.

8. Lafe, O.E. and A. H-D. Cheng, "A Perturbation Boundary Element Code for Steady State groundwater Flow in Heterogeneous Aquifers", *Water Resources Res.*, **23**(6), pp. 1079-1084, 1987.

9. Taigbenu, A.E. and J.A. Liggett, "An Integral Formulation Applied to the Diffusion and Boussinesq Equations", *Int. J. Num. Methods in Engrg.*, **23**, pp. 1057-1079, 1986.

10. Taigbenu, A.E., and O.O. Onyejekwe, "Green Element simulations of the Transient Unsaturated Flow Equation," *Applied Mathematical Modelling*, **19**, pp 675-684, 1995.

11. Richards, L.A., "Capillary Conduction of Liquids through Porous Mediums", *Physics*, **1**, pp. 318-333, 1931.

12. Brooks, R.H. and A.T. Corey, "Properties of Porous Media affecting Fluid flow", *J. Irrig. Drainage Div. ASCE*, **92**(IR2), pp. 61-68, 1966.

13. Mualem, Y., "A new model for Predicting the Hydraulic Conductivity of Unsaturated Porous Media", *Water Resources Research.*, **12**, pp. 513-522, 1976

14. Haverkamp, R., M. Vauclin, J. Touma, P. Wierenga and G. Vachaud, Comparison of Numerical Simulation Models for One-dimensional Infiltration, *Soil Sci. Soc. Am. J.*, **41**, pp. 285-294, 1977.

15. van Genuchten, M. Th., "A Closed-form Equation for Predicting the Hydraulic Conductivity of Unsaturated Soils", *Soil Sci Soc. Am. J.* **44**, 892-898, 1980.

16. Philip, J.R., "Theory of Infiltration", *Adv. Hydrosci.*, **5**, 216-291, 1969.

17. Huyakorn, P.S., S.D. Thomas, and B.M. Thompson, "Technique for Making Finite Elements Competitive in Modeling Flow in Variably Saturated Porous Media", *Water Resources Res.*, **20**, 1099-1115, 1984.

18. van Genuchten, M.T., *Moisture Transport from Disposal sites*, Report to EPA, 1976.

19. Celia M.A., E.T. Bouloutas and R.L. Zarba, "A General Mass-Conservation Numerical Solution for the Unsaturated flow Equation", *Water Resources Research*, **26**(7) pp. 1483-1496, 1990.

9 HIGHER-ORDER ELEMENTS

9.1 Introduction

In the earlier seven chapters, we solved by the Green element method a variety of engineering problems in 1-D. spatial dimensions ranging from steady to transient, linear to nonlinear, and from those which apply in homogeneous to heterogeneous media. For all these problems, line segments were used to discretize the physical domain, and functional quantities were approximated by linear interpolation polynomials within each line segment (element). These interpolating polynomials have zero-order continuity in the sense that only the functional quantity is continuous, while its first derivative is discontinuous across elements. Where the variation of the functional quantities with respect to the spatial dimension is marginal, the use of linear interpolation can be adequate in approximating these functional quantities. However, there are certain problems, earlier encountered in chapters 6, 7, and 8, whose solutions have significant spatial gradients, and for which the use of linear interpolation is inadequate. In chapter 6, such a problem is the advection-dominant transport one that exhibits the unique feature of retaining the initial concentration profile as time progresses, so that steep gradients in the initial concentration profile persist throughout the solution history. The same can be said of the momentum transport problem of chapter 7 when viscous effects are negligible. For the unsaturated flow problem which we encountered in chapter 8, we are faced with steep gradients of the soil constitutive relations and that of the solution variable. In these three cases, the use of linear interpolation polynomials in approximating the functional quantities lead to significant errors. In this chapter we shall attempt to use higher-order interpolation functions to minimize numerical errors. There are a number of interpolation functions that could be incorporated into our formulation. These range from zero-order to higher-order interpolation functions. Here we employ the *first-order cubic Hermitian polynomials* which ensure that the primary variable and its first derivative are continuous at the interzonal boundaries. Any improvement in the quality of the numerical solution from the use of this higher-order interpolation function has to be at a price of higher computing cost. The marginal gains in accuracy *vis-a-vis* computational costs are evaluated.

When the Green's function that is employed in a Green element model is complicated, it should be expected that the use of higher-order interpolating functions will be fraught with computational difficulties. In that case it becomes difficult to evaluate some of the element integrals in the integral equations. For instance, the use of Hermitian functions in conjunction with the **TGE** and **ADGE** models of chapter 6 for the transport problem will be difficult to implement numerically because of the complicated nature of their Green's functions. In contrast, the simplicity of the Green's function of the **QSGE** model makes it easy to incorporate higher-order interpolation functions.

9.2 Green Element Equations with Hermitian Interpolation Polynomials

Thus far we have concerned ourselves with interpolation functions of zero-order continuity, that is interpolation functions that provide only for the continuity of the interpolating quantity; its spatial derivative is discontinuous across elements. Here we focus on a special class of interpolation functions that are based on the Hermitian polynomials. They allow not only for the continuity of the interpolated quantity, but also its first spatial derivative. For a two-nodal one-dimensional element these conditions are satisfied by first-order continuous cubic Hermitian interpolation polynomials which give the following representation for any quantity, say $h(x,t)$

$$h(x,t) \approx \Omega_j(\zeta)h_j(t) + \hat{\Omega}_j(\zeta)\frac{\partial h_j(t)}{\partial x} \tag{9.1}$$

in which $\Omega_j(\zeta)$ and $\hat{\Omega}_j(\zeta)$ are the Hermitian basis functions which, in terms of the local coordinate ζ, are given by

$$\Omega_1 = 1 - 3\zeta^2 + 2\zeta^3 \tag{9.2a}$$

$$\Omega_2 = 3\zeta^2 - 2\zeta^3 \tag{9.2b}$$

$$\hat{\Omega}_1 = l\zeta(1 - \zeta)^2 \tag{9.2c}$$

$$\hat{\Omega}_2 = l\zeta^2(\zeta - 1) \tag{9.2d}$$

Figure 9.1 gives a graphical representation of the above functions.

The remainder of this section is used to derive the Green element equations which incorporate the Hermitian interpolation functions for the solution of the transport problem under uniform advection, the Burgers equation, and the variably saturated flow problem. These equations are applied to a number of numerical examples to assess how much enhancement in accuracy they provide. As we had earlier pointed out, the Hermitian interpolation functions are incorporated only into the **GE** formulation whose

Green's function is derived from the Laplace differential operator because of its simple nature.

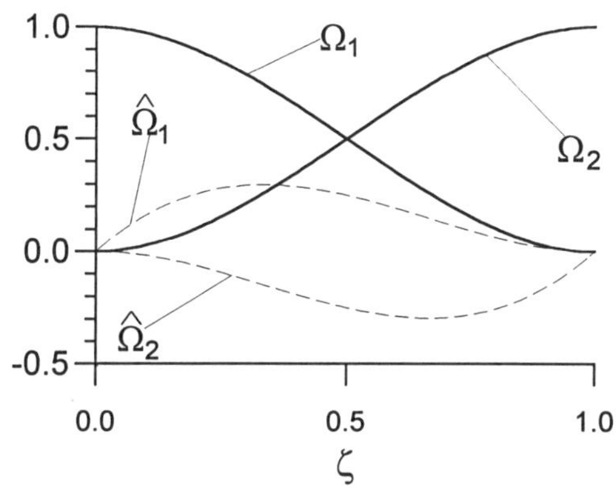

Figure 9.1 The cubic Hermitian interpolation functions

9.2.1 *Hermitian **GEM** for the Transport Equation*

It is presumed here that the procedure adopted in section 6.2 for the derivation of the **QSGE** formulation with linear elements for the transport equation is quite familiar to the reader, and as such some steps are skipped in deriving the same formulation with Hermitian interpolation functions. For this formulation, the relevant integral equation is eq. (6.3) which, for transport under uniform advection, is slightly modified to

$$\begin{aligned} D\Big[&-\lambda h(x_i,t) + G^*(x_2,x_i)h(x_2,t) - G^*(x_1,x_i)h(x_1,t) \\ &- G(x_2,x_i)\varphi(x_2,t) + G(x_1,x_i)\varphi(x_1,t)\Big] + \\ &\int_{x_1}^{x_2} G(x,x_i)\left[U\varphi(x,t) + \frac{\partial h(x,t)}{\partial t} + \mu h(x,t) + f(x,t)\right]dx = 0 \end{aligned} \tag{9.3}$$

It should be noted that eq. (9.3) is essentially the same as eq. (6.3), except that U in the first term within the square bracket in the line integral has neither spatial nor temporal dependence. To evaluate the line integral in eq. (9.3), the quantities $h(x,t)$, $\varphi(x,t)$ and $f(x,t)$ are represented by Hermitian interpolation polynomials. Since the Hermitian polynomials allow for the continuity of the spatial derivative of h, the quantity φ is approximated by the relation

$$\varphi(x,t) \equiv \frac{\partial h(x,t)}{\partial x} \approx \frac{d\Omega_j}{dx} h_j(t) + \frac{d\hat{\Omega}_j}{dx}\varphi_j(t) \tag{9.4}$$

When the relationships provided by the interpolation for h, φ and f are substituted into the integral eq. (9.3) and the line integral is evaluated, the outcome is a system of discretized equations (element equations) which apply to a typical element $[x_1, x_2]$

$$\begin{aligned} &D[R_{ij}h_j + L_{ij}\varphi_j] + [US_{ij} + \mu T_{ij}]h_j + T_{ij}\left(\frac{dh_j}{dt} + f_j\right) + \\ &[U\hat{S}_{ij} + \mu\hat{T}_{ij}]\varphi_j + \hat{T}_{ij}\left(\frac{d\varphi_j}{dt} + \frac{df_j}{dx}\right) = 0; \quad i,j=1,2 \end{aligned} \tag{9.5}$$

in which R_{ij} and L_{ij} are element matrices whose expressions are given by eqs. (2.42a) and (2.42b), respectively. The other element matrices in eq. (9.5) have the following expressions:

$$S_{ij} = \int_{x_1}^{x_2} \frac{d\Omega_j}{dx} G(x,x_i)\,dx = \frac{1}{2}\begin{bmatrix} -(2\tilde{l}+l) & 2\tilde{l}+l \\ -(2\tilde{l}+l) & 2\tilde{l}+l \end{bmatrix} \tag{9.6a}$$

$$\hat{S}_{ij} = \int_{x_1}^{x_2} \frac{d\hat{\Omega}_j}{dx} G(x,x_i)\,dx = \frac{l^2}{12}R_{ij} = \frac{l^2}{12}(-1)^{i+j-1} \tag{9.6b}$$

$$T_{ij} = \int_{x_1}^{x_2} \Omega_j G(x,x_i)\,dx = \frac{l}{20}\begin{bmatrix} 10\tilde{l}+3l & 10\tilde{l}+7l \\ 10\tilde{l}+7l & 10\tilde{l}+3l \end{bmatrix} \tag{9.6c}$$

$$\hat{T}_{ij} = \int_{x_1}^{x_2} \hat{\Omega}_j G(x,x_i)\,dx = \frac{l^2}{60}\begin{bmatrix} 5\tilde{l}+2l & -(5\tilde{l}+3l) \\ 5\tilde{l}+3l & -(5\tilde{l}+2l) \end{bmatrix} \tag{9.6d}$$

As it is readily observed, the expressions for the element matrices are simple, and that is attributable to the simple nature of the expression for $G(x,x_i)$. The temporal derivatives in eq. (9.5) are simplified by the generalized 2-level time discretization scheme which was first used in chapter 5. Simplifying dh_j/dt and $d\varphi_j/dt$ by difference expressions that are evaluated at $t = t_{1+\alpha} = t_1 + \alpha\Delta t$ (see Figure 5.2), eq. (9.5) now becomes

$$\left[\alpha\left(DR_{ij} + US_{ij} + \mu T_{ij}\right) + \frac{T_{ij}}{\Delta t}\right]h_j^{(2)} + \left[\omega\left(DR_{ij} + US_{ij} + \mu T_{ij}\right) - \frac{T_{ij}}{\Delta t}\right]h_j^{(1)} +$$

$$\left[\alpha\left(DL_{ij} + U\hat{S}_{ij} + \mu \hat{T}_{ij}\right) + \frac{\hat{T}_{ij}}{\Delta t}\right]\varphi_j^{(2)} + \left[\omega\left(DL_{ij} + U\hat{S}_{ij} + \mu T_{ij}\right) - \frac{\hat{T}_{ij}}{\Delta t}\right]\varphi_j^{(1)} + \tag{9.7}$$

$$\alpha[T_{ij}f_j^{(2)} + \hat{T}_{ij}\frac{df_j^{(2)}}{dx}] + \omega[T_{ij}f_j^{(1)} + \hat{T}_{ij}\frac{df_j^{(1)}}{dx}] = 0$$

As in earlier chapters, the bracketed superscripts 2 and 1 denote the current and previous time levels. Eq. (9.7) is assembled for all elements in the computational domain from which the solutions for $h_j^{(2)}$ and $\varphi_j^{(2)}$ are obtained at any time level.

9.2.2 *Hermitian **GEM** for the Burgers Equation*

Linear elements were incorporated into the **GEM** to obtain solutions to the Burgers equation (7.1) in chapter 7. Here Hermitian interpolation functions are employed to approximate the unknown quantities. The integral equation, obtained using the Laplace operator as complementary equation, is given by eq. (7.2) and restated below

$$\begin{aligned} &\nu\Big[-\lambda h(x_i,t) + G^*(x_2,x_i)h(x_2,t) - G^*(x_1,x_i)h(x_1,t) \\ &- G(x_2,x_i)\varphi(x_2,t) + G(x_1,x_i)\varphi(x_1,t)\Big] + \\ &\int_{x_1}^{x_2} G(x,x_i)\left[h(x,t)\varphi(x,t) + \frac{\partial h(x,t)}{\partial t}\right]dx = 0; \quad i = 1,2 \end{aligned} \tag{7.2}$$

Using the approximation provided by eq. (9.1) for $h(x,t)$, and that given by (9.4) for $\varphi(x,t)$, the line integral in eq. (7.2) can now be evaluated. The resulting element equations are given as

$$\begin{aligned} &\nu R_{ij}h_j + \nu L_{ij}\varphi_j + U_{inj}h_n h_j + \hat{U}_{inj}h_n\varphi_j + W_{inj}\varphi_n h_j + \\ &\hat{W}_{inj}\varphi_n\varphi_j + T_{ij}\frac{dh_j}{dt} + \hat{T}_{ij}\frac{d\varphi_j}{dt} = 0; \quad i,j=1,2 \end{aligned} \tag{9.8}$$

in which the element matrices R_{ij}, L_{ij}, T_{ij} and $\hat{T}_{ij}$ had earlier been defined. The matrices U_{inj}, $\hat{U}_{inj}$, W_{inj}, and $\hat{W}_{inj}$ are given as

$$U_{inj} = \int_{x_1}^{x_2} \Omega_n \frac{d\Omega_j}{dx} G(x,x_i)\,dx;$$

$$U_{1nj} = \frac{1}{70}\begin{bmatrix} -(35\tilde{l}+13l) & 35\tilde{l}+13l \\ -(35\tilde{l}+22l) & 35\tilde{l}+22l \end{bmatrix} \quad U_{2nj} = \frac{1}{70}\begin{bmatrix} -(35\tilde{l}+22l) & 35\tilde{l}+22l \\ -(35\tilde{l}+13l) & 35\tilde{l}+13l \end{bmatrix} \tag{9.9a}$$

$$\hat{U}_{inj} = \int_{x_1}^{x_2} \Omega_n \frac{d\hat{\Omega}_j}{dx} G(x,x_i)\,dx;$$

$$\hat{U}_{1nj} = \frac{l}{420}\begin{bmatrix} 42\tilde{l}-4l & -(42\tilde{l}+11l) \\ -(42\tilde{l}+31l) & 42\tilde{l}+46l \end{bmatrix} \quad \hat{U}_{2nj} = \frac{l}{420}\begin{bmatrix} 42\tilde{l}+46l & -(42\tilde{l}+31l) \\ -(42\tilde{l}+11l) & 42\tilde{l}-4l \end{bmatrix} \tag{9.9b}$$

$$W_{inj} = \int_{x_1}^{x_2} \hat{\Omega}_n \frac{d\Omega_j}{dx} G(x,x_i)\,dx;$$

$$W_{1nj} = \frac{l}{70}\begin{bmatrix} -(7\tilde{l}+3l) & 7\tilde{l}+3l \\ 7\tilde{l}+4l & -(7\tilde{l}+4l) \end{bmatrix}, \quad W_{2nj} = \frac{l}{70}\begin{bmatrix} -(7\tilde{l}+4l) & 7\tilde{l}+4l \\ 7\tilde{l}+3l & -(7\tilde{l}+3l) \end{bmatrix} \tag{9.9c}$$

$$\hat{W}_{inj} = \int_{x_1}^{x_2} \hat{\Omega}_n \frac{d\hat{\Omega}_j}{dx} G(x,x_i)\,dx;$$

$$W_{1nj} = \frac{l^2}{420}\begin{bmatrix} -2l & -(7\tilde{l}+2l) \\ 7\tilde{l}+5l & -2l \end{bmatrix}, \quad W_{2nj} = \frac{l^2}{420}\begin{bmatrix} 2l & -(7\tilde{l}+5l) \\ 7\tilde{l}+2l) & 2l \end{bmatrix} \tag{9.9d}$$

The temporal derivatives in eq. (9.8) are simplified by the 2-level time discretization scheme, while the nonlinear terms are linearized by the Picard scheme. When these simplifications are implemented, eq. (9.8) becomes

$$\left\{\alpha\left[\nu R_{ij} + U_{inj}[\alpha h_n^{(2,k)} + \omega h_n^{(1)}] + W_{inj}[\alpha\varphi_n^{(2,k)} + \omega\varphi_n^{(1)}]\right] + \frac{T_{ij}}{\Delta t}\right\} h_j^{(2,k+1)} +$$

$$\left\{\alpha\left[\nu L_{ij} + \hat{U}_{inj}[\alpha h_n^{(2,k)} + \omega h_n^{(1)}] + \hat{W}_{inj}[\alpha\varphi_n^{(2,k)} + \omega\varphi_n^{(1)}]\right] + \frac{\hat{T}_{ij}}{\Delta t}\right\} \varphi_j^{(2,k+1)} =$$

$$-\left\{\omega\left[\nu R_{ij} + U_{inj}[\alpha h_n^{(2,k)} + \omega h_n^{(1)}] + W_{inj}[\alpha\varphi_n^{(2,k)} + \omega\varphi_n^{(1)}]\right] - \frac{T_{ij}}{\Delta t}\right\} h_j^{(1)} \tag{9.10}$$

$$-\left\{\omega\left[\nu L_{ij} + \hat{U}_{inj}[\alpha h_n^{(2,k)} + \omega h_n^{(1)}] + \hat{W}_{inj}[\alpha\varphi_n^{(2,k)} + \omega\varphi_n^{(1)}]\right] - \frac{\hat{T}_{ij}}{\Delta t}\right\} \varphi_j^{(1)}$$

The linearized system of equations above has been written in such a way that the unknown quantities $h_j^{(2,k+1)}$ and $\varphi_j^{(2,k+1)}$ are on the left hand side, while the right hand side has the known quantities. Eq. (9.10) is solved at each time level for a number of times until convergence is achieved.

9.2.3 *Hermitian **GEM** for the Unsaturated Flow Equation*

In chapter 8, we addressed the unsaturated flow equation whose soil constitutive relations, for certain soils, exhibit rapid variations with respect to the primary variable. For such soils, the use of higher-order interpolation functions may be required to enhance the accuracy of the solution. The differential equation remains eq. (8.10), while its integral representation retains the form of eq. (8.11) which is restated below

$$-\lambda h(z_i,t) + G^*(z_2,z_i)h(z_2,t) - G^*(z_1,z_i)h(z_1,t)$$

$$- G(z_2,z_i)\varphi(z_2,t) + G(z_1,z_i)\varphi(z_1,t) + \tag{8.11}$$

$$\int_{z_1}^{z_2} G(z,z_i)\left\{\Psi\left(\frac{\partial h(z,t)}{\partial t} + \frac{\partial K}{\partial z}\right) - \frac{\partial \Theta}{\partial z}\frac{\partial h}{\partial z}\right\} dz = 0$$

The quantities h, K, $\Theta \equiv \ln(D)$, and $\Psi \equiv 1/D$ are approximated by expressions similar to eq. (9.1). When the approximations of those quantities are introduced into eq. (8.11), and the line integral evaluated, the outcome is the system of equations

$$
\boxed{\begin{aligned}
&R_{ij}h_j + L_{ij}\varphi_j - V_{inj}\Theta_n h_j - \hat{V}_{inj}\Theta_n\varphi_j - \hat{V}_{inj}h_n\left(\frac{d\Theta}{dh}\varphi\right)_j - \\
&\check{V}_{inj}\varphi_n\left(\frac{d\Theta}{dh}\varphi\right)_j + U_{inj}\Psi_n K_j + \hat{U}_{inj}\Psi_n\left(\frac{dK}{dh}\varphi\right)_j + W_{inj}\left(\frac{d\Psi}{dh}\varphi\right)_n K_j + \\
&\hat{W}_{inj}\left(\frac{d\Psi}{dh}\varphi\right)_n\left(\frac{dK}{dh}\varphi\right)_j + S_{inj}\Psi_n\frac{dh_j}{dt} + \hat{S}_{inj}\Psi_n\frac{d\varphi_j}{dt} + \\
&\hat{S}_{inj}\frac{dh_n}{dt}\left(\frac{d\Psi}{dh}\varphi\right)_j + \check{S}_{inj}\left(\frac{d\Psi}{dh}\varphi\right)_n\frac{d\varphi_j}{dt} = 0
\end{aligned}} \tag{9.11}
$$

in which

$$
V_{inj} = \int_{z_1}^{z_2}\frac{d\Omega_n}{dz}\frac{d\Omega_j}{dz}G(z,z_i)dz = \frac{3}{5l}\begin{bmatrix} 2\tilde{l}+l & -(2\tilde{l}+l) \\ -(2\tilde{l}+l) & 2\tilde{l}+l \end{bmatrix} = \frac{3(2\tilde{l}+l)}{5l}(-1)^{n+j} \tag{9.12a}
$$

$$
\begin{aligned}
\hat{V}_{inj} &= \int_{z_1}^{z_2}\frac{d\Omega_n}{dz}\frac{d\hat{\Omega}_j}{dz}G(z,z_i)dz; \\
\hat{V}_{1nj} &= \frac{1}{10}\begin{bmatrix} \tilde{l}+l & \tilde{l} \\ -(\tilde{l}+l) & -\tilde{l} \end{bmatrix}, \quad \hat{V}_{2nj} = \frac{1}{10}\begin{bmatrix} \tilde{l} & \tilde{l}+l \\ -\tilde{l} & -(\tilde{l}+l) \end{bmatrix}
\end{aligned} \tag{9.12b}
$$

$$
\begin{aligned}
\check{V}_{inj} &= \int_{z_1}^{z_2}\frac{d\hat{\Omega}_n}{dz}\frac{d\hat{\Omega}_j}{dz}G(z,z_i)dz; \\
\check{V}_{1nj} &= \frac{l}{60}\begin{bmatrix} 8\tilde{l}+2l & -(2\tilde{l}+l) \\ -(2\tilde{l}+l) & 8\tilde{l}+6l \end{bmatrix}, \quad \check{V}_{2nj} = \frac{l}{60}\begin{bmatrix} 8\tilde{l}+6l & -(2\tilde{l}+l) \\ -(2\tilde{l}+l) & 8\tilde{l}+2l \end{bmatrix}
\end{aligned} \tag{9.12c}
$$

$$
\begin{aligned}
S_{inj} &= \int_{z_1}^{z_2}\Omega_n\Omega_j G(z,z_i)dz; \\
S_{1nj} &= \frac{l}{140}\begin{vmatrix} 52\tilde{l}+12l & 18\tilde{l}+9l \\ 18\tilde{l}+9l & 52\tilde{l}+40l \end{vmatrix}, \quad S_{2nj} = \frac{l}{140}\begin{vmatrix} 52\tilde{l}+40l & 18\tilde{l}+9l \\ 18\tilde{l}+9l & 52\tilde{l}+12l \end{vmatrix}
\end{aligned} \tag{9.12d}
$$

$$
\begin{aligned}
\hat{S}_{inj} &= \int_{z_1}^{z_2}\Omega_n\hat{\Omega}_j G(z,z_i)dz; \\
\hat{S}_{1nj} &= \frac{l^2}{420}\begin{bmatrix} 22\tilde{l}+7l & -(13\tilde{l}+6l) \\ 13\tilde{l}+7l & -(22\tilde{l}+15l) \end{bmatrix}, \quad \hat{S}_{2nj} = \frac{l^2}{420}\begin{bmatrix} 22\tilde{l}+15l & -(13\tilde{l}+7l) \\ 13\tilde{l}+6l & -(22\tilde{l}+7l) \end{bmatrix}
\end{aligned} \tag{9.12e}
$$

$$\check{S}_{inj} = \int_{z_1}^{z_2} \hat{\Omega}_n \hat{\Omega}_j G(z,z_i)dz;$$

$$\check{S}_{1nj} = \frac{l^3}{840}\begin{bmatrix} 8\tilde{l}+3l & -(6\tilde{l}+3l) \\ -(6\tilde{l}+3l) & 8\tilde{l}+5l \end{bmatrix}, \quad \check{S}_{2nj} = \frac{l^3}{840}\begin{bmatrix} 8\tilde{l}+5l & -(6\tilde{l}+3l) \\ -(6\tilde{l}+3l) & 8\tilde{l}+3l \end{bmatrix} \tag{9.12f}$$

Eq. (9.11) is a system of discrete first-order differential equations that are also highly nonlinear. We apply the 2-level time discretization scheme to simplify the temporal derivatives, and the Picard scheme to linearize the nonlinear terms. When these steps are applied, eq. (9.11) reduces to [1]

$$\begin{aligned}
&\left\{\alpha\left[R_{ij} - V_{inj}[\alpha\Theta_n^{(2,k)} + \omega\Theta_n^{(1)}] - \hat{V}_{ijn}[\alpha\left(\frac{d\Theta}{dh}\varphi\right)_n^{(2,k)} + \omega\left(\frac{d\Theta}{dh}\varphi\right)_n^{(1)}]\right] + \right.\\
&\left.\frac{S_{inj}}{\Delta t}[\alpha\Psi_n^{(2,k)} + \omega\Psi_n^{(1)}] + \frac{\hat{S}_{ijn}}{\Delta t}[\alpha\left(\frac{d\Psi}{dh}\varphi\right)_n^{(2,k)} + \omega\left(\frac{d\Psi}{dh}\varphi\right)_n^{(1)}]\right\}h_j^{(2,k+1)} + \\
&\left\{\alpha\left[L_{ij} - \hat{V}_{inj}[\alpha\Theta_n^{(2,k)} + \omega\Theta_n^{(1)}] - \check{V}[\alpha\left(\frac{d\Theta}{dh}\varphi\right)_n^{(2,k)} + \omega\left(\frac{d\Theta}{dh}\varphi\right)_n^{(1)}]\right] + \right.\\
&\left.\frac{\hat{S}_{inj}}{\Delta t}[\alpha\Psi_n^{(2,k)} + \omega\Psi_n^{(1)}] + \frac{\check{S}_{inj}}{\Delta t}[\alpha\left(\frac{d\Psi}{dh}\varphi\right)_n^{(2,k)} + \omega\left(\frac{d\Psi}{dh}\varphi\right)_n^{(1)}]\right\}\varphi_j^{(2,k+1)} = \\
&\left\{-\omega\left[R_{ij} - V_{inj}[\alpha\Theta_n^{(2,k)} + \omega\Theta_n^{(1)}] - \hat{V}_{ijn}[\alpha\left(\frac{d\Theta}{dh}\varphi\right)_n^{(2,k)} + \omega\left(\frac{d\Theta}{dh}\varphi\right)_n^{(1)}]\right] + \right.\\
&\left.\frac{S_{inj}}{\Delta t}[\alpha\Psi_n^{(2,k)} + \omega\Psi_n^{(1)}] + \frac{\hat{S}_{ijn}}{\Delta t}[\alpha\left(\frac{d\Psi}{dh}\varphi\right)_n^{(2,k)} + \omega\left(\frac{d\Psi}{dh}\varphi\right)_n^{(1)}]\right\}h_j^{(1)} + \\
&\left\{-\omega\left[L_{ij} - \hat{V}_{inj}[\alpha\Theta_n^{(2,k)} + \omega\Theta_n^{(1)}] - \check{V}[\alpha\left(\frac{d\Theta}{dh}\varphi\right)_n^{(2,k)} + \omega\left(\frac{d\Theta}{dh}\varphi\right)_n^{(1)}]\right] + \right.\\
&\left.\frac{\hat{S}_{inj}}{\Delta t}[\alpha\Psi_n^{(2,k)} + \omega\Psi_n^{(1)}] + \frac{\check{S}_{inj}}{\Delta t}[\alpha\left(\frac{d\Psi}{dh}\varphi\right)_n^{(2,k)} + \omega\left(\frac{d\Psi}{dh}\varphi\right)_n^{(1)}]\right\}\varphi_j^{(1)} - \\
&U_{inj}[\alpha\Psi_n^{(2,k)} + \omega\Psi_n^{(1)}][\alpha K_j^{(2,k)} + \omega K_j^{(1)}] - \\
&\hat{U}_{inj}[\alpha\Psi_n^{(2,k)} + \omega\Psi_n^{(1)}][\alpha\left(\frac{dK}{dh}\varphi\right)_j^{(2,k)} + \omega\left(\frac{dK}{dh}\varphi\right)_j^{(1)}] - \\
&W_{inj}[\alpha\left(\frac{d\Psi}{dh}\varphi\right)_n^{(2,k)} + \omega\left(\frac{d\Psi}{dh}\varphi\right)_n^{(1)}][\alpha K_j^{(2,k)} + \omega K_j^{(1)}] - \\
&\hat{W}_{inj}[\left(\frac{d\Psi}{dh}\varphi\right)_n^{(2,k)} + \omega\left(\frac{d\Psi}{dh}\varphi\right)_n^{(1)}][\alpha\left(\frac{dK}{dh}\varphi\right)_j^{(2,k)} + \omega\left(\frac{dK}{dh}\varphi\right)_j^{(1)}]
\end{aligned} \tag{9.13}$$

Eq. (9.13) is assembled for all the elements, and solve iteratively at each time level.

9.3 Numerical Stability Characteristics of Hermitian GEM for the Transport Equation

In section 6.5, we outlined the procedure for evaluating the stability characteristics of three **GE** models proposed for the diffusion-advection equation. That procedure follows that which had been applied by Gray and his co-workers to a number of numerical schemes (Gray and Pinder [2], Gray and Lynch [3]). In evaluating the stability characteristics of the Hermitian **GEM** for the diffusion-advection equation, we request the reader to refer to section 6.5 to fill in the details which are herein skipped. We rewrite the element equations of the Hermitian **GEM** provided by eq. (9.7) in a form similar to eq. (6.52) which is restated below

$$P_{ij}^{(m)}\varphi_j^{(m)} + N_{ij}^{(m)}\varphi_j^{(m)} = 0, \quad i,j,m=1,2 \tag{6.52}$$

Equating coefficients of like terms we have (Taigbenu [4])

$$P_{ij}^{(2)} \equiv \alpha[DR_{ij} + US_{ij}] + \frac{T_{ij}}{\Delta t} \tag{9.14a}$$

$$N_{ij}^{(2)} \equiv \alpha[DL_{ij} + U\hat{S}_{ij}] + \frac{\hat{T}_{ij}}{\Delta t} \tag{9.14b}$$

$$P_{ij}^{(1)} \equiv \omega[DR_{ij} + US_{ij}] - \frac{T_{ij}}{\Delta t} \tag{9.14c}$$

$$N_{ij}^{(1)} \equiv \omega[DL_{ij} + U\hat{S}_{ij}] - \frac{\hat{T}_{ij}}{\Delta t} \tag{9.14d}$$

With the above relations established, the expressions for the numerical amplification g_N, the amplification factor T, and the numerical phase change after one wavelength of propagation Θ retain their earlier expressions, which are eqs. (6.63), (6.48) and (6.51), respectively. The performance of our current model depends on the values of the temporal weighting factor α, the Courant number Cr and the element Peclet number Pe used in the calculations. Having earlier established the marginal superiority of the Crank-Nicholson scheme when used in conjunction with linear **GEM** for the transport equation, that scheme is similarly incorporated into the Hermitian **GEM** for values of $Cr = 0.2$ and $Cr = 1$, $Pe = 2$ and $Pe = 50$. The comparisons are presented in Figures 9.2 and 9.3. The results in Figures 9.2a and 9.2b for the marginal advection-dominant transport ($Pe = 2$) when a Courant number of 0.2 is used in the numerical calculations show that the Hermitian **GEM** does better than linear **GEM** in propagating the amplitude of the harmonics with small wavelengths, but both models do equally well in propagating the speed of the Fourier wave components. An increase in the value of the Courant number to unity (indicative of numerical calculations with larger time steps) for the same transport mode ($Pe = 2$), which are also displayed in those figures, shows that both models correctly propagate the amplitude of the wavelengths greater than $10l$, while the Hermitian model correctly propagates the speed of the harmonics whose wavelengths are greater than $3l$, compared to the linear model which does so for wavelengths greater than $5l$. Although both models are dissipative for the short wave harmonics, they are still expected to do well for this mode of transport because steep initial concentration fronts, characterized by the small harmonics, will be smeared quickly in time by hydrodynamic dispersion.

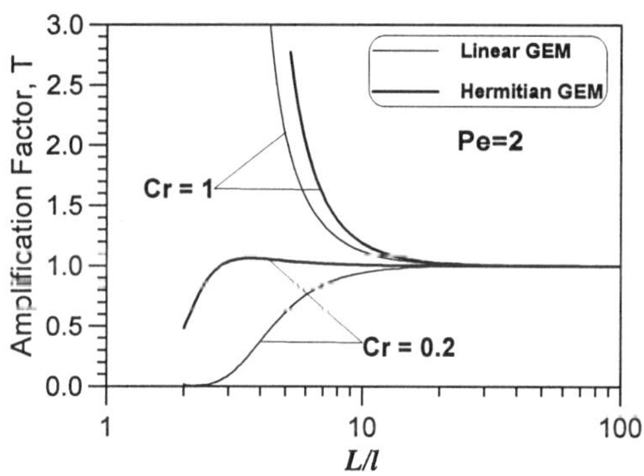

Figure 9.2a Variations in amplification factor for linear and Hermitian **GEM** for $Pe = 2$.

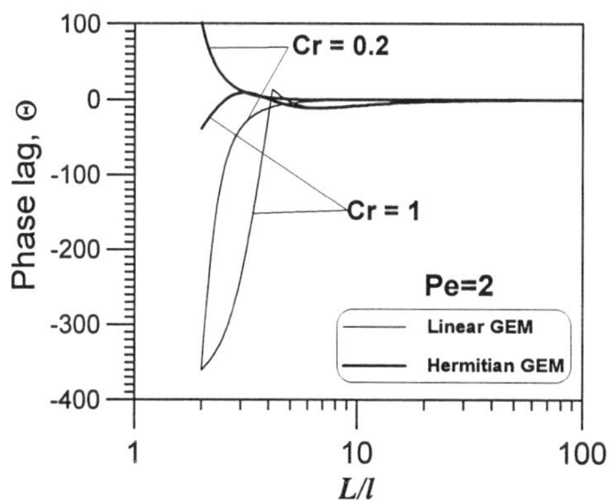

Figure 9.2b Variations in Phase lag for linear and Hermitian **GEM** for $Pe = 2$.

For the strong advection-dominant transport mode ($Pe = 50$), the Hermitian model gives a more superior result than the linear model when the Courant number has a value of 0.2 , but at $Cr = 1$ the Hermitian model has only a slight advantage (see Figures 9.3a and 9.3b). An overall assessment of both models indicates that the Hermitian model has better stability characteristics than the linear model.

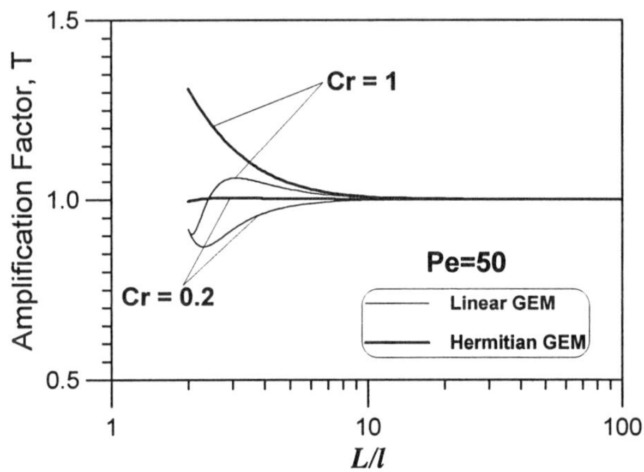

Figure 9.3a Variations in amplification factor for linear and Hermitian **GEM** for $Pe = 50$.

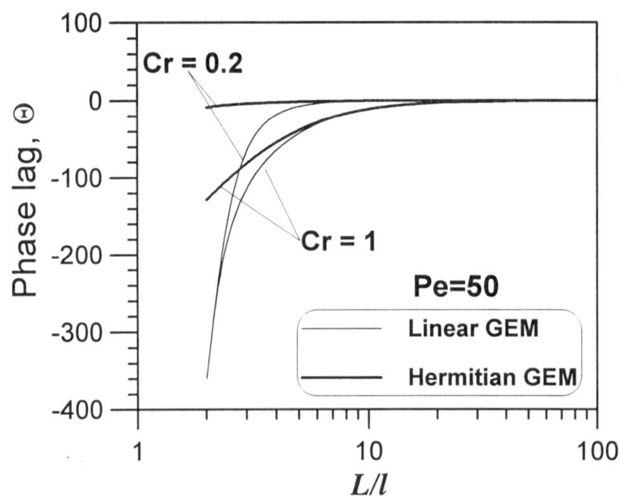

Figure 9.3b Variations in phase lag linear and Hermitian **GEM** for $Pe = 50$

.9.4 Numerical Calculations

9.4.1 *Examples of Contaminant Transport*

Two numerical examples of transient contaminant transport in 1-D. spatial dimension are used to evaluate the level of enhancement of accuracy achieved by the Hermitian Green element model described in the foregoing section. Our evaluation is based on comparison of the current model with that of the earlier model which incorporates linear interpolation functions to approximate the distribution of the concentration. The two examples had earlier been presented in chapter 6, and their exact solutions serve as bench mark results in assessing accuracy. Since one of the unique features of the solution to the transport problem is that the initial concentration profile is maintained with time when advection is dominant (large value of Peclet number), the examples have been so designed that the initial concentration profile is steep, and transport takes place when advection is dominant. The Hermitian and linear Green element calculations of the two examples are carried out with the Crank-Nicholson scheme ($\alpha = 0.5$) which had earlier been found to give optimal results from the linear model.

Example 1

This first example is the classical 1-D. transport problem with conditions: $h(0,t) = 1$, $\varphi(x=\infty,t) = 0$, and $h(x,t=0) = 0$. Since the advection-dominant case provides the stiffest test for any numerical scheme, a value of 50 is used for the element Peclet number in the numerical calculations. The simulations are done with a uniform ambient velocity $U = 1$, spatial size of each element $l = 0.025$, a uniform time step of 0.005, and the solution of the concentration front is presented at $t = 0.5$. The exact solution and those of the linear and Hermitian **GEM** are presented in Figure 9.4. The results indicate that the Hermitian model reproduced the concentration front better than the linear one, and it does so without any oscillation.

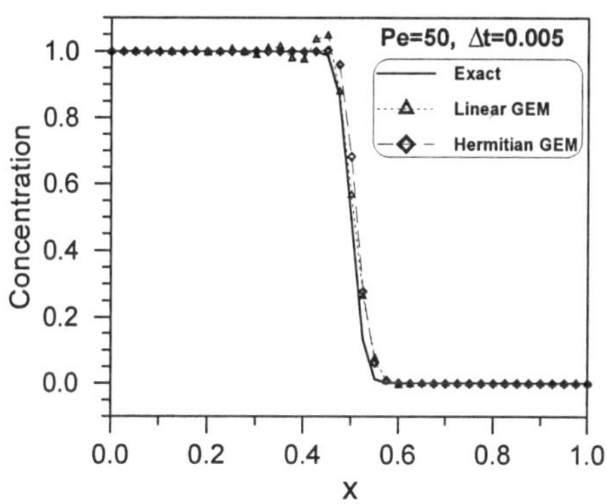

Figure 9.4 Linear and Hermitian **GEM** solutions of contaminant transport example 1

Example 2

This example, which is the third under section 6.6, is that of contaminant transport in a polluted stream that is cleaned by mechanical means of advection from a freshwater source and by biological and chemical processes that induce first-order decay of the pollutant. The initial and boundary conditions are given by eq. (6.75), and the exact solution by eq. (6.76). The following parameter values are incorporated in the numerical calculations: $U = 1$, $D = 0.0005$, $Pe = 50$, and $\mu = 0.2$. The Crank-Nicholson scheme and a uniform time step of 0.025 are used in both the linear and Hermitian Green element models. The numerical and exact solutions, obtained at times of 0.5 and 1.0, are presented in Figure 9.5. While both models did equally well in approximating the concentration front, the results of the Hermitian model are free of the oscillations that are observed in the linear model.

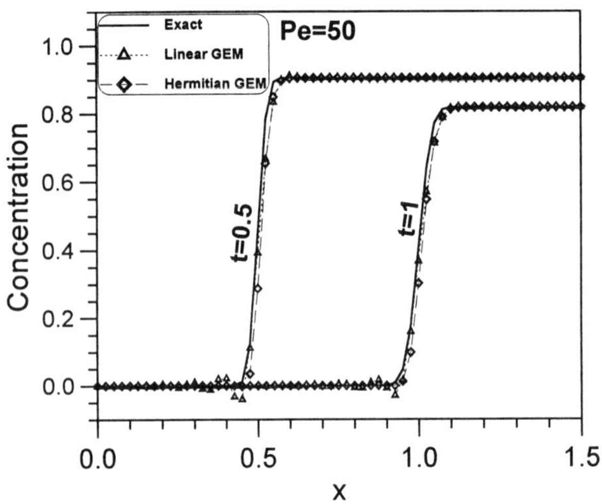

Figure 9.5 Linear and Hermitian **GEM** solutions of contaminant transport example 2.

9.4.2 *Examples of Momentum Transport (Burgers Equation)*

Two numerical examples of Burgers equation are used to evaluate the performance of the Hermitian Green element model that has been presented in section 9.2.2. The examples, which have analytic solutions, had earlier been addressed with the linear Green element model in sections 7.5 and 7.6. The results from the linear and Hermitian models are herein compared.

Example 1

The first example (example 1 in section 7.5) is an initial sinusoidal wave which is allowed to propagate and diffuse within a confined flow domain in the x-dimension. The problem has the following boundary conditions: $h(0,t)=0$, $h(1,t)=0$, and $h(x,0)=\sin\pi x$. The exact solution to this problem, in the form of an infinite series, has been provided by Cole [5].

The results of the simulation of example 1 by the Hermitian model are presented along side those of the linear model in Table 9.1 and Figure 9.6a for $\nu = 10^{-2}$, and in Figure 9.6b for $\nu = 10^{-3}$. In both models, the Crank-Nicholson scheme and a uniform time step of 0.01 are adopted, while 4 iterations are allowed at each time level. For the case of $\nu = 10^{-2}$, 50 elements are used in the Hermitian model, whereas 100 elements are used in the linear model. The numerical solutions from both models for this case are about the same, despite the fact that fewer elements are used in the Hermitian model. Computing time for Hermitian simulation is about one and a half times that of the linear simulation. When $\nu = 10^{-3}$, the solution of the linear model exhibits oscillations in the vicinity of $x = 1$, while that of the Hermitian model reflects the expected trend of the solution, which is a sharp front in the neighborhood of $x = 1$ at early times, and a smeared front at later times due to pronounced viscous action. For this case, 100 elements and a uniform time step of 0.01 are used in both models.

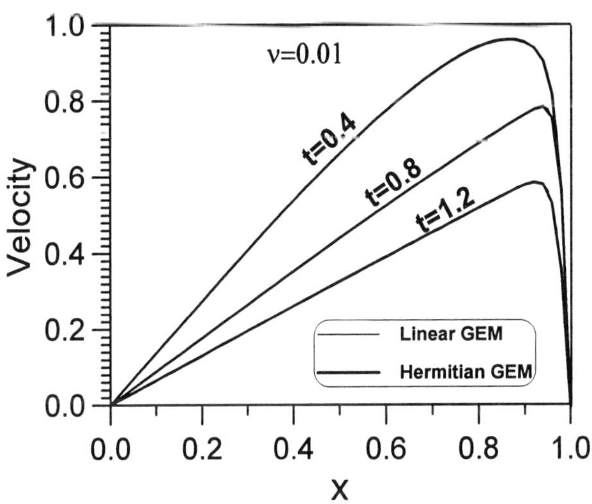

Figure 9.6a Linear **GEM** (100 elements) and Hermitian **GEM** (50 elements) solutions of example 1 of Burgers equation ($\nu = 10^{-2}$).

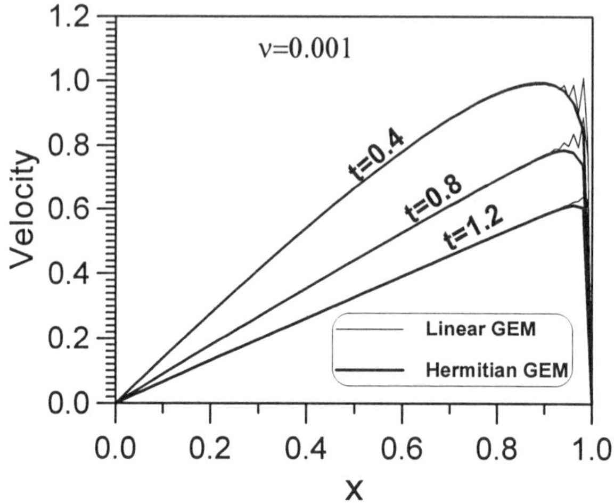

Figure 9.6b Linear **GEM** (100 elements) and Hermitian **GEM** (50 elements) solutions of example 1 of Burgers equation ($\nu = 10^{-3}$)

Table 9.1 **GEM** and Exact solutions for example 1 ($\nu = 0.01$) of Burgers equation.

Position x	$h(x,0.40)$			$h(x,0.80)$		
	Linear **GEM***	Hermitian **GEM***	Exact	Linear **GEM***	Hermitian **GEM***	Exact
0.0	0.00000	0.00000	0.00000	0.00000	0.00000	0.00000
0.2	0.27449	0.27447	0.27452	0.17734	0.17733	0.17736
0.4	0.53788	0.53790	0.53792	0.35279	0.35278	0.35282
0.6	0.77347	0.77361	0.77345	0.52398	0.52398	0.52401
0.8	0.94117	0.94157	0.94100	0.68708	0.68715	0.68708
0.9	0.95277	0.95336	0.94891	0.76311	0.76322	0.76295
0.92	0.93734	0.93794	0.93655	0.77645	0.77649	0.77586
0.94	0.90330	0.90383	0.89998	0.78319	0.78317	0.78119
0.96	0.81619	0.81692	0.81227	0.75388	0.75488	0.74905
0.98	0.56080	0.56252	0.56583	0.55927	0.56440	0.55546
1.0	0.00000	0.00000	0.00000	0.00000	0.00000	0.00000
Mean abs. error x 10^{-4}	15.618	16.773		10.427	16.2	

Table 9.1 Continued **GEM** and Exact solutions for example 1 ($\nu = 0.01$) of Burgers equation

Position x	$h(x,1.20)$			$h(x,3.0)$		
	Linear **GEM***	Hermitian **GEM***	Exact	Linear **GEM***	Hermitian **GEM***	Exact
0.0	0.00000	0.00000	0.00000	0.00000	0.00000	0.00000
0.2	0.13091	0.13090	0.13092	0.06009	0.06009	0.06009
0.4	0.26126	0.26125	0.26128	0.12016	0.12015	0.12016
0.6	0.39041	0.39040	0.39044	0.18017	0.18016	0.18018
0.8	0.51750	0.51751	0.51753	0.23867	0.23866	0.23863
0.9	0.57812	0.57808	0.57781	0.24183	0.24188	0.24159
0.92	0.58555	0.58547	0.58472	0.22636	0.22650	0.22612
0.94	0.57956	0.57965	0.57779	0.19708	0.19732	0.19690
0.96	0.52772	0.52884	0.52524	0.14919	0.14949	0.14911
0.98	0.35167	0.35422	0.35061	0.08139	0.08161	0.08139
1.0	0.00000	0.00000	0.00000	0.00000	0.00000	0.00000
Mean abs. error x 10^{-4}	5.9454	9.2727		0.7182	1.59	

* *Linear **GEM** with 100 elements, and Hermitian **GEM** with 50 elements.*

Example 2

The second example (example 2 in section 7.5) is an initial discontinuous wave form which is propagated and diffused into a continuous wave form. The initial and boundary conditions have earlier been presented and the exact solution, obtained by Cole-Hoph transform, is given by Lighthill [6].

Figure 9.7 indicates the Hermitian and linear **GEM**, and exact solutions of example 2 at different times for the case of $\nu = 10^{-2}$. Although the flow domain for this example is infinitely extensive, the numerical simulations are carried out in a finite domain of $-0.7 \le x \le 1.3$ which still ensures that the exact solutions at the end nodes satisfy the specified boundary conditions at $x = -\infty$ and $x = \infty$ throughout the simulation times. The Hermitian model, which uses half the number of elements of the linear model (50 elements), gives a better solution than the latter.

9.4.3 *Examples of Unsaturated Flow*

Simulations with Hermitian **GEM** are carried out on one-dimensional flows in unsaturated soil media for which analytical expressions of the soil constitutive relations are specified. To demonstrate the capability of the current numerical model we have chosen two numerical examples of unsaturated flow. The two examples are simulated by linear **GE** model of chapter 8, the current Hermitian **GE** model, and the finite element method. The results of the simulations are compared.

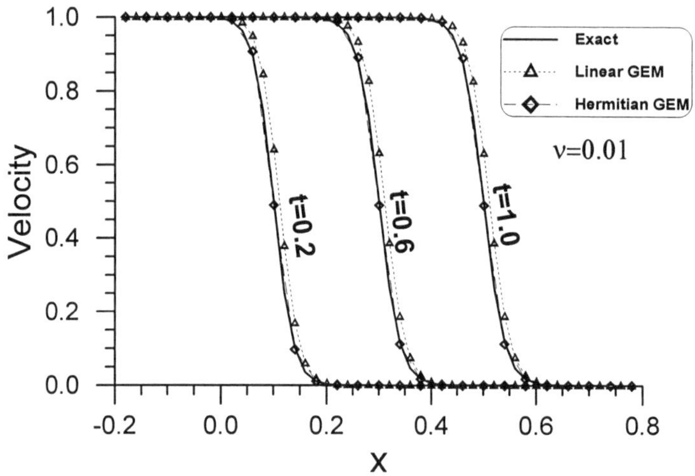

Figure 9.7 Linear **GEM** (100 elements) and Hermitian **GEM** (50 elements) solutions of example 2 of Burgers equation ($\nu = 10^{-2}$).

Example 1

In this example (also example 2 of section 8.4), water is allowed to infiltrate into a dry vertical soil column of 80cm length by maintaining a saturation condition at the top of the soil column. The soil constitutive relations were first reported by van Genuchten [7] and presented in eq. (8.20). The boundary and initial data for this example can be found in eq. (8.21).

The Hermitian model discretizes the 80cm length of flow domain into 40 uniform elements each of which has a length of 2cm. The time history of the solution follows that earlier employed in the finite element solution to this problem: a time step of 7.143×10^{-4}days is used for the time interval, $0 \le t \le 0.055$ days, and another time step of 10^{-3} days is employed within the interval $0.055 \le t \le 0.127$ days. The Hermitian Green element solutions obtained at 0.055 and 0.127days are presented in Figure 9.8 along side the linear **GEM** and the Hermitian **FEM** solutions. A difference weighting factor of 1.0 is employed in the Hermitian **GE** model. Convergence is achieved with no more than 5 iterations at any time step. The Hermitian **GEM** and **FEM** solutions are in good agreement, while that of the linear **GEM**, as earlier observed, gives less absorption of moisture by the soil than that of the other two solutions. Whereas the

FEM solution exhibits some oscillations at the tail of the advancing moisture front, no spurious oscillations are observed in the **GEM** solution.

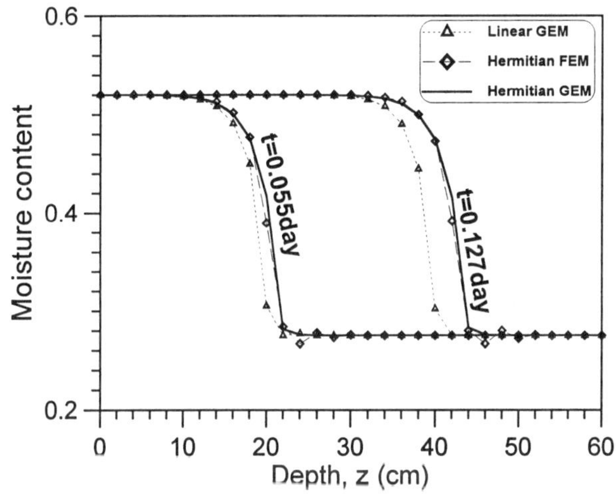

Figure 9.8 Linear and Hermitian **GEM**, and Hermitian **FEM** solutions of example 1 of unsaturated flow.

Example 2

This second example (example 3 in section 8.4) is another vertical infiltration flow into a dry soil column whose soil constitutive data was first reported by Haverkamp *et al* [8] and was used by Celia *et al.* [9] in their numerical simulations. The soil constitutive data are given by eq. (8.22), while the boundary and initial data are given by eq. (8.23).

The Hermitian Green element model simulates this example using 20 uniform elements, a time step of 30s, and a difference weighting factor of unity. A maximum of 5 iterations is allowed at each time step in the model. The accuracy of our solution is compared in Figure 9.9 to that of linear **GEM** and that of the Galerkin solution which was presented by Celia *et al.* [9]. Although we have used in the Hermitian model half the number of elements used in the linear **GEM** and **FEM** calculations, the solution from the Hermitian model compares quite favorably with that of linear **GEM** which is superior to that of linear **FEM**. In other words the Hermitian **GEM** solution with coarse discretization and the linear **GEM** solution both reproduce the numerical solution with a finer mesh discretization than the **FEM** solution. This result demonstrates the improvement in accuracy that can be offered by incorporating Hermitian interpolation functions into the Green element formulation for unsaturated flow.

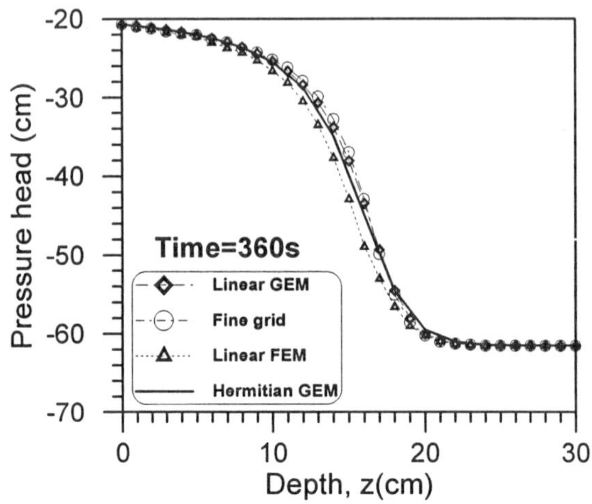

Figure 9.9 Linear and Hermitian **GEM**, and linear **FEM** solutions of example 2 of unsaturated flow.

9.5 References

1. Taigbenu, A.E., "Enhancing the accuracy of the solution to unsaturated flow by a Hermitian Green element model", *Advances in Engineering Software*, **29**(2), pp. 113-118, 1998.

2. Gray, W.G. and G.F. Pinder, "An Analysis of the Numerical Solution of the Transport Equation," *Water Resources research*, **12**, pp. 547-555, 1976.

3. Gray, W.G. and D.R. Lynch, "Time-Stepping Schemes for Finite Element Tidal Model Computations," in *Surface Flow*, ed. W.G. Gray, pp.1-14, CML Pub. Ltd., Southampton UK, 1984.

4. Taigbenu, A.E., "Numerical Stability Characteristics of a Hermitian Green element model for the transport equation",(Research Note) *Engineering Analysis with Boundary Elements*, **22**(2), pp. 161-165, 1998.

5. Cole, J.D., "On a quasi-linear parabolic equation occurring in aerodynamics", *Quarterly of Applied mathematics*, **23**, pp. 225-236 1951.

6. Lighthill, M.J., "Viscosity effects in sound waves of finite amplitude", in *Surveys in Mechanics*, Ed.: G.K. Batchelor and R.M. Davis, cambridge Univ. Press, 1956.

7. van Genuchten, M.T., *Moisture Transport from disposal sites*, Report to Environmental Protection Agency, 1976.

8. Haverkamp, R., M. Vauclin, J. Touma, P. Wierenga, and G. Vachaud, Comparison of Numerical Simulation Models for One-dimensional Infiltration, *Soil Sci. Soc. Am. J.*, **41**, pp. 285-294, 1977.

9. Celia, M.A., E.T. Bouloutas, and R.L. Zarba, "A General Mass-Conservation Numerical Solution for the Unsaturated flow Equation", *Water Resources Research*, **26**(7) pp. 1483-1496,` 1990.

10 STEADY TWO-DIMENSIONAL PROBLEMS

10.1 Introduction

Two dimensional problems present an additional spatial dimension to one-dimensional problems, and their Green element calculations essentially follow the procedure which we had adopted for 1-D. problems in the earlier chapters. For every differential equation, there has to be found an appropriate complimentary or auxiliary differential equation to which the fundamental solution is obtained. Green's second identity in two dimensions provides the tool to transform the governing differential equation into an integral one which is discretized by appropriate 2-D. elements such as triangles and rectangles. The resulting discretized integral equation constitutes the element equations which are assembled to form the global matrix equation that is solved to obtain the nodal unknowns. In situations where the differential equation is nonlinear, the global matrix equation has to be linearized and solved by either the Picard or Newton-Raphson algorithm. In the remaining chapters of this text, we shall apply the **GEM** to steady, transient, linear and nonlinear problems in 2-D. domains which are either homogeneous or heterogeneous.

10.2 Steady Second-Order Differential Equation

We consider the steady problem governed by the second-order partial differential equation applicable to a two dimensional domain Λ

$$\nabla.(\boldsymbol{K}\nabla h) = f(x,y) \quad \text{on } \Lambda \tag{10.1}$$

in which $h = h(x,y)$ is the dependent or primary variable, $\nabla = \boldsymbol{i}\partial/\partial x + \boldsymbol{j}\partial/\partial y$ is the 2-D. gradient operator, x and y are the spatial coordinates, $f(x,y)$ is a known external forcing function that accounts for either point or distributed externally imposed inputs, and $\boldsymbol{K}$ is a parameter which, in general, is a second-order ranked tensor given by the expression

$$\boldsymbol{K} = \begin{bmatrix} K_x & 0 \\ 0 & K_y \end{bmatrix} \tag{10.2}$$

Eq. (10.1) is encountered in many engineering applications — heat transfer, groundwater flow, vibration and acoustics problems, *etc*. When the parameter $\boldsymbol{K}$ is a constant then the medium is said to be homogeneous and isotropic, when it has components in the x and y directions then the medium is anisotropic, when its components have spatial dependence then the medium is heterogeneous, and when $\boldsymbol{K}$ depends on the primary variable, the problem is then a nonlinear one. We shall restrict the application of **GEM** in the subsequent section to the case when the medium is isotropic, that is $\boldsymbol{K} \equiv K$, though the method can accommodate anisotropy.

10.3 Boundary Conditions

Conditions that are similar to those prescribed for 1-D. spatial problems in section 2.2 have to be specified along with the governing differential equation in order to obtain its unique solutions. As with 1-D. problems, three types of boundary conditions are normally encountered. They are:

i. the 1st-type or Dirichlet condition which specifies the value of the primary variable h on a part of the boundary Γ_1 (see Figure 10.1)

$$h = h_1(x,y) \text{ on } \Gamma_1 \tag{10.3a}$$

ii. the 2nd-type or Neumann condition which specifies the normal flux on another part of the boundary Γ_2

$$K\nabla h.\boldsymbol{n} = \varphi_2(x,y) \text{ on } \Gamma_2 \tag{10.3b}$$

where $\boldsymbol{n}$ is the unit outward-pointing normal vector (Figure 10.1).

iii. the 3rd-type or Cauchy condition which specifies the value for a linear combination of the primary variable and normal flux on a part of the boundary Γ_3

$$\beta_1 h + \beta_2 K\nabla h.\boldsymbol{n} = v_3(x,y) \text{ on } \Gamma_3 \tag{10.3c}$$

in which β_1 and β_2 are known coefficients, and $\Gamma_1 + \Gamma_2 + \Gamma_3 = \Gamma$ is the boundary of the computational domain.

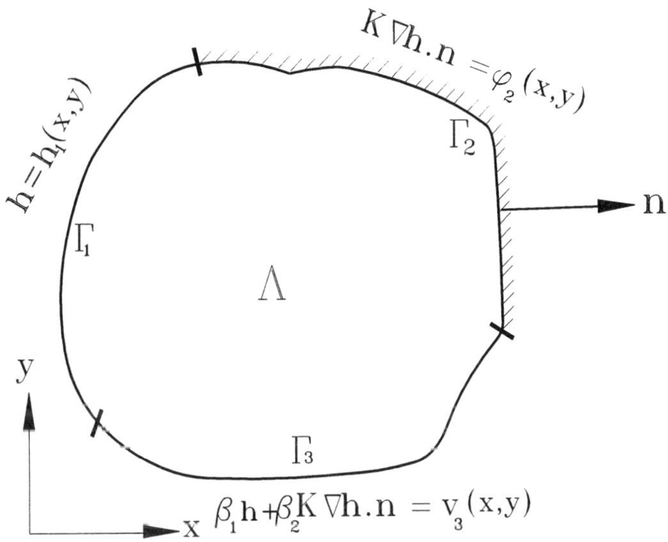

Figure 10.1 Definition sketch of a 2-D. domain with boundary conditions.

10.4 Integral Equations

The governing partial differential equation expressed by eq. (10.1) in an isotropic medium is given by

$$\nabla.[K(x,y,h)\nabla h] = f(x,y) \text{ on } \Lambda \tag{10.4}$$

which can be rewritten as

$$\nabla^2 h = -\nabla\Theta.\nabla h + \Psi f \text{ on } \Lambda \tag{10.5}$$

in which, as in the 1-D. problem, $\Theta \equiv \ln K$ and $\Psi \equiv 1/K$. We propose a complimentary differential equation from which we obtain the free-space Green's function. That complimentary differential equation, which applies to an infinitely extensive domain, is

$$\nabla^2 G = \delta(r - r_i) \tag{10.6}$$

in which $r = (x,y)$ is considered the coordinate of the field node, and $r_i = (x_i, y_i)$ is the coordinate of the source node. The solution to eq. (10.6) is[1]

$$G = \ln(r - r_i) \tag{10.7}$$

In two dimensions, Green's second identity for any two functions U and V which are at least twice differentiable is given by

$$\iint_{\Lambda} [U\nabla^2 V - V\nabla^2 U]\, dA = \int_{\Gamma} [U\nabla V - V\nabla U].\boldsymbol{n}\, ds \tag{10.8}$$

In essence eq. (10.8) transforms a domain integral into a boundary one. If we set U to h and V to G, and then apply Green's identity to eqs. (10.5) and (10.6) we have the integral equation

$$\iint_{\Lambda} [h\nabla^2 G - G\nabla^2 h] dA = \int_{\Gamma} [h\nabla G - G\nabla h].\boldsymbol{n}\, ds \tag{10.9a}$$

We substitute the terms $-\nabla\Theta.\nabla h + \Psi f$ for $\nabla^2 h$ and $\delta(r - r_i)$ for $\nabla^2 G$ into eq. (10.9a) to have

$$\iint_{\Lambda} [h\delta(r - r_i) - G(-\nabla\Theta.\nabla h + \Psi f)] dA = \int_{\Gamma} [h\nabla G - G\nabla h].\boldsymbol{n}\, ds \tag{10.9b}$$

It can be shown that in two dimensions

$$\iint_{\Lambda} h\delta(r - r_i) dA = \lambda h(x_i, y_i) = \lambda h(r_i) \tag{10.10}$$

in which λ is the nodal angle at the source point r_i. That is

$\lambda = 2\pi$ if r_i is within Λ,

$\lambda = \pi$ if r_i is on Γ and Γ is straight at r_i

$\lambda = \theta$ (nodal angle at r_i) if r_i is on Γ and Γ is not straight at r_i

With the above relations, eq. (10.9b) can now be expressed as

$$-\lambda h(r_i) + \int_\Gamma (h\frac{\partial G}{\partial n} - G\frac{\partial h}{\partial n})\, ds + \iint_\Lambda [G(-\nabla\Theta.\nabla h + \Psi f)]\, dA = 0 \tag{10.11}$$

Eq. (10.11) is the integral equation of the differential equation, and it is based on the singular boundary integral theory. It is a Fredholm-type integral equation of the second kind whose kernel is singular. In conventional boundary element applications, the same integral equation is arrived at. It may be necessary to point out that it comprises a boundary and a domain integral. In Green element calculations, evaluating the domain integral should pose no unusual difficulties since the domain is discretized into simple regular polygonal elements and domain integrations always involve source and field nodes being on the same element.

10.5 Green Element Models

The Green element method requires that the computational domain be discretized by suitable polygonal elements which, collectively, closely represent the shape of the domain. We reserve discussions on some of these elements to a subsequent section. When eq. (10.11) is discretized, it becomes

$$-\lambda^{(e)} h_i^{(e)} + \sum_{e=1}^{M} \left[\int_{\Gamma^{(e)}} \left(h\frac{\partial G}{\partial n} - G\frac{\partial h}{\partial n} \right) ds + \iint_{\Lambda^{(e)}} G(-\nabla\Theta.\nabla h + \Psi f)\, dA \right] = 0 \tag{10.12}$$

in which the superscript e denotes a typical element, $h_i^{(e)}$ stands for the value of h at the source point r_i in the element e, M is the total number of polygonal elements that are used to discretize the domain, and $\Gamma^{(e)}$ and $\Lambda^{(e)}$ are, respectively, the boundary and domain of the typical element. Over each element h, $\varphi \equiv \partial h/\partial n$, Θ, Ψ, and f are approximated by basis functions whose expressions are in terms of the coordinates of the vertices of the element. When that is done, any of the functional quantities, say h, is approximated by the relation

$$h(x,y) \approx \Omega_j(x,y)\, h_j \tag{10.13}$$

in which $\Omega_j(x,y)$ is the basis function of the Lagrange family with respect to node j of the element. The expressions for these interpolating functions for some polygonal elements are presented later. When eq. (10.13) is substituted into eq. (10.12) for each functional quantity, we have the element equations for the first model of the 2-D Poisson equation.

$$\boxed{\sum_{e=1}^{M} \left(R_{ij}^{(e)} h_j + L_{ij}^{(e)} \varphi_j - V_{ijl}^{(e)} \Theta_j h_l + U_{ijl}^{(e)} \Psi_j f_l \right) = 0} \tag{10.14}$$

in which the elemental matrices are given

$$R_{ij}^{(e)} = \int_{\Gamma^{(e)}} \frac{\partial G(r,r_i)}{\partial n} \Omega_j \, ds - \delta_{ij} \lambda^{(e)} \tag{10.15a}$$

$$L_{ij}^{(e)} = - \int_{\Gamma^{(e)}} G(r,r_i) \Omega_j \, ds \tag{10.15b}$$

$$V_{ijl}^{(e)} = \iint_{\Lambda^{(e)}} G(r,r_i) [\frac{\partial \Omega_j}{\partial x}\frac{\partial \Omega_l}{\partial x} + \frac{\partial \Omega_j}{\partial y}\frac{\partial \Omega_l}{\partial y}] \, dA \tag{10.15c}$$

$$U_{ijl}^{(e)} = \iint_{\Lambda^{(e)}} G(r,r_i) \Omega_j \Omega_l \, dA \tag{10.15d}$$

It is worth noting the symmetry of the element matrix $U_{ijl}^{(e)} = U_{ilj}^{(e)}$. The expressions for these element matrices depend on the type of element that is used in discretizing the domain. In a later section, the procedure for evaluating these element matrices for triangular and rectangular elements is presented. Because the Green's function that we are dealing with is simple, the above element matrices can be evaluated analytically for simple polygonal elements. That assures the accuracy of the numerical scheme.

If we relate the expression for the element equation (10.14) to that earlier obtained in modelling the 1-D. Poisson equation in a heterogeneous medium (section 3.7), we observe some semblance with the **GESNH-2** model. In a similar vein, a 2-D. version of the **GESNH-1** model comes by assuming that K is uniform in each element, but can be allowed to vary from element to element. In that case, $K \approx \bar{K}^{(e)}$, where $\bar{K}^{(e)}$ is some mean value of K within the element. By this approximation the term having ∇K becomes zero. This approximation essentially treats the entire domain as a piece-wise homogeneous medium, and with it, the differential equation reduces to

$$\nabla^2 h = \Psi f(x,y) \tag{10.16}$$

Applying Green's identity, the integral representation of eq. (10.16) becomes

$$-\lambda h(r_i) + \int_{\Gamma} (h\frac{\partial G}{\partial n} - G\frac{\partial h}{\partial n})\, ds + \int\int_{\Lambda} G(r,r_i)\bar{\Psi} f(r)\, dA = 0 \tag{10.17}$$

The integral equation is discretized by spatial elements, and functional quantities are approximated by interpolation functions over those elements to have

$$\sum_{e=1}^{M} \left(R_{ij}^{(e)} h_j + L_{ij}^{(e)} \varphi_j + \bar{\Psi}^{(e)} T_{ij}^{(e)} f_j \right) = 0 \tag{10.18}$$

in which

$$T_{ij}^{(e)} = \int\int_{\Lambda^{(e)}} G(r,r_i)\Omega_j\, dA \tag{10.19}$$

Eq. (10.18) is the system of discretized element equations of the second model for the 2-D. Poisson equation.

In one dimension, we had 2 degrees of freedom at each node. For two-dimensional problems, we may wish to still retain 2 degrees of freedom at each node which means solving for the primary variable and the normal flux. If we proceed in this way, it puts considerable demand on computing resources. To illustrate the additional demand on computations when we solve for both h and φ at every node, we consider the node i which is shared by 5 elements. In that case there are 6 unknown quantities that need to be calculated. These unknowns comprise one value of h and 5 values of φ which are indicated in Figure 10.2a by the arrows. We can now generalize that when a node is common to n elements, calculations have to be done for $m = n+1$ unknowns. We can, however, reduce the number of unknowns at each node to one if we adopt the procedure outlined below. It involves evaluating h or φ at nodes that are on external segments (either h or φ is specified at these nodes), while at nodes on internal segments, the normal flux is expressed in terms of the primary variable so that only h needs to be computed. Although expressing φ in terms of h will lead to some errors, it is expected that the error will be of the order of the spatial size of the element. To obtain φ in terms of h, eq. (10.13) is differentiated with respect to the normal direction

$$\varphi = \frac{\partial h}{\partial n} \approx \frac{\partial \Omega_j}{\partial n} h_j \tag{10.20}$$

It is necessary to point out that the continuity requirement across an internal segment shared by two elements should also be imposed to assure the accuracy of the numerical method. In such a case (see Figure 10.2b), the continuity requirements are that

$$h^{(e)} = h^{(f)} \tag{10.21a}$$

$$K^{(e)} \nabla h . \boldsymbol{n}^{(e)} = -K^{(f)} \nabla h . \boldsymbol{n}^{(f)} \quad \text{or} \quad K^{(e)} \varphi^{(e)} = -K^{(f)} \varphi^{(f)} \tag{10.21b}$$

As indicated in Figure 10.2b, the relationship $\boldsymbol{n}^{(e)} = -\boldsymbol{n}^{(f)}$ holds. If there is no sudden change in the value of K across the internal segment, then eq. (10.21b) can be modified to

$$\varphi^{(e)} = -\varphi^{(f)} \tag{10.22}$$

When dealing with the normal fluxes, their coefficients come from the element matrix L_{ij}. When it is computed by traversing element (e) in the counterclockwise direction, L_{ij} has the same value as when it is computed by traversing element (f) in the clockwise direction, as long as the nodes i and j are on the same internal segment. Thus, when the contributions from the elements (e) and (f) are assembled in the global coefficient matrix, the terms $L_{ij}\varphi^{(e)}$ and $L_{ij}\varphi^{(f)}$ add to zero. For this reason, it is advisable not to evaluate in the computer program the element matrix L_{ij} when dealing with an internal segment which has the nodes i and j as source and field nodes (see also Taigbenu [2]).

Furthermore, the element matrix R_{ij} resulting from the integration of $\nabla G(r,r_i).\boldsymbol{n}\Omega_j$ on the boundary has zero value whenever the nodes i and j are on the same segment, and that is due to the fact that the normal distance from the source node to the boundary segment is zero. Thus R_{ij} should only be evaluated on a segment which does not have the source node i.

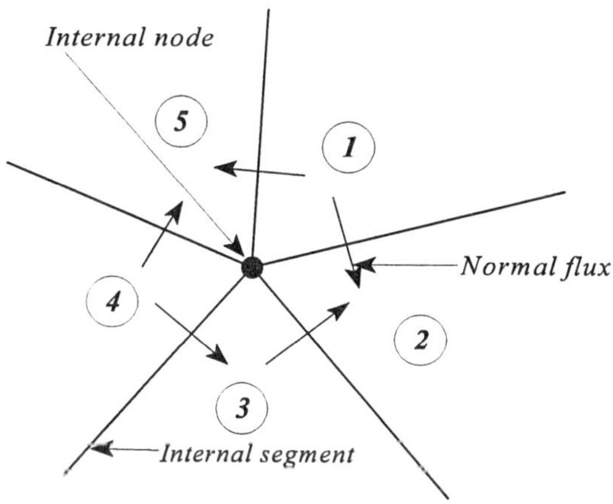

Figure 10.2a Indication of normal fluxes at a node shared by 5 elements.

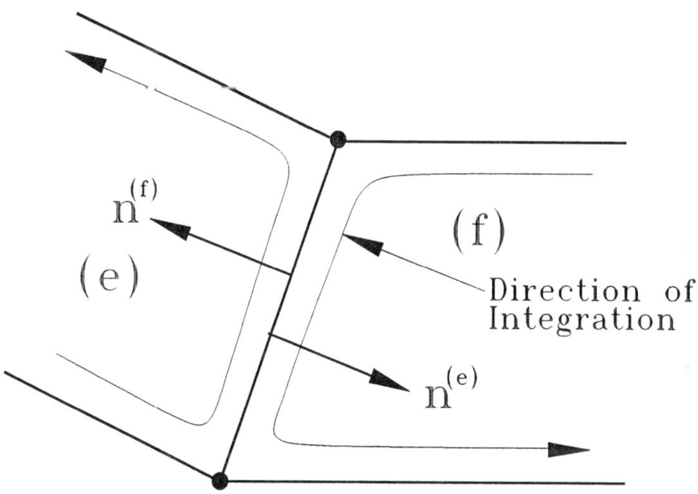

Figure 10.2b An internal segment shared by elements (e) and (f).

10.6 Polygonal Elements and their Interpolation Functions

As we have observed from the preceding sections, the Green element method involves spatial discretization of the computational domain into elements over which the distributions of certain quantities are prescribed with the aid of interpolating functions. For one-dimensional elements, we obtained the interpolation functions by the Lagrange interpolation method which provided a unique relationship between the number of nodes in the element and the degree of the interpolation polynomial. Such a unique relationship between the number of nodes in the element and the degree of the polynomial which is used for interpolation does not exist in 2-D. elements. It is, however, possible to derive interpolation functions that are polynomials which reproduce the values of the interpolating quantity at specified nodes on the element (Lagrange-type interpolation). That is

$$\Omega_j(x_i,y_i) = \delta_{ij} \tag{10.23}$$

There are a number of texts which give procedures for the derivation of these interpolation functions for triangular and rectangular elements. One of such texts is that by Reddy [3]. Here we shall only state the final expressions of the linear interpolation functions for those elements.

10.6.1 *Bilinear Rectangular Element*

The linear interpolation polynomial functions of the Lagrange family for the 4-node rectangular element shown in Figure 10.3a are

$$\Omega_1 = (1 - \zeta)(1 - \eta) \tag{10.24a}$$

$$\Omega_2 = \zeta(1 - \eta) \tag{10.24b}$$

$$\Omega_3 = \zeta\eta \tag{10.24c}$$

$$\Omega_4 = (1 - \zeta)\eta \tag{10.24d}$$

in which $\zeta = x/a$ and $\eta = y/b$, where a and b are the dimensions of the sides of the rectangular element. These functions are schematically represented in Figure 10.3b.

10.6.2 *Linear Triangular Element*

The linear interpolation functions for the three-node triangular element shown in Figure 10.4a correspond to the expressions for the area coordinates They are given as

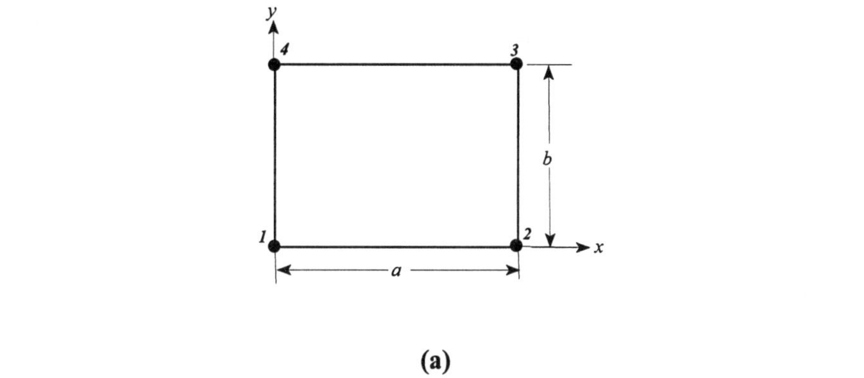

(b)

Figure 10.3 Linear rectangular element — **(a)** element, **(b)** interpolation functions.

$$\Omega_1 \equiv L_1 = 1 - \eta \tag{10.25a}$$

$$\Omega_2 \equiv L_2 = \mu\eta - \zeta \tag{10.25b}$$

$$\Omega_3 \equiv L_3 = 1 - \Omega_1 - \Omega_2 \tag{10.25c}$$

in which $\eta = x/l$, $\zeta = y/p$, and $\mu = d/p$, where the dimensions l, p, and d are indicated in Figure 10.4a. These interpolation functions are schematically presented in Figure 10.4b.

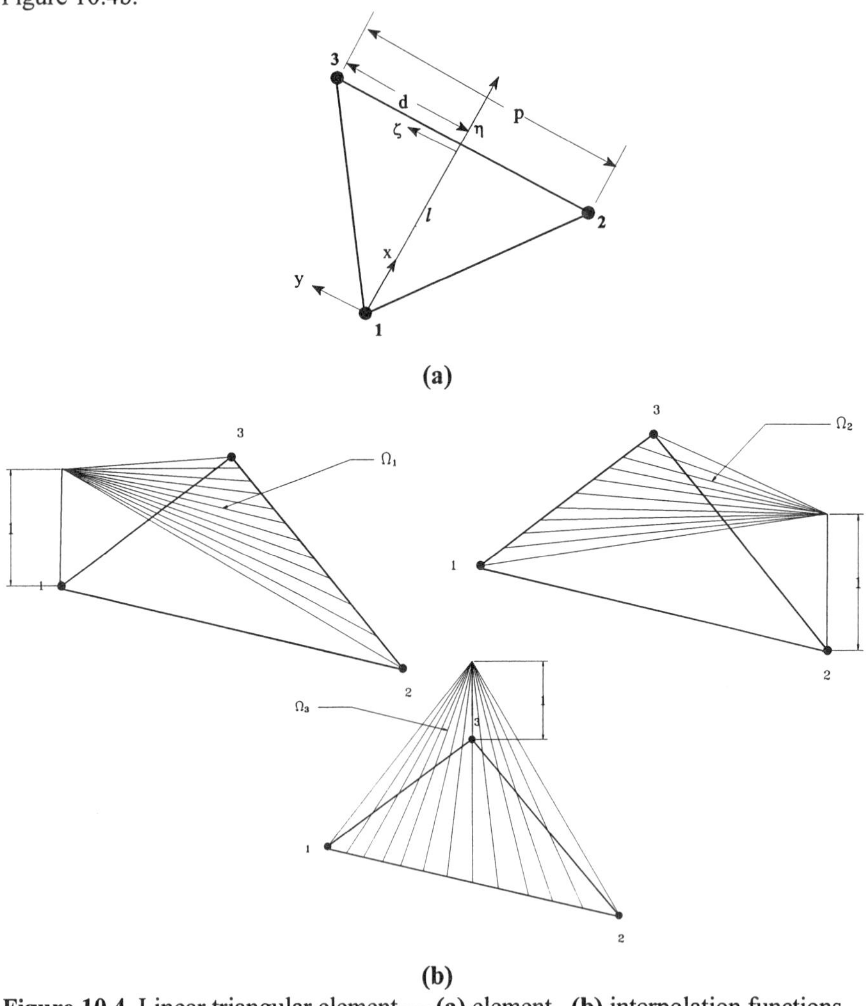

Figure 10.4 Linear triangular element — **(a)** element, **(b)** interpolation functions.

10.7 Boundary Integrations

The proper choice of coordinate system enhances the evaluation of the boundary integrals. A good choice is one in which one axis of the coordinate system is along the boundary element, while the other is normal to it[4]. In that case, the elemental boundary integrals involve integrations of combination of powers of the variable ζ (axis along the boundary element) and either $G(r,r_i)$ or $\partial G(r,r_i)/\partial n$ (see Figure 10.5). The highest power of ζ depends on the degree of the polynomial function used to interpolate quantities over the boundary elements.

The boundary integrals that have to be evaluated are

$$B1_m = \int_{\zeta_1}^{\zeta_2} \zeta^m \frac{\partial G}{\partial n} d\zeta \tag{10.26}$$

and

$$B2_m = \int_{\zeta_1}^{\zeta_2} \zeta^m G \, d\zeta \tag{10.27}$$

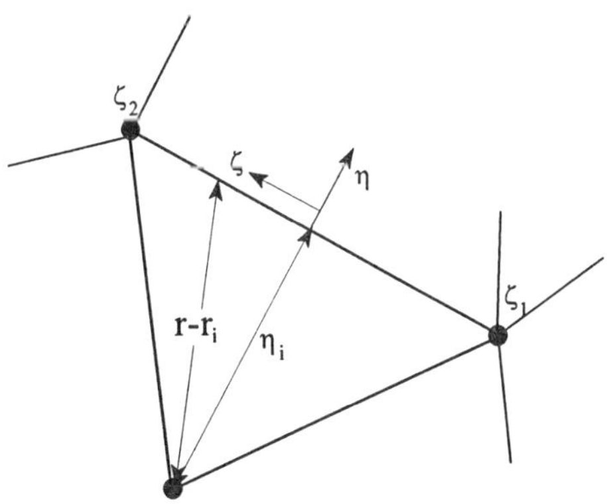

Figure 10.5 Local coordinates for line integrations.

To evaluate the above integrals we need to know the expressions for G and $\partial G/\partial n$ in terms of ζ and η. From Figure 10.5, we have that

$$r - r_i = \sqrt{\eta_i^2 + \zeta^2} \tag{10.28}$$

Thus,

$$G(r,r_i) = \ln(r - r_i) = \frac{1}{2}\ln(\eta_i^2 + \zeta^2) \tag{10.29}$$

and

$$\frac{\partial G}{\partial n} = \frac{1}{r - r_i}\frac{\partial(r - r_i)}{\partial n} = \frac{1}{r - r_i}\frac{\partial(r - r_i)}{\partial \eta_i} = \frac{\eta_i}{(r - r_i)^2} = \frac{\eta_i}{\eta_i^2 + \zeta^2} \tag{10.30}$$

As we had earlier pointed out, it is readily observed from eq. (10.30) that when the source node is on a segment, then $\partial G(r,r_i)/\partial n$ is zero on that segment. Introducing the expressions for G and $\partial G/\partial n$ from eqs. (10.29) and (10.30) into eqs. (10.26) and (10.27), respectively, those boundary integrals can now be evaluated. They are presented below:

$$B1_0 = \left[\tan^{-1}\left(\frac{\zeta_2}{\eta_i}\right) - \tan^{-1}\left(\frac{\zeta_1}{\eta_i}\right)\right] \tag{10.31a}$$

$$B1_1 = \frac{\eta_i}{2}\ln\left[\frac{\eta_i^2 + \zeta_2^2}{\eta_i^2 + \zeta_1^2}\right] = \eta_i \ln\left[\frac{r_2}{r_1}\right] \tag{10.31b}$$

$$B1_2 = \zeta_2 - \zeta_1 - \eta_i\left[\tan^{-1}\left(\frac{\zeta_2}{\eta_i}\right) - \tan^{-1}\left(\frac{\zeta_1}{\eta_i}\right)\right] \tag{10.31c}$$

$$B2_0 = \zeta_2 \ln r_2 - \zeta_1 \ln r_1 - (\zeta_2 - \zeta_1) + \eta_i\left[\tan^{-1}\left(\frac{\zeta_2}{\eta_i}\right) - \tan^{-1}\left(\frac{\zeta_1}{\eta_i}\right)\right] \tag{10.31d}$$

$$B2_1 = \frac{1}{4}\left\{r_2^2[\ln r_2^2 - 1] - r_1^2[\ln r_1^2 - 1]\right\} \tag{10.31e}$$

$$B2_2 = \frac{1}{6}\left\{\zeta_2^3 \ln r_2^2 - \zeta_1^3 \ln r_1^2 - \frac{2}{3}(\zeta_2^3 - \zeta_1^3) + 2\eta_i^2(\zeta_2 - \zeta_1) - 2\eta_i^3\left[\tan^{-1}\left(\frac{\zeta_2}{\eta_i}\right) - \tan^{-1}\left(\frac{\zeta_1}{\eta_i}\right)\right]\right\} \tag{10.31f}$$

The above results are used in deriving the expressions for the element matrices $R_{ij}^{(e)}$ and $L_{ij}^{(e)}$ for linear rectangular and triangular elements, and these are presented in sections B.1 and B.2 of Appendix B.

10.8 Domain Integrations

The domain integrations that are evaluated depend on both the kind of element used in discretizing the computational domain, the interpolation function, and the location of the source node relative to all other nodes on the element. When linear rectangular elements are employed in discretizing the computational domain, the source node is always at the corner of the element. In that case the domain integrations that are evaluated are of the form

$$D1_{ij} = \frac{1}{2}\int_0^b\int_0^a \left(\frac{x}{a}\right)^i \left(\frac{y}{b}\right)^j \ln(x^2 + y^2)\, dx\, dy \tag{10.32}$$

The exact expressions for the integral given by eq. (10.32) for various powers of x and y have been obtained. They are presented below:

$$D1_{00} = \frac{ab}{2}[\ln(a^2+b^2) + \frac{b}{a}\tan^{-1}(\frac{a}{b}) + \frac{a}{b}\tan^{-1}(\frac{b}{a}) - 3] \tag{10.33a}$$

$$D1_{10} = \frac{b}{12}[3a\ln(a^2+b^2) + \frac{b^2}{a}\ln(\frac{a^2+b^2}{b^2}) + \frac{4a^2}{b}\tan^{-1}(\frac{b}{a}) - 7a] \tag{10.33b}$$

$$DI_{01} = \frac{a}{12}\left[3b\ln(a^2+b^2) + \frac{a^2}{b}\ln(\frac{a^2+b^2}{a^2}) + \frac{4b^2}{a}\tan^{-1}(\frac{a}{b}) - 7b\right] \tag{10.33c}$$

$$DI_{11} = \frac{1}{16}\left[2ab\ln(a^2+b^2) + \frac{b^3}{a}\ln(\frac{a^2+b^2}{b^2}) + \frac{a^3}{b}\ln(\frac{a^2+b^2}{a^2}) - 3ab\right] \tag{10.33d}$$

$$DI_{22} = \frac{1}{18}\left[ab\ln(a^2+b^2) - \frac{a^4}{b^2}\tan^{-1}(\frac{b}{a}) - \frac{b^4}{a^2}\tan^{-1}(\frac{a}{b}) - ab + \frac{a^3}{b} + \frac{b^3}{a}\right] \tag{10.33e}$$

$$\begin{aligned} DI_{21} = \frac{1}{180}\Bigg[& 15ab\ln(a^2+b^2) + 9\frac{a^3}{b}\ln(\frac{a^2+b^2}{a^2}) - \\ & 12\frac{b^4}{a^2}\tan^{-1}(\frac{a}{b}) + 12\frac{b^3}{a} - 19ab\Bigg] \end{aligned} \tag{10.33f}$$

$$\begin{aligned} DI_{12} = \frac{1}{180}\Bigg[& 15ab\ln(a^2+b^2) + 9\frac{b^3}{a}\ln(\frac{a^2+b^2}{b^2}) - \\ & 12\frac{a^4}{b^2}\tan^{-1}(\frac{b}{a}) + 12\frac{a^3}{b} - 19ab\Bigg] \end{aligned} \tag{10.33g}$$

$$DI_{20} = \frac{1}{6}\left[ab\ln(a^2+b^2) + \frac{3}{2}a^2\tan^{-1}(\frac{b}{a}) - \frac{b^4}{2a^2}\tan^{-1}(\frac{a}{b}) - \frac{13}{6}ab + \frac{b^3}{2a^2}\right] \tag{10.33h}$$

$$DI_{02} = \frac{1}{6}\left[ab\ln(a^2+b^2) + \frac{3}{2}b^2\tan^{-1}(\frac{a}{b}) - \frac{a^4}{2b^2}\tan^{-1}(\frac{b}{a}) - \frac{13}{6}ab + \frac{a^3}{2b^2}\right] \tag{10.33i}$$

The above expressions are employed in evaluating the element matrices presented in section B.1 of Appendix B.

In the case of linear triangular elements, the source node is usually at one of the three vertices, so that the kind of domain integration that is encountered is of the form

$$D2_{ij} = \frac{1}{2l^{i+j}}\int_0^l x^i dx \int_{\beta x}^{\mu x} y^j \ln(x^2+y^2)\, dy \tag{10.34}$$

in which $\mu = d/l$ and $\beta = (d-p)/l$ (see Figure 10.4a). The exact integration of eq. (10.34) for various powers of x and y have been obtained, and they are presented below (see Liggett [5]):

$$D2_{i0} = \frac{l^{i+2}}{i+2}[(\mu-\beta)(\ln l - \frac{1}{i+2} - 1) + \frac{\mu\ln(1+\mu^2)}{2} - \frac{\beta\ln(1+\beta^2)}{2} + \tan^{-1}\mu - \tan^{-1}\beta \tag{10.35a}$$

$$D2_{i1} = \frac{l^{i+3}}{4(i+3)}[(\mu^2-\beta^2)(2\ln l - 1 - \frac{2}{(i+3)}) + (\mu^2+1)\ln(1+\mu^2) - (\beta^2+1)\ln(1+\beta^2)] \tag{10.35b)}$$

$$D2_{i2} = \frac{l^{i+4}}{6(i+4)}[2(\mu^3-\beta^3)(\ln l - \frac{1}{(i+4)} - \frac{1}{3}) + \mu^3\ln(1+\mu^2) - \beta^3\ln(1+\beta^2) + 2(\mu - \beta - \tan^{-1}\mu + \tan^{-1}\beta)] \tag{10.35c}$$

The above expressions are employed in deriving the expressions for the element matrices with linear triangular elements that can be found in section B.2 of Appendix B.

It is important to note that the ease with which we have obtained analytical expressions for the element matrices for the linear rectangular and triangular elements is due to the fact that the source and field nodes always belong to the same element. That cannot be said to apply when domain integrations are involved in boundary element formulations. In that case numerical quadrature provides the most viable approach in evaluating the domain integrals.

The contribution of the recharge or mass generation term $f(x,y)$ is handled at the level of the integral equation by a domain integration. Depending on the model, it is either the term $U_{ijl}\Psi_j f_l$ or $\overline{\Psi} T_{ij} f_j$. As in the 1-D. analysis, the evaluation of that term must take cognisance of whether it is a point recharge or a distributed one. For computational purposes each of them is treated differently. The total recharge is expressed as a sum of point sources/sinks and distributed recharge.

$$f(x,y) = f_p(x,y) + f_d(x,y) \tag{10.36}$$

where the point sources/sinks are represented mathematically as

$$f_p(x,y) = \sum_{m=1}^{N_p} Q_m \delta(r - r_m) \tag{10.37}$$

in which Q_m is the strength of the m-th source or sink located at $r_m = (x_m, y_m)$, and N_p is the total number of sources/sinks. The property of the Dirac delta function is utilised in evaluating the domain integral associated with the contribution of point sources and sinks. When dealing with the first model, the recharge term with point sources/sinks has the expression

$$\begin{aligned} U_{ijl}^{(e)} \Psi_j f_l &= \int\int_{\Lambda^{(e)}} G(r,r_i) \Psi(r) f_p(r) \, dA = \\ \int\int_{\Lambda^{(e)}} G(r,r_i) \Psi(r) \Big[\sum_{m=1}^{N_p} Q_m \delta(r-r_m)\Big] \, dA &= \sum_{m=1}^{N_p} Q_m \Psi(r_m) \ln(r_m - r_i) \end{aligned} \tag{10.38a}$$

while that for the second model is

$$\begin{aligned} \bar{\Psi} T_{ij}^{(e)} f_j &= \bar{\Psi} \int\int_{\Lambda^{(e)}} G(r,r_i) f_p(r) \, dA = \\ \bar{\Psi} \int\int_{\Lambda^{(e)}} G(r,r_i) \Big[\sum_{m=1}^{N_p} Q_m \delta(r-r_m)\Big] \, dA &= \bar{\Psi} \sum_{m=1}^{N_p} Q_m \ln(r_m - r_i) \end{aligned} \tag{10.38b}$$

Figure 10.6 illustrates the distance $r_m - r_i$ for a source/sink located at r_m in a triangular element. It should be noted that only nodes which belong to an element in which is located a source/sink that directly receive its contribution.

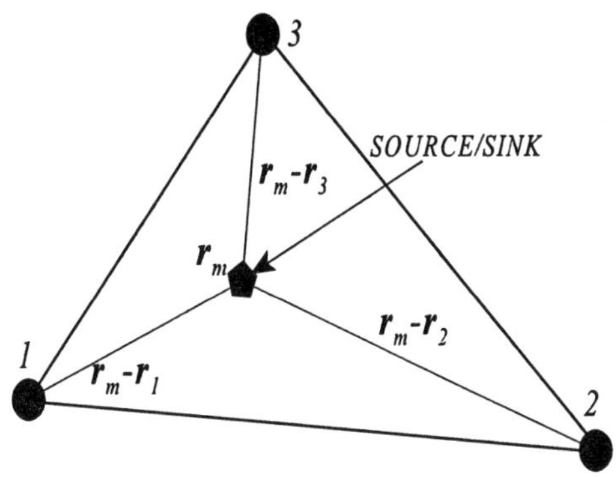

Figure 10.6 A source/sink at $r_m = (x_m, y_m)$ in a triangular element.

For distributed recharge, it is usual in practical situations that a functional representation for $f(x,y)$ will not be available, but rather its values at a number of points. A functional representation is then obtained from those discrete values. It is advisable to use simple functional expressions so that the domain integration is easily carried out. Because of the simplicity of the Green's function, the domain integration can be carried out analytically. Numerical quadrature integration can also be applied, and in that case the domain integration takes the general form

$$\begin{aligned} U_{ijl}^{(e)} \Psi_j f_l &= \iint_{\Lambda^{(e)}} G(r,r_i)\Omega_j(r)\Psi_j\Omega_l(r)f_l\,dA \\ &\approx A^{(e)}\Psi_j f_l \sum_{n=1}^{N_g} w_n G(r_n,r_i)\Omega_j(r_n)\Omega_l(r_n) \end{aligned} \tag{10.39a}$$

for model 1 and

$$\begin{aligned} \bar{\Psi} T_{ij}^{(e)} f_j &= \bar{\Psi} \iint_{\Lambda^{(e)}} G(r,r_i)\Omega_j(r) f_j\,dA \\ &\approx A^{(e)}\bar{\Psi} f_j \sum_{n=1}^{N_g} w_n G(r_n,r_i)\Omega_j(r_n) \end{aligned} \tag{10.39b}$$

for model 2. Where $A^{(e)}$ is the area of the element, w_n and r_n are the weights and sampling coordinates of the quadrature rule, N_g is the number of such sampling points, and the Ω's are the interpolation polynomial functions. It is observed that the integrand in these domain integrals consists of a combination of polynomials and the logarithmic function. The literature is dotted with various numerical quadrature on rectangular and triangular elements, but they are mainly for integrands consisting of polynomials. The 1-D. Gaussian quadrature presented in chapter 2 can be appropriately extended to rectangular elements. For triangular elements, the Gauss-type quadrature formula of Cowper [6] can be applied. It evaluates a function $F(x,y)$ over a triangular element using the formula

$$\int\int_{\Lambda^{(e)}} F(x,y)\, dA = A^{(e)} \sum_{m=1}^{N_g} w_m F(\xi_m, \eta_m) \tag{10.40}$$

where the weights and sampling points for the 1-, 3- and 4-point quadrature are below

1-point rule

$$w_1 = 1.0, \quad \{\xi_1 \quad \eta_1\} = [1/3 \quad 1/3 \quad 1/3] \begin{Bmatrix} x_1 & y_1 \\ x_2 & y_2 \\ x_3 & y_3 \end{Bmatrix} \tag{10.41a}$$

3-point rule

$$w_1 = w_2 = w_3 = 1/3, \quad \begin{Bmatrix} \xi_1 & \eta_1 \\ \xi_2 & \eta_2 \\ \xi_3 & \eta_3 \end{Bmatrix} = \begin{bmatrix} 2/3 & 1/6 & 1/6 \\ 1/6 & 2/3 & 1/6 \\ 1/6 & 1/6 & 2/3 \end{bmatrix} \begin{Bmatrix} x_1 & y_1 \\ x_2 & y_2 \\ x_3 & y_3 \end{Bmatrix} \tag{10.41b}$$

4-point rule

$$w_1 = -27/48, \; w_2 = w_3 = w_4 = 25/48,$$

$$\begin{Bmatrix} \xi_1 & \eta_1 \\ \xi_2 & \eta_2 \\ \xi_3 & \eta_3 \\ \xi_4 & \eta_4 \end{Bmatrix} = \begin{bmatrix} 1/3 & 1/3 & 1/3 \\ 0.6 & 0.2 & 0.2 \\ 0.2 & 0.6 & 0.2 \\ 0.2 & 0.2 & 0.6 \end{bmatrix} \begin{Bmatrix} x_1 & y_1 \\ x_2 & y_2 \\ x_3 & y_3 \end{Bmatrix} \tag{10.41c}$$

In the formulas above (x_i, y_i), $i=1,3$ represents the coordinates of the vertices of the triangular element, and (ξ_m, η_m) are the coordinates of the quadrature rule which are indicated in Figure 10.7. As can be observed, the sampling points are symmetric about the three vertices of the triangle. In the alternative, the quadrature rule later developed by Reddy and Shippy [7] can also be applied.

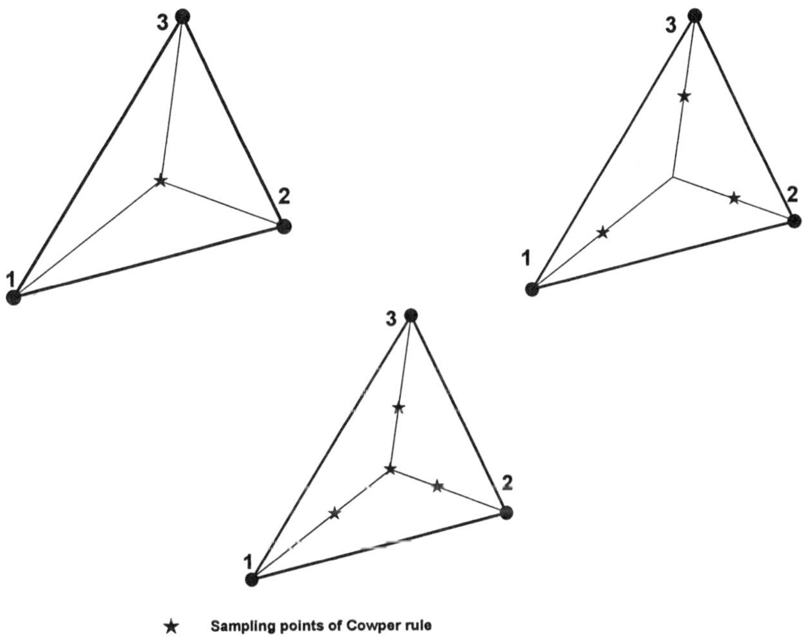

Figure 10.7 1-, 3-, and 4-point sampling points of Cowper [6].

10.9 Numerical Examples

Three examples of steady flow in 2-D. spatial dimensions have been designed to validate the models discussed in the foregoing sections. Computer programs were coded on the basis of the theoretical developments and tested on the examples. The examples cover linear and nonlinear steady problems. The second example is solved manually using 4 linear rectangular elements. The hand calculations are done deliberately with intent of aiding our comprehension of the computational features arising from the theoretical developments. Most practitioners in the field on computational methods seem to agree that such hand calculations are a powerful pedagogical tool.

Example 1

Our first steady state example in 2-D. spatial dimension is governed by the Laplace equation with the following prescribed conditions

$$\begin{aligned} h(0,y) &= 0; \quad \frac{\partial h}{\partial x}(0.5,y) = 0; \quad 0 \le y \le 0.5 \\ h(x,0) &= 0; \quad h(x,0.5) = \sin\pi x \quad 0 \le x \le 0.5 \end{aligned} \tag{10.42}$$

The exact solution to this example is given by

$$h(x,y) = \frac{\cosh(\pi y)\sin(\pi x)}{\cosh(\pi/2)} \tag{10.43}$$

Since there is no recharge term, coupled with the fact that medium is homogeneous, the two numerical models are essentially the same. The absence of the recharge term means that domain integrations can be avoided. The Green element calculations are carried out on a 0.5×0.5 computational domain that is discretized uniformly into 16 rectangular elements. The exact and **GEM** solutions are presented in Table 10.1. The results show that the numerical solution reflects the trend of the exact solution.

Example 2

We consider the linear form of the Poisson equation which is described by

$$\nabla^2 h = -1 \text{ in } \Lambda = \{(x,y):\ 0 \le (x,y) \le 1\} \tag{10.44}$$

with the boundary data

$$h(1,y) = h(x,1) = 0, \text{ and } \frac{\partial h(0,y)}{\partial x} = \frac{\partial h(x,0)}{\partial y} = 0 \tag{10.45}$$

The exact solution is well known, and it is stated below

$$\begin{aligned} h(x,y) = \frac{1}{2}\Bigg[&(1-y^2) + \\ &\frac{32}{\pi^3}\sum_{n=0}^{\infty}\frac{(-1)^n\cos[(2n-1)\pi y/2]\cosh[(2n-1)\pi x/2]}{(2n-1)^3\cosh[(2n-1)\pi/2]}\Bigg] \end{aligned} \tag{10.46}$$

We now proceed to carry out the Green element calculation of this example by hand using a uniform grid of four linear rectangular elements (see Figure 10.8a). Because all the four elements are identical, it is necessary to compute the element matrices for only one element, and these are given below. From eqs. (B.4) through (B.7) we obtain the expressions for the element matrix R_{ij} (note that $a = 0.5$ and $b = 0.5$).

Table 10.1 Exact and **GEM** solutions for example 1

Position (x,y)	Solution type	Exact solution	Linear **GEM**	Absolute relative error
(0,0)	$\partial h/\partial x$	1.2530	1.2149	0.0304
(0.125,0)	h	0.1525	0.1464	0.0400
(0.25,0)	h	0.2818	0.2706	0.0397
(0.375,0)	h	0.3590	0.3564	0.0072
(0.5,0)	h	0.3985	0.3989	0.0010
(0,0.125)	$\partial h/\partial x$	1.3498	1.3788	0.0215
(0.125,0.125)	h	0.1644	0.1636	0.0049
(0.25,0.125)	h	0.3038	0.3022	0.0053
(0.375,0.125)	h	0.3970	0.3979	0.0023
(0.5,0.125)	h	0.4297	0.4451	0.0358
(0,0.25)	$\partial h/\partial x$	1.6585	1.7026	0.0266
(0.125,0.25)	h	0.2020	0.2039	0.0094
(0.25,0.25)	h	0.3733	0.3758	0.0067
(0.375,0.25)	h	0.4877	0.4942	0.0133
(0.5,0.25)	h	0.5279	0.5520	0.0457
(0,0.375)	$\partial h/\partial x$	2.2262	2.3268	0.0452
(0.125,0.375)	h	0.2712	0.2757	0.0166
(0.25,0.375)	h	0.5011	0.5084	0.0146
(0.375,0.375)	h	0.6547	0.6641	0.0144
(0.5,0.375)	h	0.7086	0.7393	0.0435
(0,0.5)	$\partial h/\partial y$	3.1416	3.1810	0.0125
(0.125,0.5)	$\partial h/\partial y$	1.1026	1.1605	0.0525
(0.25,0.5)	$\partial h/\partial y$	2.0374	2.1458	0.0532
(0.375,0.5)	$\partial h/\partial y$	2.6620	2.9960	0.1255
(0.5,0.5)	$\partial h/\partial y$	2.8813	2.4880	0.1365

$$R_{ij} = \begin{bmatrix} -1.5708 & 0.4388 & 0.6931 & 0.4388 \\ 0.4388 & -1.5708 & 0.4388 & 0.6931 \\ 0.6931 & 0.4388 & -1.5708 & 0.4388 \\ 0.4388 & 0.6931 & 0.4388 & -1.5708 \end{bmatrix}$$

The expressions for the element matrix L_{ij} are obtained from eqs. (B.8) through (B.15). The superscripts b and f that are attached to L_{ij} represent the coefficients for the flux *before* and *after* the node j, respectively, as the sides of the element are traversed in an anti-clockwise manner. With this representation, the element matrices L_{ij}^{b} and L_{ij}^{f} for each of the linear rectangular elements become

$$L_{ij}^{b} = \begin{bmatrix} 0.5483 & 0.2983 & 0.1250 & 0.1556 \\ 0.1556 & 0.5483 & 0.2983 & 0.1250 \\ 0.1250 & 0.1556 & 0.5483 & 0.2983 \\ 0.2983 & 0.1250 & 0.1556 & 0.5483 \end{bmatrix}$$

and

$$L_{ij}^{f} = \begin{bmatrix} 0.5483 & 0.1556 & 0.1250 & 0.2983 \\ 0.2983 & 0.5483 & 0.1556 & 0.1250 \\ 0.1250 & 0.2983 & 0.5483 & 0.1556 \\ 0.1556 & 0.1250 & 0.2983 & 0.5483 \end{bmatrix}$$

The domain integration involving the external forcing term $f(x,y) = -1$ is obtained from the expression for DI_{00} in eq. (10.33a)

$$F_i = -DI_{00} = -0.2653, \quad i = 1,2,3,4$$

The element matrices are assembled taking cognisance of the fact that at nodes 1, 2, 4, and 5 h is calculated, while φ is calculated at the other nodes. (It should be noted that the boundary data at node 9 are over specified in the sense that the primary variable and the normal flux are known. However, we have treated the normal flux as the unknown in our calculations.) The procedure for achieving the global matrix equation is outlined below. A distinction is made between the global and local node numbers in Figure 10.9 — the global node numbers are circled, whereas the local node numbers are not. A row of the global matrix comes from the positioning of the source node, while the column entries come from the field nodes that share an element with the source node. The column entries for the 9 rows are given below.

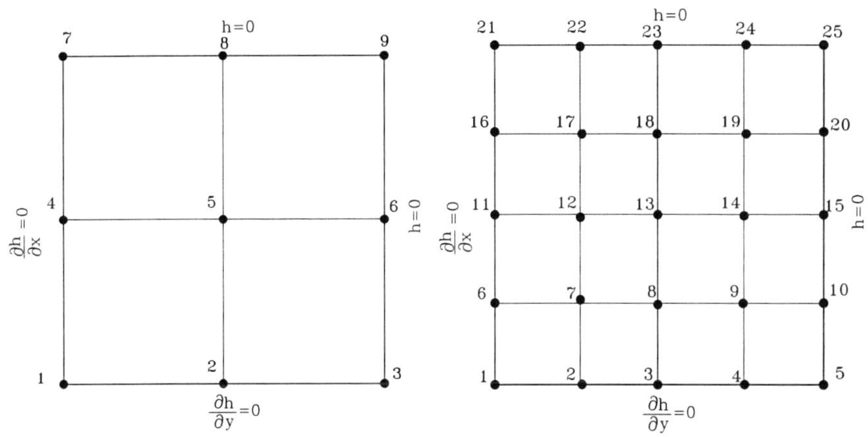

(a) Four elements **(b)** sixteen elements

Figure 10.8 Discretization of domain of example 2.

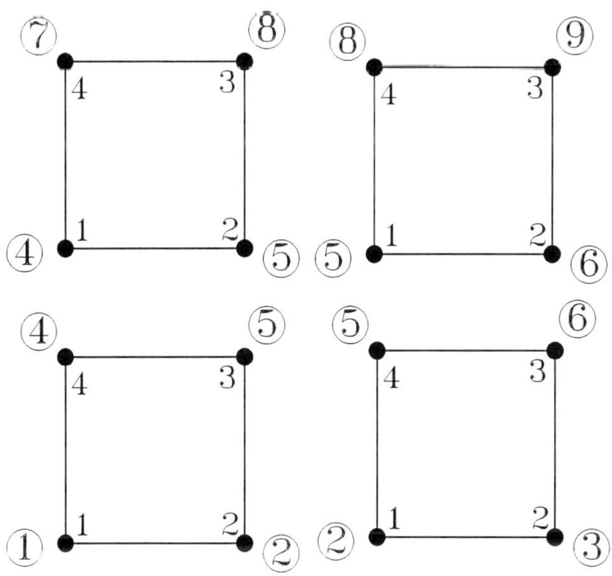

Figure 10. 9 Local and global node numbers of 4-element discretizion of example 2.

Row 1 or source node at 1

Column **1**: $R_{11} - (L_{12}^{f} + L_{14}^{b})/0.5 = -1.5708 - 4\times0.1556 = -2.1932$

Column **2**: $R_{12} + (L_{12}^{f} - L_{13}^{f})/0.5 = 0.4388 + 2\times(0.1556 - 0.125) = 0.5$

Column **4**: $R_{14} + (L_{14}^{b} - L_{13}^{b})/0.5 = 0.4388 + 2\times(0.1556 - 0.125) = 0.5$

Column **5**: $R_{13} + (L_{13}^{b} + L_{13}^{f})/0.5 = 0.6931 + 2\times(0.125 + 0.125) = 1.1931$

All other column entries are zero.

RHS: $$F_1 - L_{11}^{f}\frac{\partial h_1}{\partial y} - L_{11}^{b}\frac{\partial h_1}{\partial x} - L_{12}^{b}\frac{\partial h_2}{\partial y} - L_{14}^{f}\frac{\partial h_4}{\partial x} = -0.2653$$

Row 2 or source node at 2

Column **1**: $R_{21} - L_{24}^{b}/0.5 = 0.4388 - 2\times0.125 = 0.1888$

Column **2**: $R_{22} + R_{11} - (L_{23}^{f} + L_{14}^{b})/0.5 =$
$2\times(-1.5708 - 0.1556 - 0.1556) = -3.764$

Column **3**: $L_{12}^{f} = 0.1556$

Column **4**: $R_{24} + L_{24}^{b}/0.5 = 0.6931 + 2\times0.125 = 0.9431$

Column **5**: $R_{23} + R_{14} + (L_{23}^{f} + L_{14}^{b})/0.5 = 2\times(0.4388 + 0.1556 + .1556) = 1.5$

Column **6**: $L_{13}^{b} = 0.125$

All other column entries are zero.

RHS:
$$F_1 + F_2 - L_{21}^f \frac{\partial h_1}{\partial y} - L_{21}^b \frac{\partial h_1}{\partial x} - L_{22}^b \frac{\partial h_2}{\partial y} - L_{11}^f \frac{\partial h_2}{\partial y} - L_{12}^b \frac{\partial h_3}{\partial y} -$$
$$(R_{12} - \frac{L_{13}^f}{0.5}) h_3 - (R_{13} + \frac{L_{13}^f}{0.5}) h_6 - L_{24}^f \frac{\partial h_4}{\partial x} = -2\times 0.2653 = -0.5306$$

Row 3 or source node at ***3***

Column **2**: $R_{21} + (L_{21}^b - L_{24}^b)/0.5 = 0.4388 + 2\times(0.1556 - 0.125) = 0.5$

Column **3**: $L_{22}^f = 0.5483$

Column **5**: $R_{24} + (L_{24}^b - L_{24}^f)/0.5 = 0.6931 + 2\times(0.125 + 0.125) = 1.1931$

Column **6**: $L_{23}^b = 0.2983$

All other column entries are zero.

RHS:
$$F_2 - L_{21}^f \frac{\partial h_2}{\partial y} - L_{22}^b \frac{\partial h_3}{\partial y} - (R_{22} - \frac{L_{21}^b + L_{23}^f}{0.5}) h_3 -$$
$$(R_{23} - \frac{L_{23}^f - L_{24}^f}{0.5}) h_6 = -0.2653$$

Row 4 or source node at ***4***

Column **1**: $R_{41} - L_{43}^f/0.5 = 0.4388 - 2\times 0.125 = 0.1888$

Column **2**: $R_{42} + L_{42}^f/0.5 = 0.6931 + 2\times 0.125 = 0.9431$

Column **4**: $R_{44} + R_{11} - (L_{43}^b + L_{12}^f)/0.5 =$
$2\times(-1.5708 - 0.1556 - 0.1556) = -3.764$

Column **5**: $R_{43} + R_{12} + (L_{43}^b + L_{12}^f)/0.5 = 2\times(0.4388 + 0.1556 + .1556) = 1.5$

Column **7**: $L_{14}^b = 0.1556$

Column **8**: $L_{13}^f = 0.125$

All other column entries are zero.

RHS: $$F_1 + F_4 - L_{41}^f \frac{\partial h_1}{\partial y} - L_{41}^b \frac{\partial h_1}{\partial x} - L_{42}^b \frac{\partial h_2}{\partial y} - (R_{13} + \frac{L_{13}^b}{0.5}) h_8 - (R_{14} - \frac{L_{13}^b}{0.5}) h_7 - L_{14}^f \frac{\partial h_7}{\partial x} - (L_{11}^b + L_{44}^f) \frac{\partial h_4}{\partial x} = -2 \times 0.2653 = -0.5306$$

Row 5 or source node at ***5***

Column **1**: $R_{31} = 0.6931$

Column **2**: $R_{32} + R_{41} = 0.4388 + 0.4388 = 0.8776$

Column **3**: $L_{42}^f = 0.125$

Column **4**: $R_{34} + R_{21} = 0.4388 = 0.4388$

Column **5**: $R_{33} + R_{44} + R_{22} + R_{11} = -4 \times 1.5708 = -6.2838$

Column **6**: $L_{43}^b + L_{12}^f = 2 \times 0.1556 = 0.3112$

Column **7**: $L_{24}^f = 0.125$

Column **8**: $L_{23}^f + L_{14}^b = 2 \times 0.1556 = 0.3112$

Column **9**: $L_{13}^b + L_{13}^f = 2 \times 0.125 = 0.25$

RHS: $$F_3 + F_4 + F_2 + F_1 - L_{31}^f \frac{\partial h_1}{\partial y} - L_{31}^b \frac{\partial h_1}{\partial x} - (L_{32}^b + L_{41}^f) \frac{\partial h_2}{\partial y} - L_{42}^b \frac{\partial h_3}{\partial y} - R_{42} h_3 - (R_{43} + R_{12}) h_6 - R_{13} h_9 - (R_{14} + R_{23}) h_8 - R_{24} h_7 - L_{24}^f \frac{\partial h_7}{\partial x} - (L_{21}^b + L_{34}^f) \frac{\partial h_4}{\partial x} = -4 \times 0.2653 = -1.0612$$

Row 6 or source node at ***6***

Column **2**: $R_{31} + L^b_{31}/0.5 = 0.6931 + 2\times 0.125 = 0.9431$

Column **3**: $L^f_{32} = 0.2983$

Column **5**: $R_{34} + R_{21} + (L^f_{34} + L^b_{21})/0.5 = 2\times(0.4388 + 0.1556 + .1556) = 1.5$

Column **6**: $L^b_{33} + L^f_{22} = 2\times 0.5483 = 1.0966$

Column **8**: $L^b_{24} = 0.125$

Column **9**: $L^f_{23} + L^f_{23} = 0.2983 + 0.1556 = 0.4539$

All other column entries are zero.

RHS:
$$F_3 + F_2 - L^f_{31}\frac{\partial h_2}{\partial y} - L^b_{32}\frac{\partial h_3}{\partial y} - (R_{32} - \frac{L^b_{31}}{0.5})h_3 - (R_{33} + R_{22} - \frac{L^f_{34} + L^b_{21}}{0.5})h_6 -$$
$$(R_{13} - \frac{L^f_{24}}{0.5})h_9 - (R_{24} + \frac{L^f_{24}}{0.5})h_8 = -2\times 0.2653 = -0.5306$$

Row 7 or source node at 7

Column **4**: $R_{41} + (L^f_{41} - L^f_{42})/0.5 = 0.4388 + 2\times(0.1556 - 0.125) = 0.5$

Column **5**: $R_{42} + (L^f_{42} + L^b_{42})/0.5 = 0.6931 + 2\times(0.125 + 0.125) = 1.1931$

Column **7**: $L^b_{44} = 0.5483$

Column **8**: $L^f_{43} = 0.2983$

All other column entries are zero.

$$\text{RHS: } F_4 - (R_{43} - \frac{L_{42}}{0.5})h_8 - (R_{24} - \frac{L^f_{41} + L^b_{43}}{0.5})h_7 - L^f_{44}\frac{\partial h_7}{\partial x} = -0.2653$$

*Row 8 or source node at **8***

Column **4**: $R_{31} + L_{31}^{f}/0.5 = 0.6931 + 2\times0.125 = 0.9431$

Column **5**: $R_{32} + R_{41} + (L_{32}^{b} + L_{41}^{f})/0.5 = 2\times(0.4388 + 0.1556 + .1556) = 1.5$

Column **6**: $L_{42}^{f} = 0.125$

Column **7**: $L_{32}^{b} = 0.2983$

Column **8**: $L_{33}^{f} + L_{44}^{b} = 2\times0.5483 = 1.0966$

Column **9**: $L_{43}^{b} + L_{43}^{f} = 0.1556 + 0.2983 = 0.4539$

All other column entries are zero.

RHS:
$$F_3 + F_4 - (R_{42} + \frac{L_{42}^{b}}{0.5})h_6 - (R_{43} - \frac{L_{42}^{b}}{0.5})h_9 - (R_{33} + R_{44} - \frac{L_{32}^{b} + L_{41}^{b}}{0.5})h_8 -$$
$$(R_{34} - \frac{L_{31}^{f}}{0.5})h_7 - L_{44}^{f}\frac{\partial h_7}{\partial x} - L_{44}^{f}\frac{\partial h_4}{\partial x} = -2\times0.2653 = -0.5305$$

*Row 9 or source node at **9***

Column **5**: $R_{31} + (L_{31}^{b} + L_{31}^{f})/0.5 = 0.6931 + 2\times(0.125 + 0.125) = 1.1931$

Column **6**: $L_{32}^{f} = 0.5483$

Column **8**: $L_{34}^{b} = 0.2983$

Column **9**: $L_{33}^{b} + L_{33}^{f} + 2\times0.5483 = 1.0966$

All other column entries are zero.

RHS:

$$F_3 - (R_{32} - \frac{L_{31}}{0.5})h_6 - (R_{33} - \frac{L_{32}^b + L_{34}^f}{0.5})h_9 -$$
$$(R_{34} - \frac{L_{31}^f - L_{34}^f}{0.5})h_8 = -0.2653$$

The results above are summarized in the matrix equation

$$\begin{bmatrix} -2.19 & 0.5 & 0 & 0.5 & 1.19 & 0 & 0 & 0 & 0 \\ 0.18 & -3.77 & 0.16 & 0.94 & 1.5 & 0.13 & 0 & 0 & 0 \\ 0 & 0.5 & 0.55 & 0 & 1.19 & 0.30 & 0 & 0 & 0 \\ 0.19 & 0.94 & 0 & -3.77 & 1.5 & 0 & 0.16 & 0.13 & 0 \\ 0.69 & 0.88 & 0.13 & 0.88 & -6.28 & 0.31 & 0.13 & 0.31 & 0.25 \\ 0 & 0.94 & 0.30 & 0 & 1.5 & 1.10 & 0 & 0.13 & 0.45 \\ 0 & 0 & 0 & 0.5 & 1.19 & 0 & 0.55 & 0.30 & 0 \\ 0 & 0 & 0 & 0.94 & 1.5 & 0.13 & 0.30 & 1.1 & 0.45 \\ 0 & 0 & 0 & 0 & 1.19 & 0.55 & 0 & 0.30 & 1.10 \end{bmatrix} \begin{Bmatrix} h_1 \\ h_2 \\ \varphi_3 \\ h_4 \\ h_5 \\ \varphi_6 \\ \varphi_7 \\ \varphi_8 \\ \varphi_9 \end{Bmatrix} = - \begin{Bmatrix} 0.27 \\ 0.53 \\ 0.27 \\ 0.53 \\ 1.06 \\ 0.53 \\ 0.27 \\ 0.53 \\ 0.27 \end{Bmatrix}$$

The solution to the global matrix equation is presented in Table 10.2a. It is noted that the solution is symmetric about the north-east axis of the domain. Another simulation is carried out on this same example with the computational domain discretized into 16 uniform rectangular elements (Figure 10.8b). The solution from this simulation and the exact solution are presented in Table 10.2b. The solutions are in good agreement.

Table 10.2a Exact and **GEM** (with 4 rectangular element-grid) solutions for example 2.

Node	Solution Variable	Exact solution	**GEM** solution	Absolute Error
1	h	0.29469	0.32378	0.02909
2	h	0.22934	0.23494	0.00560
3	φ	-0.67531	-0.74808	0.07277
5	h	0.18114	0.17589	0.00525
6	φ	-0.56276	-0.61170	0.04894

Table 10.2b Exact and **GEM** (with 16 rectangular element-grid) solutions for example 2.

Location (x,y)	Solution Variable	Exact solution	**GEM** solution	Absolute Error
(0,0)	h	0.29469	0.31126	0.01657
(0.25,0)	h	0.27888	0.28736	0.00848
(0.5,0)	h	0.22934	0.23240	0.00306
(0.75,0)	h	0.13973	0.13719	0.00254
(1,0)	φ	-0.67531	-0.72100	0.04569
(0.25,0.25)	h	0.26415	0.26579	0.00164
(0.5,0.25)	h	0.21780	0.21584	0.00196
(0.75,0.25)	h	0.13333	0.12869	0.00517
(1,0.25)	φ	-0.64875	-0.67798	0.02923
(0.5,0.5)	h	0.18114	0.17704	0.00410
(0.75,0.5)	h	0.11274	0.10670	0.00604
(1,0.5)	φ	-0.56276	-0.59029	0.02753
(0.75,0.75)	h	0.07282	0.06662	0.00620
(1,0.75)	φ	-0.39183	-0.40926	0.01743

Example 3

This example is a steady nonlinear flow problem in a variably saturated multiply zoned porous medium which is shown in Figure 10.10. The problem is governed by

$$\nabla.(K\nabla h) + \nabla K.\boldsymbol{k} = 0 \tag{10.47}$$

in which h is the pressure head, $K = K(h)$ is the hydraulic conductivity of the medium, and $\boldsymbol{k}$ is the unit vector in the y-direction (positive in the upward vertical direction). The computational domain, which is shown in Figure 10.10, is half the size of the actual flow domain because of the symmetry of the flow about the $x = 0$ line. The boundary conditions for h and the flux are specified along side eq. (10.47). They are:

$$\begin{gathered} h(x,y{=}0) = 0, \quad h(x,y{=}15.2m) = -7.5m \\ -K\frac{\partial h(x{=}0,y)}{\partial x} = -K\frac{\partial h(x{=}9.14m,y)}{\partial x} = 0 \end{gathered} \tag{10.48}$$

For this example we have adopted the soil parameters and spatial discretization employed by Huyakorn *et al.* [8] in their finite element model. We observe that the predominant alluvial sand formation has lenses of clay and loam which are of much lower hydraulic conductivities. The computational challenge posed by this example arises not only from the highly nonlinear soil constitutive relations, but also from the abrupt changes in soils of contrasting properties.

Comparing eq. (10.47) to (10.4) which we started with in this chapter, we observe that the term $f(x,y)$ on the right hand side of eq. (10.4) can be treated in a manner similar to the second term $\nabla K.\boldsymbol{k}$ on the left hand side of eq. (10.47).

With that, the element equations for the differential operator is (the reader should attempt to verify this)

$$R_{ij}^{(e)}h_j + L_{ij}^{(e)}\varphi_j - V_{ijl}^{(e)}\Theta_j h_l - P_{ij}^{(e)}\Theta_j \equiv g_i^{(e)} = 0 \tag{10.49}$$

in which

$$P_{ij}^{(e)} = \int\int_{A^{(e)}} G(r,r_i)\nabla\Omega_j.\boldsymbol{k}\,dA \tag{10.50}$$

Treating the medium as piecewise homogeneous, that is $K \approx \bar{K}^{(e)} = K^{(e)}(\bar{h})$, where $\bar{h}$ is the average of h within the element e, then the term ∇K is zero, and the element equations based on this assumption are similar to eq. (10.18) which earlier was presented. That is

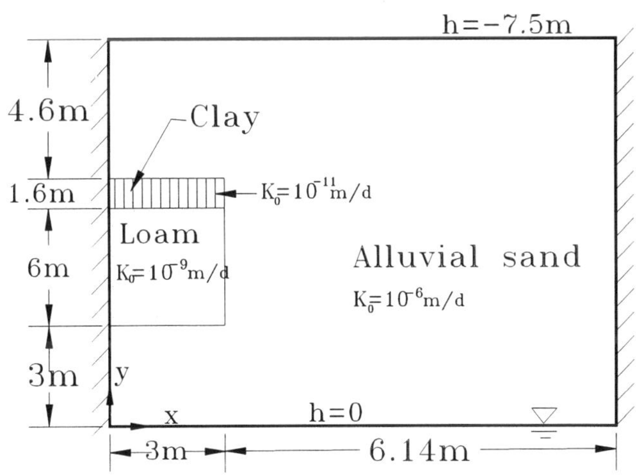

Figure 10.10 Flow domain for example 3.

$$R_{ij}^{(e)} h_j + L_{ij}^{(e)} \varphi_j \equiv g_i^{(e)} = 0 \tag{10.51}$$

The model based on eq. (10.51) is referred to as model 2, while that based on (10.49) as model 1.

The Newton-Raphson algorithm is used to linearize both models. This algorithm was earlier described in chapter 3, so it will not be repeated here. Essentially we seek to solve for the increment of the mixed vector of unknowns $\Delta u_j^{(k+1)}$ to a known estimate $u_j^{(k)}$. That is

$$u_j^{(k+1)} = \begin{Bmatrix} h_j^{(k+1)} \\ \varphi_j^{(k+1)} \end{Bmatrix} = u_j^{(k)} + \Delta u_j^{(k+1)} = \begin{Bmatrix} h_j^{(k)} + \Delta h_j^{(k+1)} \\ \varphi_j^{(k)} + \Delta \varphi_j^{(k+1)} \end{Bmatrix} \tag{10.52}$$

The increment $\Delta u_j^{(k+1)}$ is obtained from the matrix equation

$$J_{ij}^{(e,k)} \Delta u_j^{(k+1)} = -g_i^{(e,k)} \tag{10.53}$$

in which $g_i^{(e,k)}$ is the right hand side of either eq. (10.49) or eq. (10.51). The expression for the Jacobian matrix for model 1 is given by

$$J_{ij}^{(e,k)} = \begin{cases} \left.\dfrac{\partial g_i^{(e)}}{\partial h_j}\right|_{h_j = h_j^{(k)}} = R_{ij}^{(e)} - V_{inl}^{(e)} \Theta_n^{(k)} \delta_{lj} - V_{inl}^{(e)} h_l^{(k)} \dfrac{d\Theta_n}{dh_j} - P_{in}^{(e)} \dfrac{d\Theta_n}{dh_j} \\ \left.\dfrac{\partial g_i^{(e)}}{\partial \varphi_j}\right|_{\varphi_j = \varphi_j^{(k)}} = L_{ij}^{(e)} \end{cases} \tag{10.54}$$

and that for model 2 is

$$J_{ij}^{(e,k)} = \begin{cases} \left.\dfrac{\partial g_i^{(e)}}{\partial h_j}\right|_{h_j = h_j^{(k)}} = R_{ij}^{(e)} \\ \left.\dfrac{\partial g_i^{(e)}}{\partial \varphi_j}\right|_{\varphi_j = \varphi_j^{(k)}} = L_{ij}^{(e)} \end{cases} \tag{10.55}$$

The second model is programmed and employed to carry out a one-stage steady state solution on the computational domain which is discretized into 135 rectangular elements according to the data in Table 10.3. Five iterations are allowed to achieve convergence of the solution. The soil constitutive relations, which were presented graphically by Huyakorn *et al.* [8], are approximated in the least square sense by the functional relations

$$K = K_0 10^{(a_1 R + a_2 R^2 + a_3 R^3)} \tag{10.56a}$$

$$S = (1 + b_1 h^{b_2})^{b_3} \tag{10.56b}$$

$$R = S - 1 \tag{10.56c}$$

in which K_0 is the hydraulic conductivity at saturation and its values for three soils are found in Table 10.4, and S is the degree of saturation. The coefficients a_i and b_i $(i = 1,2,3)$ are presented in Table 10.5 for the three soils. The solutions for the pressure head at $x = 0$ and $x = 7.52$ cm, and degree of saturation at $x = 0.375$ and $x = 7.52$ cm from the second Green element model are presented in Figures 10.11a and 10.11b in the form that Huyakorn *et al.* [8] displayed their finite element solutions. Their finite element model also treated the medium as piece-wise homogeneous. The solutions from both models are in good agreement. We have also presented contours of pressure head and degree of saturation from the solutions of the Green element model in Figures 10.12a and 10.12b. The contour plots reflect the phenomenon of capillary perching in which, at the interface of soils of different conductivities, a buildup of soil moisture and steep gradient in pressure head occurs.

Table 10.3 Spatial discretization data for example 3

Δx_i; $i=1,...,9$ (m)	0.75, 0.75, 0.75, 0.75, 0.8, 1.2, 1.2, 1.34, 1.6
Δy_j; $j=1,...,15$ (m)	1.2, 1.0, 0.8, 1.2, 1.4, 1.4, 1.0, 1.0, 0.6, 0.4, 0.6 1.2, 1.2, 1.2, 1.0

Table 10.4 Hydraulic conductivity values at saturation for soils in example 3.

Soil	Alluvial sand	Loam	Clay
Hydraulic conductivity at saturation K_0(m/day)	35955	36047	36108

Table 10.5 Values of coefficients in soil constitutive relations

Soil	a_1	a_2	a_3	b_1	b_2	b_3
Alluvial sand	2.918	3.730	7.754	33.64	2.270	-0.269
Loam	4.252	2.210	6.065	2.130	2.140	-0.259
Clay	10.567	12.190	12.269	0.354	0.918	-0.431

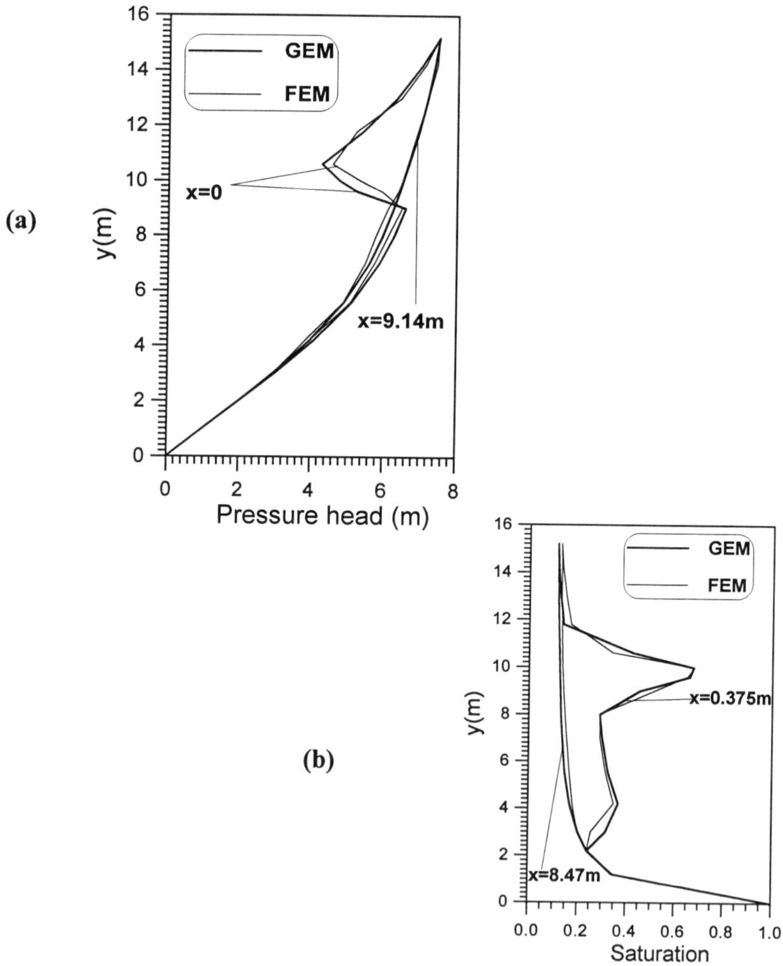

Figure 10.11 **GEM** and **FEM** solutions of example 3 — **(a)** Pressures curves, **(b)** Saturation curves.

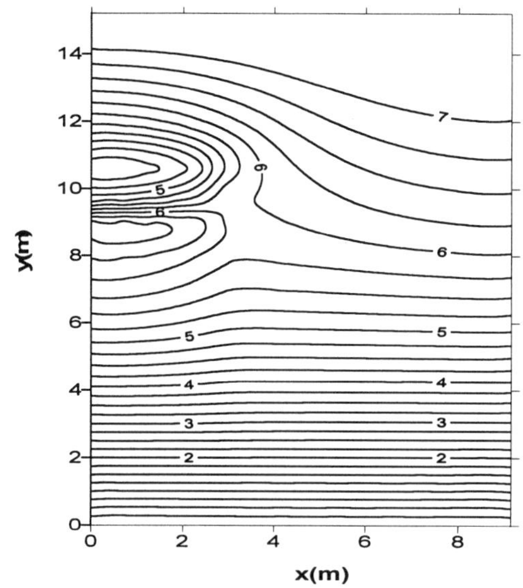

(a)

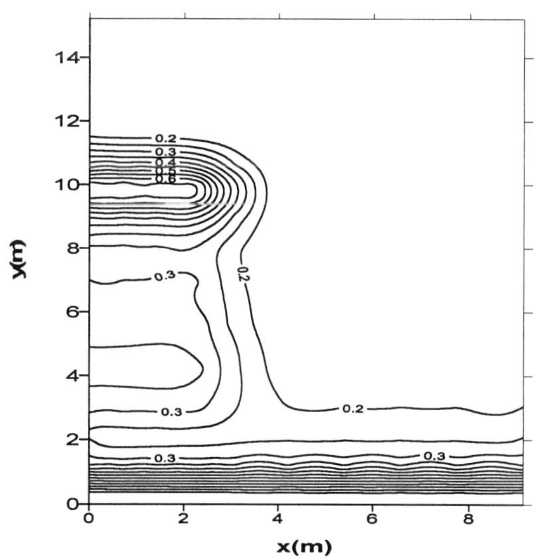

(b)

Figure 10.12 Contour plots of steady solution of example 3 — **(a)** Pressure head, **(b)** Degree of saturation.

10.10 Remarks

The Green element formulations for two-dimensional steady flows have been derived by essentially following the derivation procedure for the 1-D. formulations, and that has involved constructing the integral representation of the differential equation on the basis of Green's second identity, discretizing the computational domain by suitable elements, aggregating the discretized element equations to form the global matrix equation, and solving it to obtain the nodal unknowns. Where the flow equations are nonlinear, the Picard and Newton-Raphson algorithms have been quite valuable in linearizing the discrete element equations. Two formulations have been pursued here — one allows medium and flow parameters to be approximated by appropriate interpolating functions, while the second considers the medium parameters to be uniform within each element but provides for the interpolation of the dependent variable. The second formulation is simpler to handle. We have considered linear rectangular and triangular elements in discretizing the computational domain, and provided expressions for the element matrices. However, we have made a modification to the implementation of the 2-D. formulation primarily to allow for only one degree of freedom at the internal nodes. That modification provides savings on computing resources. Linear and nonlinear steady problems have been solved with considerable success.

10.11 References

1. Greenberg, M.D., *Applications of Green's functions in Science and Engineering*, Prentice-Hall, Inc. Englewood Cliffs, NJ., 1971.

2. Taigbenu, A.E., "The Green Element Method," *Int. J. for Numerical Methods in Engineering*, **38**, pp 2241-2263, 1995.

3. Reddy, J.N., *An Introduction to the Finite Element Method*, McGraw-Hill, Inc., New York, 1984.

4. Liggett, J.A. and L-F. Liu, *The Boundary Integral Equation Method for Porous Media Flow*, Allen & Unwin, London, UK., 1983.

5. Liggett, J.A., “Singular Cubature over Triangles”, *Int. J. Num. Meth. Eng.*, **18**, pp. 1375-1384, 1982.

6. Cowper, G.R., “Gaussian Quadrature Formular for Triangles”, 7(3), pp. 405-408, 1973.

7. Reddy, C.T. and D.J. Shippy, “Alternate Integration Formulae for Triangular Finite Elements”, *Int. J. Num. Meth. Eng.*, **17**, pp. 133-139, 1981.

8. Huyakorn, P.S., S.D. Thomas, and B.M. Thompson, "Technique for Making Finite Elements Competitive in Modeling Flow in Variably Saturated Porous Media", *Water Resources Res.*, **20**, pp. 1099-1115, 1984.

11 UNSTEADY TWO-DIMENSIONAL PROBLEMS

11.1 Introduction

The additional dimension of time which is incorporated into the two-dimensional steady problems addressed in the earlier chapter allows for solutions that provide information on the time history of the behaviour of the primary scalar or vector variable. Such solutions do, in most cases, give better representation of the behaviour of engineering systems which are usually under dynamic forces that alter equilibrium states from time to time.

As indicated in chapter 5, there is a long history of applications of the boundary element theory to transient 2-D. problems dating back to the early 60's. A sizeable portion of those applications had been directed at the diffusion equation which is one of the simplest linear transient differential equations, and they have proceeded along three approaches: direct, Laplace transform, and dual-reciprocity methods. However, in this chapter, the boundary integral theory will be implemented in the Green element sense, and shall not only be directed at the diffusion equation, but also at the contaminant transport and variably saturated flow problems. The nonlinear nature of the latter problem has posed computational difficulties for conventional **BEM**, but it is demonstrated here how those difficulties are eliminated by **GEM**. The ease with which **GEM** solves the nonlinear variably saturated flow problem further strengthens our motivation for the method.

11.2 Flow Equation

There are a number of transient problems that are addressed in this chapter, and they are expressed in the general form

$$\nabla.(K\nabla h) + \nabla K.\boldsymbol{k} = \nabla.(\boldsymbol{u}h) + \rho\frac{\partial h}{\partial t} + f(x,y,t) \tag{11.1}$$

in which $\boldsymbol{k}$ is the unit vector in the vertical direction, and $\boldsymbol{u} = \boldsymbol{i}u_x + \boldsymbol{j}u_y$ is the velocity vector. Depending on the problem being addressed, a number of terms of eq. (11.1) could be eliminated, while, for nonlinear problems, the parameters of the equation depend on the primary variable whose solution is desired. For instance, the simplest version of eq. (11.1), which is the linear diffusion or heat conduction equation, considers only the first term on the LHS and the last two terms on the RHS of eq. (11.1).

As we have always indicated in earlier chapters, appropriate boundary and initial data have to be specified in order to obtain unique solutions of the differential equation. The 1st-, 2nd-, and 3rd-type boundary conditions can be specified on different parts of the boundary, while initial data are specified for the primary variable throughout the solution domain at the start of the solution history.

11.3 Green Element Modeling

11.3.1 *Quasi-Steady Models*

Our first model for the transient problem uses the Green's function that we had employed for the steady problems in chapter 10. Our complementary differential equation is eq. (10.6) and its fundamental solution is eq. (10.7). We can rearrange eq. (11.1) to give

$$\nabla^2 h + \nabla\Theta.\nabla h + \nabla\Theta.\boldsymbol{k} = \Psi\nabla.(\boldsymbol{u}h) + \sigma\frac{\partial h}{\partial t} + \Psi f(x,y,t) \tag{11.2}$$

in which $\sigma = \rho/K$, $\Theta = \ln(K)$, and $\Psi = 1/K$. We are now in a position to apply Green's second identity, given by eq. (10.8), to eqs. (10.6) and (11.2) which yields

$$\begin{aligned} &-\lambda h(r_i,t) + \int_\Gamma \left(h\frac{\partial G}{\partial n} - G\frac{\partial h}{\partial n} \right) ds + \\ &\iint_A G\left[-\nabla\Theta.\nabla h - \nabla\Theta.\boldsymbol{k} + \Psi(\nabla.(\boldsymbol{u}h) + f) + \sigma\frac{\partial h}{\partial t} \right] dA = 0 \end{aligned} \tag{11.3}$$

Eq. (11.3) is the singular integral representation of the differential equation. The modelling of the differential equation by the Green element method takes off at this point. It involves discretizing the computational domain by suitable polygonal elements over which is prescribed the distribution of functional quantities. We shall not repeat this procedure here since it was extensively treated in chapter 10. We are at liberty to choose the type of element to employ for the discretization of the computational domain and as well as the type of interpolation functions to use to approximate functional quantities. We, however, advise that the choice produces simple elements and interpolation functions so that the mathematical calculations are easy to explain and apply. When these steps of discretization and interpolation are carried out, eq. (11.3) gives the system of discrete equations for a typical element. That is

$$\boxed{\begin{aligned} &R_{ij}h_j + L_{ij}\varphi_j - V_{ijl}\Theta_j h_l - P_{ij}\Theta_j + \\ &(\bar{u}_x S^x_{ijl} + \bar{u}_y S^y_{ijl})\Psi_j h_l + U_{ijl}\left[\sigma_j \frac{dh_l}{dt} + \Psi_j f_l\right] = 0 \end{aligned}} \tag{11.4}$$

Although eq. (11.4) is written for a typical element, we have omitted the superscript e for convenience. However, whenever there arises a mix up between the global and element equations, the superscript will be reintroduced. Some of the element matrices in eq. (11.4) had earlier been defined in eqs. (10.15a), (10.15b), (10.15c), (10.15d), and (10.48), while the remaining ones that are now newly encountered have the following expressions

$$S^x_{ijl} = \iint_{\Lambda^{(e)}} G(r,r_i)\Omega_j \frac{\partial \Omega_l}{\partial x} dA \tag{11.5a}$$

$$S^y_{ijl} = \iint_{\Lambda^{(e)}} G(r,r_i)\Omega_j \frac{\partial \Omega_l}{\partial y} dA \tag{11.5b}$$

The quantities $\bar{u}_x$ and $\bar{u}_y$ in eq. (11.4) represent the centroidal values of the velocity vector in the x and y directions within a typical element. The system of discrete element equations of eq. (11.4) has been arrived at in a manner similar to that of model 1 in chapter 10 for the steady state problem — medium and flow variables have been approximated by interpolation functions prescribed over each element. It is herein referred to as model 1 for the transient problem.

Eq. (11.4) is, however, still a system of first-order equations in time. To simplify the temporal derivative, the generalized 2-level time discretization scheme which was employed in chapter 5 is again applied. The temporal derivative is approximated by a difference expression at $t = t_{1+\alpha} = t_1 + \alpha\Delta t$, $0 \leq \alpha \leq 1$ (see Fig. 5.2) which simplifies eq. (11.4) to

$$\alpha\Big\{R_{ij} - V_{ilj}\Theta_l + (\bar{u}_x S^x_{ilj} + \bar{u}_y S^y_{ilj})\Psi_l\Big\}h_j^{(2)} +$$
$$\omega\Big\{R_{ij} - V_{ilj}\Theta_l + (\bar{u}_x S^x_{ilj} + \bar{u}_y S^y_{ilj})\Psi_l\Big\}h_j^{(1)} + L_{ij}(\alpha\varphi_j^{(2)} + \omega\varphi_j^{(1)}) -$$
$$P_{ij}\Theta_j + U_{ijl}\left[\sigma_j\frac{h_l^{(2)} - h_l^{(1)}}{\Delta t} + \Psi_j(\alpha f_l^{(2)} + \omega f_l^{(1)})\right] = 0$$

or

$$\left\{\alpha\Big[R_{ij} - V_{ilj}\Theta_l + (\bar{u}_x S^x_{ilj} + \bar{u}_y S^y_{ilj})\Psi_l\Big] + \sigma_j\frac{U_{ijl}}{\Delta t}\right\}h_j^{(2)} +$$
$$\left\{\omega\Big[R_{ij} - V_{ilj}\Theta_l + (\bar{u}_x S^x_{ilj} + \bar{u}_y S^y_{ilj})\Psi_l\Big] - \sigma_j\frac{U_{ijl}}{\Delta t}\right\}h_j^{(1)} + \tag{11.6}$$
$$L_{ij}(\alpha\varphi_j^{(2)} + \omega\varphi_j^{(1)}) - P_{ij}\Theta_j + U_{ijl}\Big[\Psi_j(\alpha f_l^{(2)} + \omega f_l^{(1)})\Big] = 0$$

in which $\omega = 1 - \alpha$. In arriving at eq. (11.6), it is assumed that the medium parameters K and ρ neither depend on time nor on the primary variable. Later we shall derive an equation similar to eq. (11.6) when these parameters depend on time and the primary variable. As in earlier chapters, the bracketed superscript 2 and 1 denote the current and previous time levels.

For nonlinear problems where the medium parameters depend on the primary variable, the system of element equations (11.4) can be solved either by the Picard algorithm or the Newton-Raphson algorithm. In that case when the generalized 2-level time discretization scheme is applied to eq. (11.4), it yields

$$\alpha\Big\{R_{ij} - V_{ilj}[\alpha\Theta_l^{(2)} + \omega\Theta_l^{(1)}] + (\bar{u}_x S^x_{ilj} + \bar{u}_y S^y_{ilj})(\alpha\Psi_l^{(2)} + \omega\Psi_l^{(1)})\Big\}h_j^{(2)} +$$
$$\omega\Big\{R_{ij} - V_{ilj}(\alpha\Theta_l^{(2)} + \omega\Theta_l^{(1)}) + (\bar{u}_x S^x_{ilj} + \bar{u}_y S^y_{ilj})(\alpha\Psi_l^{(2)} + \omega\Psi_l^{(1)})\Big\}h_j^{(1)} +$$
$$L_{ij}(\alpha\varphi_j^{(2)} + \omega\varphi_j^{(1)}) - P_{ij}(\alpha\Theta_j^{(2)} + \omega\Theta_j^{(1)}) +$$
$$U_{ijl}\left[(\alpha\sigma_j^{(2)} + \omega\sigma_j^{(1)})\frac{h_l^{(2)} - h_l^{(1)}}{\Delta t} + (\alpha\Psi_j^{(2)} + \omega\Psi_l^{(1)})(\alpha f_l^{(2)} + \omega f_l^{(1)})\right] \equiv g_i = 0$$

or arranged differently

$$\left\{\alpha\left[R_{ij} - V_{ilj}[\alpha\Theta_l^{(2)}+\omega\Theta_l^{(1)}] + (\bar{u}_x S^x_{ilj}+\bar{u}_y S^y_{ilj})(\alpha\Psi_l^{(2)}+\omega\Psi_l^{(1)})\right] + \frac{(\alpha\sigma_j^{(2)}+\omega\sigma_j^{(1)})}{\Delta t}U_{ijl}\right\}h_j^{(2)} +$$

$$\left\{\omega\left[R_{ij} - V_{ilj}(\alpha\Theta_l^{(2)}+\omega\Theta_l^{(1)}) + (\bar{u}_x S^x_{ilj}+\bar{u}_y S^y_{ilj})(\alpha\Psi_l^{(2)}+\omega\Psi_l^{(1)}) - \frac{(\alpha\sigma_j^{(2)}+\omega\sigma_j^{(1)})}{\Delta t}U_{ijl}\right]\right\}h_j^{(1)} +$$

$$L_{ij}(\alpha\varphi_j^{(2)}+\omega\varphi_j^{(1)}) - P_{ij}(\alpha\Theta_j^{(2)}+\omega\Theta_j^{(1)}) + U_{ijl}\left[(\alpha\Psi_j^{(2)}+\omega\Psi_l^{(1)})(\alpha f_l^{(2)}+\omega f_l^{(1)})\right] \equiv g_i = 0 \tag{11.7}$$

To solve eq. (11.7) by the Newton-Raphson algorithm, it suffices to give the expression for the Jacobian matrix. It is given by

$$J_{ij}^{(2,k)} = \begin{cases} \left.\dfrac{\partial g_i}{\partial h_j}\right|_{h_j = h_j^{(2,k)}} = & \alpha\Big[R_{ij} - V_{ilj}(\alpha\Theta_l^{(2,k)} + \omega\Theta_l^{(1)}) + \\ & (\bar{u}_x S^x_{ilj} + \bar{u}_y S^y_{ilj})(\alpha\Psi_l^{(2,k)} + \omega\Psi_l^{(1)})\Big] + \\ & \alpha\left[-V_{iln}\dfrac{d\Theta_l^{(2,k)}}{dh_j} + (\bar{u}_x S^x_{iln} + \bar{u}_y S^y_{iln})\dfrac{d\Psi_l^{(2,k)}}{dh_j}\right](\alpha h_n^{(2,k)} + \omega h_n^{(1)}) - \\ & \alpha\left[P_{il}\dfrac{d\Theta_l^{(2,k)}}{dh_j} - U_{inl}\dfrac{d\Psi_n^{(2,k)}}{dh_j}(\alpha f_l^{(2)} + \omega f_l^{(1)})\right] + \\ & \dfrac{1}{\Delta t}\left[U_{ilj}(\alpha\sigma_l^{(2,k)} + \omega\sigma_l^{(1)}) + \alpha U_{iln}\dfrac{d\sigma_l^{(2,k)}}{dh_j}(h_n^{(2,k)} - h_n^{(1)})\right] \\ \left.\dfrac{\partial g_i}{\partial \varphi_j}\right|_{\varphi_j = \varphi_j^{(2,k)}} = \alpha L_{ij} & \end{cases} \tag{11.8}$$

where the superscript k stands for the iteration number.

A simplified formulation of Model 1 can be achieved in a manner similar to what was done in chapter 10 by making the assumption that the medium is piecewise homogeneous. This second model considers K and σ to be uniform within each element, that is $K \approx \bar{K}$ and $\sigma \approx \bar{\sigma}$. With this assumption, the differential equation is modified from eq. (11.2) to

$$\bar{K}\nabla.(\nabla h) = \nabla.(\bar{\boldsymbol{u}}h) + \bar{\rho}\frac{\partial h}{\partial t} + f(x,y,t) \tag{11.9}$$

whose integral representation according to the quasi-steady formulation is given by

$$\begin{aligned} &-\lambda h(r_i,t) + \int_\Gamma \left(h\frac{\partial G}{\partial n} - G\frac{\partial h}{\partial n} \right) ds + \\ &\int\int_\Lambda G\left[\bar{\Psi}(\nabla.(\bar{\boldsymbol{u}}h) + f) + \bar{\sigma}\frac{\partial h}{\partial t} \right] dA = 0 \end{aligned} \tag{11.10a}$$

or in discrete form, after discretization and interpolation, as

$$\boxed{R_{ij}h_j + L_{ij}\varphi_j + (\bar{u}_x Q_{ij} + \bar{u}_y P_{ij})\bar{\Psi}h_j + T_{ij}\left[\bar{\sigma}\frac{dh_j}{dt} + \bar{\Psi}f_j\right] = 0} \tag{11.10b}$$

in which

$$Q_{ij} = \int\int_{\Lambda^{(e)}} G(r,r_i)\frac{\partial \Omega_j}{\partial x} dA \tag{11.11}$$

The time derivative in eq. (11.10b) is simplified by the generalized 2-level time discretization scheme to give

$$\begin{aligned} &\alpha\left\{R_{ij} + (\bar{u}_x Q_{ij} + \bar{u}_y P_{ij})(\alpha\bar{\Psi}^{(2)} + \omega\bar{\Psi}^{(1)})\right\}h_j^{(2)} + \\ &\omega\left\{R_{ij} + (\bar{u}_x Q_{ij} + \bar{u}_y P_{ij})(\alpha\bar{\Psi}^{(2)} + \omega\bar{\Psi}^{(1)})\right\}h_j^{(1)} + \\ &L_{ij}(\alpha\varphi_j^{(2)} + \omega\varphi_j^{(1)}) + T_{ij}\left[(\alpha\bar{\sigma}^{(2)} + \omega\bar{\sigma}^{(1)})\frac{h_l^{(2)} - h_l^{(1)}}{\Delta t} + \right. \\ &\left.(\alpha\bar{\Psi}^{(2)} + \omega\bar{\Psi}^{(1)})(\alpha f_j^{(2)} + \omega f_j^{(1)})\right] \equiv g_i = 0 \end{aligned}$$

or arranged differently as

$$
\begin{aligned}
&\left\{\alpha\left[R_{ij} + (\bar{u}_x Q_{ij} + \bar{u}_y P_{ij})(\alpha\bar{\Psi}^{(2)} + \omega\bar{\Psi}^{(1)})\right] + \frac{(\alpha\bar{\sigma}^{(2)} + \omega\bar{\sigma}^{(1)})}{\Delta t} T_{ij}\right\} h_j^{(2)} + \\
&\left\{\omega\left[R_{ij} + (\bar{u}_x Q_{ij} + \bar{u}_y P_{ij})(\alpha\bar{\Psi}^{(2)} + \omega\bar{\Psi}^{(1)})\right] - \frac{(\alpha\bar{\sigma}^{(2)} + \omega\bar{\sigma}^{(1)})}{\Delta t} T_{ij}\right\} h_j^{(1)} + \\
&L_{ij}(\alpha\varphi_j^{(2)} + \omega\varphi_j^{(1)}) + T_{ij}\left[(\alpha\bar{\Psi}^{(2)} + \omega\bar{\Psi}^{(1)})(\alpha f_j^{(2)} + \omega f_j^{(1)})\right] \equiv g_i = 0
\end{aligned}
\tag{11.12}
$$

It should be noted that we have considered $\bar{K}$ and $\bar{\sigma}$ to be dependent on time and the primary variable φ (nonlinear case). The solution to the system of nonlinear discrete equations (11.12) is achieved by the Newton-Raphson algorithm whose Jacobian matrix is given by

$$
J_{ij}^{(2,k)} = \begin{cases}
\left.\dfrac{\partial g_i}{\partial h_j}\right|_{h_j = h_j^{(2,k)}} = \begin{aligned}[t]
&\alpha\left\{R_{ij} + (\bar{u}_x Q_{ij} + \bar{u}_y P_{ij})(\alpha\bar{\Psi}^{(2,k)} + \omega\bar{\Psi}^{(1)})\right\} + \\
&\alpha(\bar{u}_x Q_{il} + \bar{u}_y P_{il})(\alpha h_l^{(2,k)} + \omega h_l^{(1)})\frac{d\bar{\Psi}^{(2,k)}}{dh_j} + \\
&\alpha T_{il}\frac{d\bar{\Psi}^{(2,k)}}{dh_j}(\alpha f_l^{(2)} + \omega f_l^{(1)}) + \\
&\frac{1}{\Delta t}\left\{T_{ij}[\alpha\bar{\sigma}^{(2,k)} + \omega\bar{\sigma}^{(1)}] + \alpha T_{in}\frac{d\bar{\sigma}^{(2,k)}}{dh_j}(h_n^{(2,k)} - h_n^{(1)})\right\}
\end{aligned} \\[2ex]
\left.\dfrac{\partial g_i}{\partial \varphi_j}\right|_{\varphi_j = \varphi_j^{(2,k)}} = \alpha L_{ij}
\end{cases}
\tag{11.13}
$$

11.3.2 *Transient Model*

This model can only applied to the form of the original differential equation which is given by eq. (11.9) because its Green's function comes from the linear heat conduction equation

$$\nabla^2 G + \bar{\sigma}\frac{\partial G}{\partial t} = \delta(r-r_i)\delta(t_2-t); \quad -\infty<r<\infty, \quad t_2>t \tag{11.14}$$

whose solution (free-space Green's function) is (Morse and Feshbach [1])

$$G = G(r,t;r_i,t_2) = \frac{\bar{\sigma}H(t_2-t)}{(t_2-t)}\exp\left[-\frac{\bar{\sigma}(r-r_i)^2}{4(t_2-t)}\right] \tag{11.15}$$

The singular integral equation that arises from this approach has been presented by other investigators.[2, 3, 4] By applying Green's second identity to eqs. (11.9) and (11.14) and performing a time integration within the temporal element $[t_1, t_2]$, we have

$$\begin{aligned}\int_{t_1}^{t_2}\iint_{\Lambda}\left[h\left(\delta(r-r_i)\delta(t_2-t)-\bar{\sigma}\frac{\partial G}{\partial t}\right)-G\left(\bar{\Psi}\nabla.(\bar{\boldsymbol{u}}h)+\bar{\sigma}\frac{\partial h}{\partial t}+\bar{\Psi}f(r,t)\right)\right]dA\,dt \\ = \int_{t_1}^{t_2}\int_{\Gamma}\left[h\nabla G - G\nabla h\right].\boldsymbol{n}\,dS\,dt\end{aligned} \tag{11.16}$$

We employ the relationships outlined below to simply the above equation

$$\int_{t_1}^{t_2}\iint_{\Lambda} h(r,t)\delta(r-r_i)\delta(t_2-t)dA\,dt = 2\lambda h(r_i,t_2) \tag{11.17a}$$

$$\begin{aligned}-\bar{\sigma}\int_{t_1}^{t_2}\iint_{\Lambda}\left(h\frac{\partial G}{\partial t}+G\frac{\partial h}{\partial t}\right)dA\,dt \\ = -\bar{\sigma}\int_{t_1}^{t_2}\iint_{\Lambda}\frac{\partial}{\partial \tau}(hG)dA\,dt = \bar{\sigma}\iint_{\Lambda}G(r,t_1;r_i,t_2)h(r,t_1)dA\end{aligned} \tag{11.17b}$$

in which λ is the nodal angle at the source point which has been defined in chapter 10. It should be noted that we have used the property of the Green's function that $G(r,t_2;r_i,t_2) = 0$ to simplify eq. (11.17b). When we introduce the above two relationships into eq. (11.16) we have the integral equation

$$
\begin{aligned}
&-2\lambda h(r_i,t) + \int_\Gamma\int_{t_1}^{t_2}[h\frac{\partial G(r,t;r_i,t_2)}{\partial n} - G(r,t;r_i,t_2)\frac{\partial h(r,t)}{\partial n}]\,dS\,dt + \\
&\int\int_\Lambda h(r,t_1)\,G(r,t_1;r_i,t_2)dA + \\
&\int\int_\Lambda\int_{t_1}^{t_2}\bar{\Psi}[\nabla.(\bar{\boldsymbol{u}}h(r,t)) + f(r,t)]G(r,t;r_i,t_2)dtdA = 0
\end{aligned}
\tag{11.18}
$$

where

$$
\frac{\partial G(r,t;r_i,t_2)}{\partial n} = -\frac{\bar{\sigma}\eta_i}{2(t_2-t)}G(r,t;r_i,t_2) \tag{11.19}
$$

where η_i is the normal distance from the source point to the boundary over which integration is being done. The functional quantities are approximated in space and time by linear interpolation functions and the integrations in time and space are implemented to reduce eq. (11.18) to

$$
\boxed{
\begin{aligned}
&R_{ij}^{(m)}h_j^{(m)} + L_{ij}^{(m)}\varphi_j^{(m)} + W_{ij}^{(1)}h_j^{(1)} + \\
&\bar{\Psi}\Big[(\bar{u}_x Q_{ij}^{(m)} + \bar{u}_y P_{ij}^{(m)})h_j^{(m)} + T_{ij}^{(m)}f_j^{(m)}\Big] = 0; \quad m=1,2
\end{aligned}
}
\tag{11.20}
$$

in which

$$
\begin{aligned}
R_{ij}^{(2)} &= -2\delta_{ij}\lambda + \int_\Gamma\int_0^{\Delta t}\Omega_j(r)\Omega^{(2)}(\tau)\frac{\partial G(r,\tau;r_i,\Delta t)}{\partial n}ds\,d\tau \\
&= -2\delta_{ij}\lambda + \frac{\eta_i}{2}\int_\Gamma\Omega_j(r)\left\{\bar{\sigma}E_1\left[\frac{\bar{\sigma}(r-r_i)^2}{4\Delta t}\right] - \frac{4}{(r-r_i)^2}\exp\left[\frac{-\bar{\sigma}(r-r_i)^2}{4\Delta t}\right]\right\}ds
\end{aligned}
\tag{11.21a}
$$

$$\begin{aligned} R_{ij}^{(1)} &= \int_{\Gamma}\int_{0}^{\Delta t}\Omega_j(r)\Omega^{(1)}(\tau)\frac{\partial G(r,\tau;r_i,\Delta t)}{\partial n}\,ds\,d\tau \\ &= -\frac{\eta_i}{2}\int_{\Gamma}\Omega_j(r)\bar{\sigma}E_1\left[\frac{\bar{\sigma}(r-r_i)^2}{4\Delta t}\right]ds \end{aligned} \tag{11.21b}$$

$$L_{ij}^{(2)} = -\int_{\Gamma}\int_{0}^{\Delta t}\Omega_j(r)\Omega^{(2)}(\tau)\,G(r,\tau;r_i,\Delta t)\,dS\,d\tau = -\int_{\Gamma}\Omega_j(r)M_i^{(2)}ds \tag{11.21c}$$

$$L_{ij}^{(1)} = -\int_{\Gamma}\int_{0}^{\Delta t}\Omega_j(r)\Omega^{(1)}(\tau)G(r,\tau;r_i,\Delta t)\,ds\,d\tau = \int_{\Gamma}\Omega_j(r)M_i^{(1)}ds \tag{11.21d}$$

$$Q_{ij}^{(2)} = \iint_{\Lambda}\int_{0}^{\Delta t}\frac{\partial\Omega_j(r)}{\partial x}\Omega^{(2)}(\tau)G(r,\tau;r_i,\Delta t)ds\,d\tau = \iint_{\Lambda}\frac{\partial\Omega_j(r)}{\partial x}M_i^{(2)}ds \tag{11.21e}$$

$$P_{ij}^{(2)} = \iint_{\Lambda}\int_{0}^{\Delta t}\frac{\partial\Omega_j(r)}{\partial y}\Omega^{(2)}(\tau)G(r,\tau;r_i,\Delta t)ds\,d\tau = \iint_{\Lambda}\frac{\partial\Omega_j(r)}{\partial y}M_i^{(2)}ds \tag{11.21f}$$

$$Q_{ij}^{(1)} = \iint_{\Lambda}\int_{0}^{\Delta t}\frac{\partial\Omega_j(r)}{\partial x}\Omega^{(1)}(\tau)\,G(r,\tau;r_i,\Delta t)ds\,d\tau = \iint_{\Lambda}\frac{\partial\Omega_j(r)}{\partial x}M_i^{(1)}ds \tag{11.21g}$$

$$P_{ij}^{(1)} = \iint_{\Lambda}\int_{0}^{\Delta t}\frac{\partial\Omega_j(r)}{\partial y}\Omega^{(1)}(\tau)G(r,\tau;r_i,\Delta t)ds\,d\tau = \iint_{\Lambda}\frac{\partial\Omega_j(r)}{\partial y}M_i^{(1)}\,ds \tag{11.21h}$$

$$T_{ij}^{(2)} = \iint_{A}\int_{0}^{\Delta t} \Omega_j(r)\Omega^{(2)}(\tau)G(r,\tau;r_i,\Delta t)ds d\tau = \iint_{A} \Omega_j(r)M_i^{(2)}ds \tag{11.21i}$$

$$T_{ij}^{(1)} = \iint_{A}\int_{0}^{\Delta t} \Omega_j(r)\Omega^{(2)}(\tau)G(r,\tau;r_i,\Delta t)ds d\tau = \iint_{A} \Omega_j(r)M_i^{(1)}ds \tag{11.21j}$$

$$W_{ij}^{(1)} = \iint_{A} \Omega_j(r)\,G(r,0;r_i,\Delta t)\,dA \tag{11.21k}$$

where

$$M_i^{(2)} = [1 + \frac{\bar{\sigma}(r-r_i)}{4\Delta t}]\,E_1\left[\frac{\bar{\sigma}(r-r_i)^2}{4\Delta t}\right] - \exp\left[\frac{-\bar{\sigma}(r-r_i)^2}{4\Delta t}\right] \tag{11.21l}$$

$$M_i^{(1)} = -\frac{\bar{\sigma}(r-r_i)}{4\Delta t}E_1\left[\frac{\bar{\sigma}(r-r_i)^2}{4\Delta t}\right] + \exp\left[\frac{-\bar{\sigma}(r-r_i)^2}{4\Delta t}\right] \tag{11.21m}$$

In the above equations, $E_1(z) = \int_z^{\infty} \exp(-s)/s\,ds$ is the exponential integral whose expressions in the form of rational approximations are found in Abramowitz and Stegun [5]. As in the 1-D. transient model, the expressions for the coefficient matrices are also complicated, and that is due solely to the complexity of the Green's function. Few of the above integrals in the spatial dimension can be implemented analytically and as such require numerical treatment along the lines of our discussions in section 10.8.

We may need to state that the transient model can still be applied if $\bar{K}$ and $\bar{\sigma}$ are functions of the primary variable (nonlinear case). In that case, we need only recognize that all the element coefficients become functions of the primary variable which have to be re-evaluated at each iteration in each simulation interval. That certainly is expected to be computationally messy.

11.4 Numerical Examples

Four examples of transient flows in 2-D. spatial dimensions are employed to evaluate the capabilities of the Green element formulations which were presented in the foregoing sections. The examples have been designed in such a way that they cover a wide range of transient problems ranging from linear to nonlinear which are of profound engineering significance. Only the models that are derived from the fundamental solution to the Laplacian operator have been applied here because of the computational ease in coding the element equations with this solution. Where analytical solutions of these examples are available, they have been employed as bench mark results, but where such solutions do not exist, results from simulations by other numerical methods are utilized in validating our numerical solutions.

Example 1

Our first example is a nonlinear heat conduction problem which is governed by the differential equation

$$\nabla.(K\nabla h) = C\frac{\partial h}{\partial t} \tag{11.22}$$

This example is solved as a one-dimensional problem so that the results from the 2-D. model can be compared to the 1-D. exact solution that is available for the linear case. To this end a rectangular domain 1×0.1 is used as computational region with the top and bottom sides serving as no-flux boundaries and other sides having the boundary conditions:

$$h(0,y,t) = 1; \quad h(1,y,t) = 0 \tag{11.23}$$

At initial time, the primary variable h has zero value everywhere. The parameter C takes a value of unity, and K has the expression

$$K = (1 + \beta h)^{\gamma} \tag{11.24}$$

In the Green element calculations, two pairs of values of β and γ are considered. They are: (0,1) and (3,2). The first pair of values represents the linear heat conduction problem for which we had the exact solution given by eq. (5.72). Our 2-D. simulation of this linear case is compared to that exact solution.

The 2-D. **GEM** calculations are carried out on the computational domain that is discretized into 20 uniform rectangular elements. Model 1 of the quasi-steady formulation is implemented in solving this example. A uniform time step of 0.025 and a difference weighting factor of unity are adopted in the calculations. For the nonlinear case, a maximum of 4 iterations is allowed at each time step. The exact and **GEM** solutions are presented at times of 0.1, 0.3 and 0.8 in Figure 11.1 for the linear case, and we observe good agreement between both solutions. The **GEM** solutions at those same times for the nonlinear case are presented in Figure 11.2.

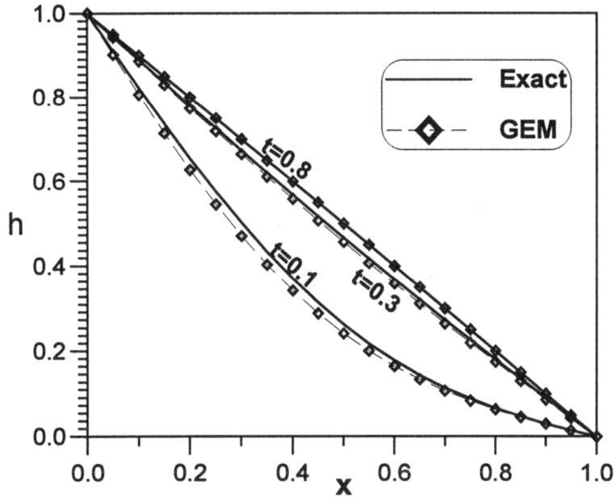

Figure 11.1 Exact and **GEM** solutions of example 1 (linear case)

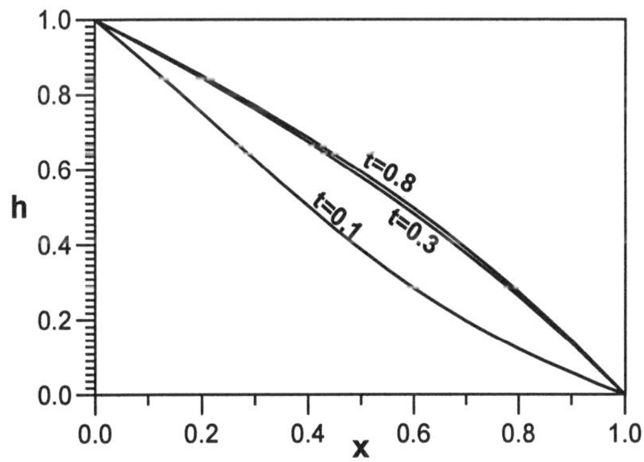

Figure 11.2 **GEM** solution of example 1 (nonlinear case).

Example 2

There are two parts to this example which occupies a significant place in the groundwater literature on well hydraulics. The first part deals with confined flow to a well discharging at a constant rate of Q_w located in an extensive confined aquifer of uniform thickness b.

$$T\nabla^2 h = S\frac{\partial h}{\partial t} \tag{11.25}$$

in which S and $T = bK$ are the storativity and transmissivity of the aquifer, where K is the hydraulic conductivity. Because of the radial symmetry of the problem, the solution varies only in the radial direction from the well.
The boundary and initial conditions are

$$2\pi T\frac{\partial h(r\to 0,t)}{\partial r} = Q_w, \quad h(\infty,t) = h_0; \quad h(r,0) = h_0 \tag{11.26}$$

Its solution in terms of the drawdown s is well known (Theis [6]).

$$h_0 - h = s = \frac{Q_w}{4\pi T}E_1(u) \tag{11.27}$$

where $u = r^2 S/(4Tt)$. In the groundwater literature, the exponential integral is denoted as the well function, that is $W(u) \equiv E_1(u)$. Using the following parameters values: $S = 10^{-5}$ and $T = 10^5\,\text{m}^2/\text{day}$, the problem is simulated on a the domain shown in Figure 11.3 which approximates a $\pi/18$ sector of a circle. The domain is discretized into 29 triangular elements with a total of 31 nodes that are all on the boundary and arranged along a radius r from the well. (The node arrangement along the radius from the well is as follows: $r =$ 0, 5, 10, 50, 10^2, 500, 10^3, 2000, 3000, 4000, 5000, 6000, 7000, 8000, 9000, 10^4m.) In the numerical calculations, the well is placed at a small distance of 10^{-25}m from the node at $r = 0$ where the well ought to have been located. The second model of the quasi-steady formulation is programmed for the calculations, and a difference weighting of 0.75 is incorporated in the program. The time history is discretized as follows: $\Delta t = 10^{-5}$ for $0 \le t \le 10^{-4}$, $\Delta t = 10^{-3}$ for $10^{-4} < t \le 10^{-2}$, and, $\Delta t = 10^{-2}$ for $10^{-2} < t \le 1$. The dimension of time is day. The results of the Green element method in terms of the dimensionless drawdown $s* = 4\pi Ts/Q_W$ versus the dimensionless time $t* = 0.25/u$ are plotted along with the analytic solution (Theis curve) in Figure 11.4. The numerical and analytic solutions are in good agreement. Three dimensional plots of the drawdown in the confined aquifer at times of 10^{-4} and 10^{-3}day, displayed in Figure 11.5, show, as expected, a decrease in drawdown from the well at any time and an increase in drawdown at any radius with time.

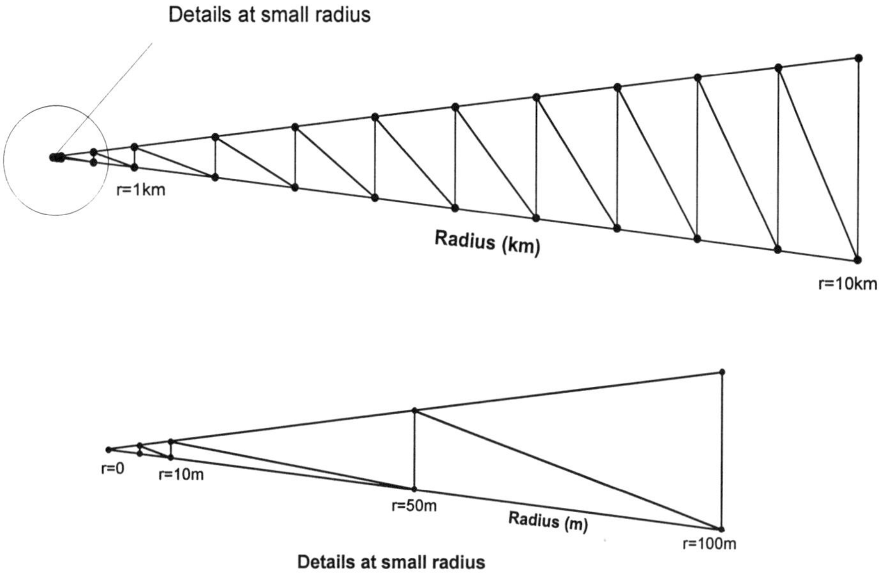

Figure 11.3 Computational domain of example 2.

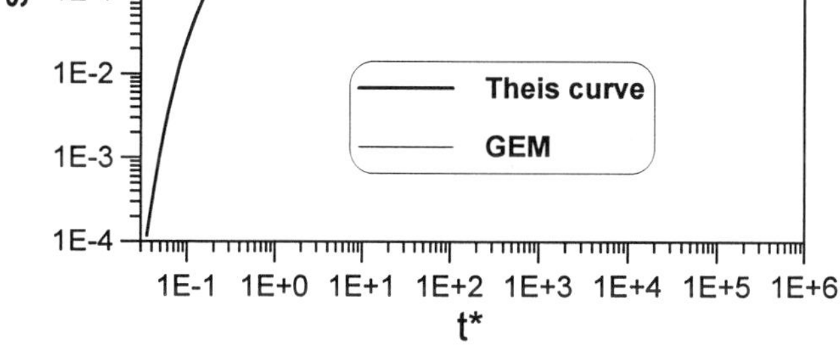

Figure 11.4 Exact and **GEM** solutions for confined flow example.

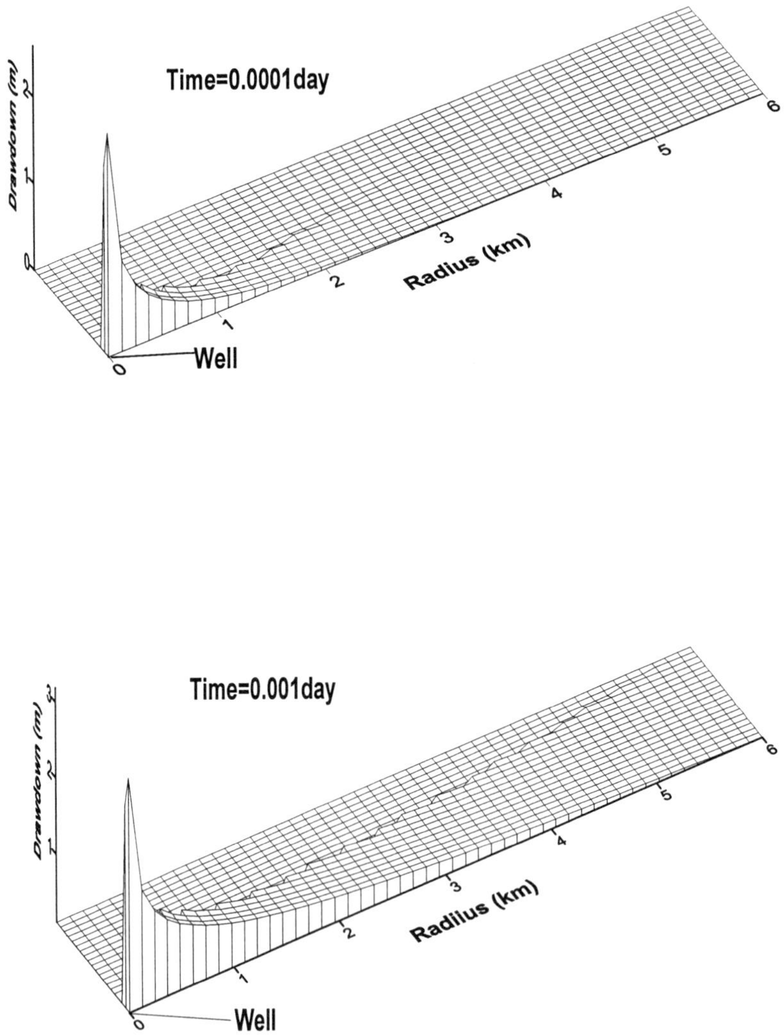

Figure 11.5 Solid plots of drawdown at $t = 10^{-4}$ day and $t = 10^{-3}$ day.

The second part of this example examines the influence on leakage flux from another aquifer on the solution earlier obtained in the first part. It considers the situation of flow to a constant-discharging well in a confined aquifer overlain by a leaky stratum of thickness b_L and hydraulic conductivity K_L. The boundary and initial conditions remain eq. (11.26), while the differential equation now becomes

$$T\nabla^2 h = S\frac{\partial h}{\partial t} - \frac{K_L}{b_L}(h_0 - h) \tag{11.28}$$

The analytical solution, provided by Jacob [7] , is

$$h_0 - h = s = \frac{Q_w}{4\pi T}\int_u^\infty \frac{1}{z}\exp[-(z + \frac{r^2}{4B^2 z})]\,dz = \frac{Q_w}{4\pi T}W(u,r/B) \tag{11.29}$$

where $B^2 = b_L T/K_L$. The integral denoted $W(u,r/B)$ is tabulated in Bear [8].

Using the same time discretization of the solution history and computational domain as the first example, the analytic and **GEM** solutions in terms of the drawdown $s* = 4\pi Ts/Q_w$ versus the dimensionless time $t* = 0.25/u$ are displayed in Figure 11.6, and they are in good agreement.

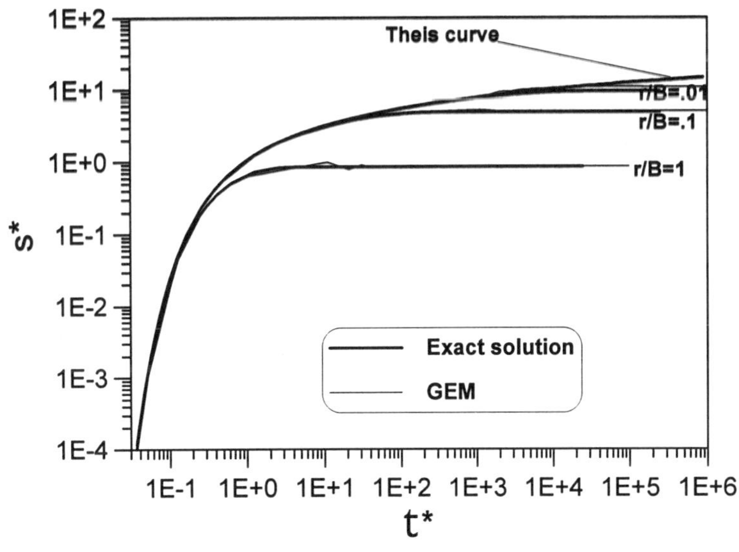

Figure 11.6 Exact and **GEM** solutions for leaky confined flow example.

Example 3

This example is a transient contaminant transport problem in 2-D. for which Cleary [9] derived an exact solution. The pollutant, which is continuously injected into the flow medium from a line source, is dispersed in the longitudinal and transverse directions and advected by a uniform velocity field along the longitudinal direction. The transport problem is described mathematically as

$$\nabla.(D\nabla h) = U\frac{\partial h}{\partial x} + \frac{\partial h}{\partial t}$$

with the boundary and initial data

$$h(0,y,t) = 1.0, \quad 2 \le y \le 3 \tag{11.50a}$$

$$h(0,y,t) = 0, \quad 0 \le y < 2 \text{ and } 3 < y \le 5 \tag{11.50b}$$

$$\frac{\partial h(x,0,t)}{\partial y} = \frac{\partial h(x,5,t)}{\partial y} = 0; \quad x > 0 \tag{11.50c}$$

$$\frac{\partial h(\infty,y,t)}{\partial x} = 0, \quad 0 \le y \le 5 \tag{11.50d}$$

$$h(x,y,0) = 0 \tag{11.50e}$$

Finite element simulations of this example were carried out by Yeh [10], and the parameters employed in those simulations are also used here. Although the domain is semi-infinite, a finite domain of 15 x 2.5, which is discretized into 150 uniform rectangular elements each of 0.5×0.5, is used for the calculations. Following Yeh [10], three transport modes are examined: the case of dispersion-dominant transport ($Pe = 0.05$), equal dominance of dispersion and advection ($Pe = 1$), and the advection-dominant transport ($Pe = 50$). A uniform time step of 0.2 is employed in all the calculations, and the breakthrough concentration fronts at $t = 4$ along $y = 2.5$ are presented in Figures 11.7a through 11.7c for the three cases. The finite element solutions are obtained by orthogonal weighting scheme, and the results show that the **GEM** solutions are better than the **FEM** solutions for all the cases. For the advection-dominant transport, the **GEM** solutions are presented in the form of 3-D. plots at time values of 2 and 4 in Figures 11.8a and 11.8b which show an advancing concentration front whose shape is maintained fairly the same at different times.

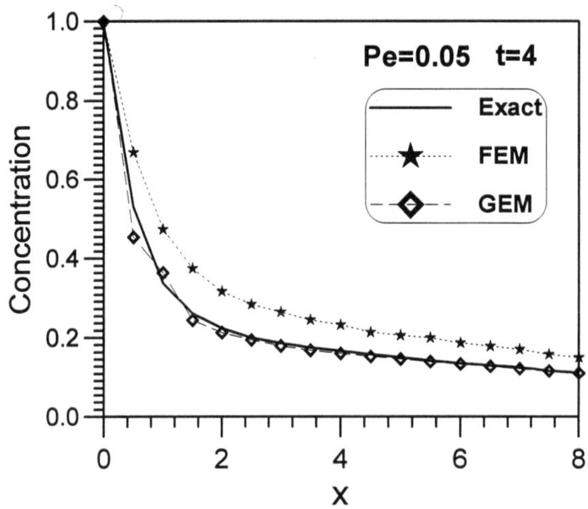

Figure 11.7a Exact, **GEM**, and **FEM** solutions of example 2 (dispersion-dominant, $Pe = 0.05$).

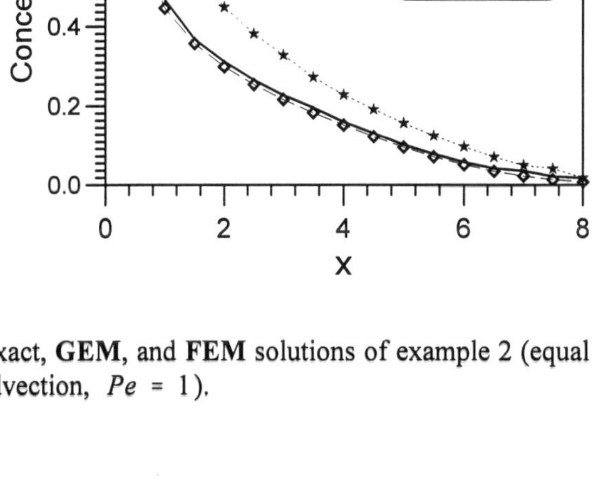

Figure 11.7b Exact, **GEM**, and **FEM** solutions of example 2 (equal dominance of dispersion and advection, $Pe = 1$).

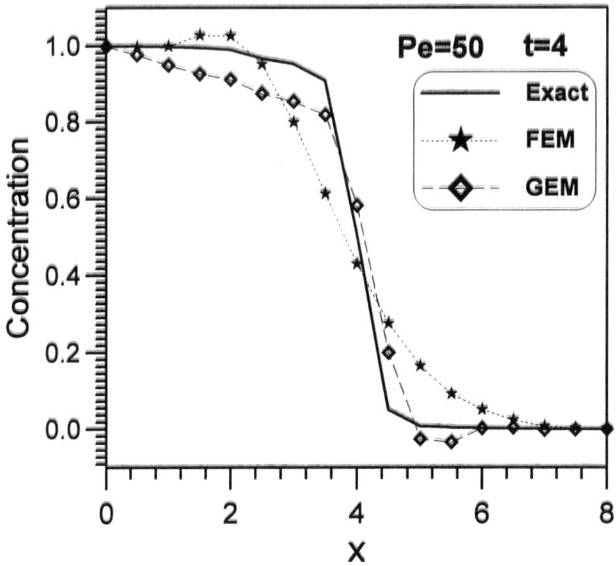

Figure 11.7c Exact, **GEM**, and **FEM** solutions of example 2 (advection-dominant, $Pe = 50$).

Example 4

This example is a 2-D. infiltration flow problem which was taken from Warrick and Lomen [11] who presented a semianalytic solution to it. It is governed by a form of the differential eq. (11.1) which is given by

$$\nabla . [D(h)\nabla h] + \frac{\partial K(h)}{\partial y} = \frac{\partial h}{\partial t} \tag{11.51}$$

in which $D(h)$ is the unsaturated diffusivity, h is the moisture content, and $K(h)$ is the unsaturated hydraulic conductivity. In this example flow of water takes place from a constant-flux strip source into a dry rectangular soil 75cm x 85cm which initially has a moisture content of 2.3×10^{-5} (or suction pressure head ψ of 250cm) throughout the soil medium. The constant-flux strip source, which is located on the top left corner of the boundary, is 5cm wide and has a strength $q_y = -0.0278$ cm/min ($q_y = -D\partial h/\partial y - K$). The remainder of the top boundary and the entire left boundary are no flux boundaries, while the bottom and right boundaries are Dirichlet boundaries where the moisture content is prescribed to have a value of 2.3×10^{-5} or suction head ψ of 250cm (see Figure 11.9).

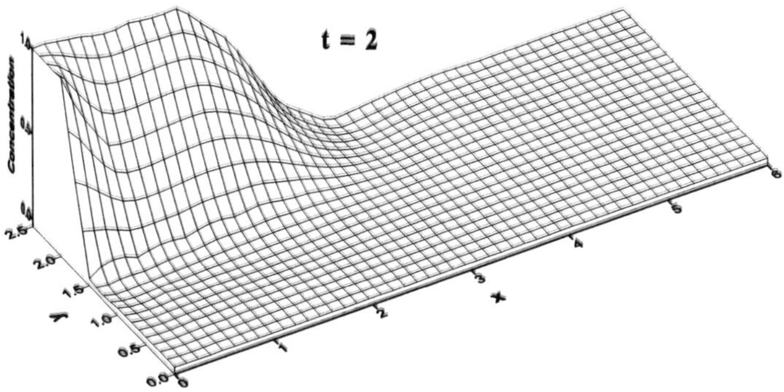

(a)

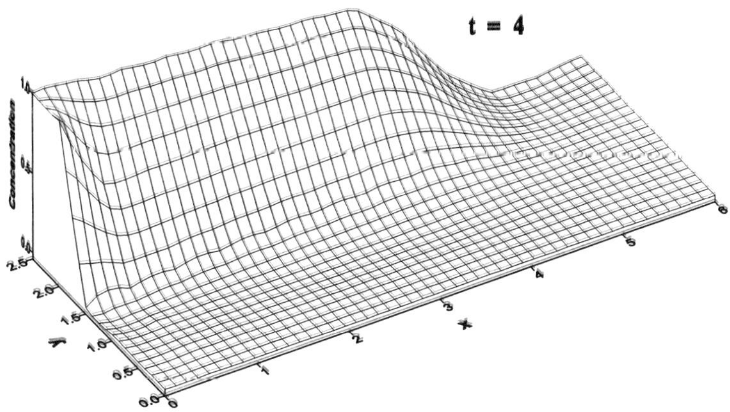

(b)

Figure 11.8 Solid plots of concentration for example 3. — (a) solution at $t = 2$, (b) solution at $t = 4$.

The soil constitutive relations for the fourth example are expressed as

$$K(h) = A_0 h \text{ cm/min} \tag{11.19a}$$

$$\psi(h) = \frac{1}{\gamma}\left[\ln\left(\frac{A_0}{K_0}\right) + \ln h\right] \text{ cm} \tag{11.19b}$$

$$\frac{dh}{d\psi} = \gamma h \tag{11.19c}$$

in which K_0, γ and A_0 are empirical constants whose values for the Panoche clay loam are $K_0 = 0.0694$ cm/min, $\gamma = 0.04$, and $A_0 = 0.1388$ cm/min [11].

To make a meaningful comparison between our results and those from other numerical methods, we have used the same spatial and time discretizations used in those numerical simulations. A uniform time step of 4 minutes is employed in the simulations, and 120 nonuniform linear rectangular elements are used to discretize the domain. The spatial discretization taken from Tsai *et al.* [12] corresponds to 143 nodal points which are defined by the grid spacings in Table 11.1.

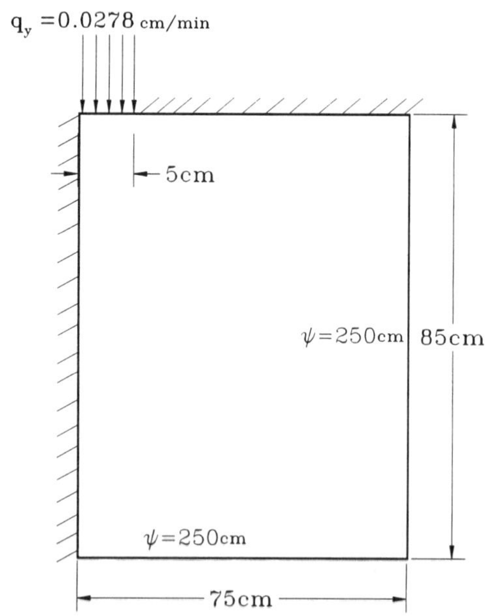

Figure 11.9 Computational domain of example 4.

Table 11.1 Spatial discretization data for example 4

Δx_i; $i=1,\ldots,9$	1 3.9 0.2, 9.9, 10, 10, 10, 10, 10, 10
Δy_j; $j=1,\ldots,15$	10, 10, 10, 10, 10, 10, 10, 10, 2.4, 0.4, 0.4, 0.2

At each time step a convergence tolerance of 10^{-5} based on the mean absolute error is imposed. Figures 11.10a and 11.10b show a comparison of the **GEM** solutions against the **FEM** solutions at times of 36 and 72 minutes obtained from the computer code FEMWATER of Yeh [13] and presented by Tsai *et al.* [12], the finite analytic solutions of Tsai *et al.* [12], and the semi-analytic solutions of Warrick and Lomen [11]. The results indicate that the Green element solutions reflect the general trend of the semi-analytic solution [14].

11.5 Remarks

Transient problems in 2-D. spatial dimensions have essentially been solved by two Green element formulations which arise from the quasi-steady Green's function and that from the transient one. The simple nature of the Green's function of the former greatly enhances its computational capability that is observed in the relative ease in which linear and nonlinear problems are solved. We may adopt it as a general rule that when a number of Green's functions are applicable to a problem, the most mathematically simple of them should be used in the numerical implementation. Our experience with the transient 1-D. diffusion-advection problem in chapter 6 where three Green's functions were applicable to the problem confirms the earlier statement. Furthermore, the first formulation for the problems addressed in this chapter is more appealing than the second because its Green's function applies both to steady and unsteady problems, and as such can be used in one program to solve both types of problems.

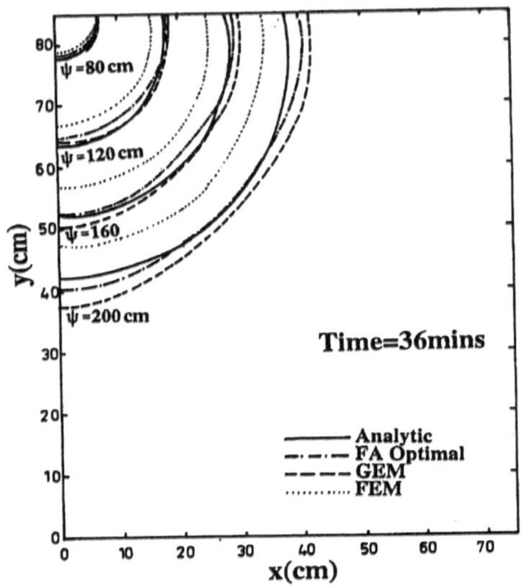

(a)

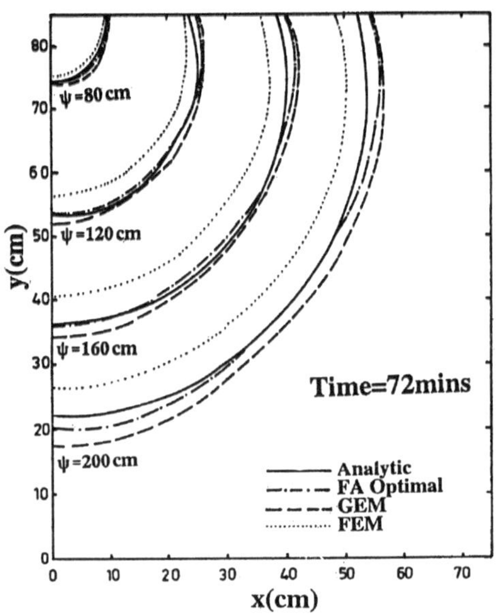

(b)

Figure 11.7 Contour plots of numerical solutions of example 4. — **(a)** Solutions at time = 36 minutes **(b)** Solutions at time = 72 minutes.

11.6 References

1. Morse, P.M. and H. Feshbach, *Methods of Theoretical Physics*, Vol. 1, McGraw-Hill, NY., 1953.

2. Shaw, R.P., "An Integral Equation Approach to Diffusion", *Int. J. Heat and mass Transfer*, **17**, 693-699, 1974.

3. Banerjee, P.K. and R. Butterfield, *Boundary Element Methods in Engineering Science*, McGraw-Hill, London UK, 1981.

4. Taigbenu, A.E. and J.A. Liggett, "Boundary Element calculations of Diffusion Equation", *J. Engineering Mech.*, **111**(3), pp. 311-328, 1985.

5. Abramowitz, M. and I.A. Stegun, *Handbook of Mathematical Functions With Formulas, Graphs, and Mathematical Tables*, National Bureau of Standards, 1970.

6. Theis, C.V., "The relation between the lowering of the Piezometric surface and the rate and duration of discharge of a well using groundwater storage", *Trans. Amer. Geophysical Union*, **16**, pp. 519-524, 1935.

7. Jacob,, C.E., "Radial flow in a leaky artesian aquifer", *Trans. Amer. Geophysical Union*, **27**(2), pp. 198-208, 1946.

8. Bear, J., *Hydraulics of Groundwater*, McGraw-Hill, 1979.
9. Cleary, R.W., *Groundwater Pollution and Hydrology: Mathematical models and computer programs*, Report No. 78-WR-15, Water resources Program, Princeton University, 1978.

10. Yeh, G.T., "Solution of groundwater flow equations using an orthogonal finite element scheme", *Proc. of Int. AMSE Conf. on modelling and Simulations*, Tassin, France, pp. 219-351, 1983.

11. Warrick, A.W. and D.O. Lomen, "Time-dependent linearized infiltration: III. Strip and Disc sources, *Soil Sci. Soc. Am. J.*, **40**, pp. 639-643, 1976.

12. Tsai, W.F., C-I. Chen and H-C. Tien, "Finite analytic numerical solutions for unsaturated flow with irregular boundaries, *ASCE J. Hydr. Eng.*, **119**(11), pp. 1274-1298, 1993.

13. Yeh, G.T., *FEMWATER: A Finite-element model of water flow through saturated-unsaturated porous media*, 1st. Rev. Rep. ONRL-5567/R1, Oak Ridge Mat. Lab, Oak Ridge, TN, 1987.

14. Taigbenu, A.E., and O.O. Onyejekwe, "Green Element simulations of the Transient Unsaturated Flow Equation," *Applied Mathematical Modelling*, **19**, pp. 675-684, 1995.

12 FURTHER CONSIDERATIONS

12.1 Introduction

Earlier chapters provided the theoretical basis of the Green element formulation for steady and unsteady problems in one and two spatial dimensions, and its solution capabilities were demonstrated on various examples which are of interest in many fields of engineering. Hereafter, we may wish to refine the numerical aspects of the method to achieve higher accuracy and/or savings in computing resources, and extend the applicability of the method to three-dimensional problems and those that were not addressed in this text.

In this chapter, we shall discuss incorporating higher-order and isoparametric elements into the 2-D. Green element model. We shall make mention of a three-dimensional Green element model.

12.2 Quadratic Rectangular and Triangular Elements

In chapter 10, we had prescribed a linear distribution for h and φ over rectangular and triangular elements, and evaluated the relevant boundary and domain integrations in the discretized element equations. In general, however, h and φ can have higher-order variation, and their individual distributions could be of different order. Here, we present rectangular and triangular elements over which quadratic variation is assumed for h and φ. Furthermore, we derive the analytic expressions for the elements equations arising from boundary and domain integrations for these elements. As we observed with linear elements, the exact integrations have been achieved because the source and field nodes belong to the same element. This is always the case in Green element calculations.

12.2.1 *Quadratic Rectangular Element*

The quadratic rectangular element in which the interpolating variable takes its value at each node (element of the Lagrange family) has 9 nodes as indicated in Fig. 12.1. The interpolation functions, using the node numbering in Fig. 12.1, are found in many texts on computational methods. They are repeated here, and expressed in terms of the local coordinates ζ and η whose origin is at node 1 ($0 \le \zeta, \eta \le 1$).

$$\Omega_1 = 4\zeta^2\eta^2 - 6\zeta^2\eta + 2\zeta^2 - 6\zeta\eta^2 + 9\zeta\eta - 3\zeta + 2\eta^2 - 3\eta + 1 \tag{12.1a}$$

$$\Omega_2 = 4(-2\zeta^2\eta^2 + 3\zeta^2\eta + 2\zeta\eta^2 - 3\zeta\eta + \zeta - \zeta^2) \tag{12.1b}$$

$$\Omega_3 = 4\zeta^2\eta^2 - 6\zeta^2\eta + 2\zeta^2 - 2\zeta\eta^2 + 3\zeta\eta - \zeta \tag{12.1c}$$

$$\Omega_4 = 4(-2\zeta^2\eta^2 + 2\zeta^2\eta + \zeta\eta^2 - \zeta\eta) \tag{12.1d}$$

$$\Omega_5 = 4\zeta^2\eta^2 - 2\zeta^2\eta - 2\zeta\eta^2 + \zeta\eta \tag{12.1e}$$

$$\Omega_6 = 4(-2\zeta^2\eta^2 + \zeta^2\eta + 2\zeta\eta^2 - \zeta\eta) \tag{12.1f}$$

$$\Omega_7 = 4\zeta^2\eta^2 - 2\zeta^2\eta + 2\eta^2 - 6\zeta\eta^2 + 3\zeta\eta - \eta \tag{12.1g}$$

$$\Omega_8 = 4(-2\zeta^2\eta^2 + 2\zeta^2\eta - \eta^2 + 3\zeta\eta^2 - 3\zeta\eta + \eta) \tag{12.1h}$$

$$\Omega_9 = 16(\zeta^2\eta^2 - \zeta^2\eta - \zeta\eta^2 + \zeta\eta) \tag{12.1i}$$

in which $\zeta = x/a$ and $\eta = y/b$

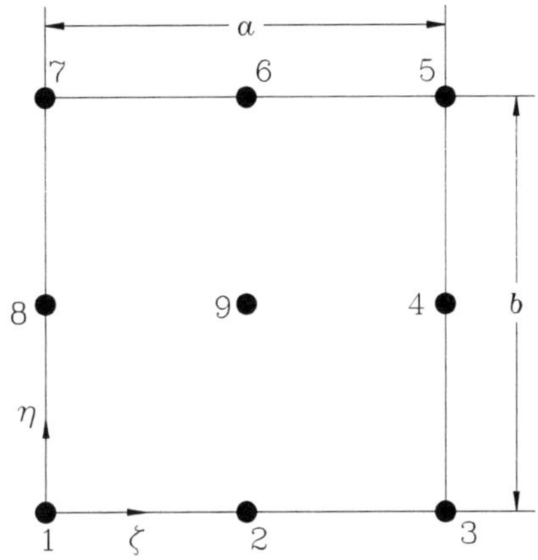

Fig. 12.1 9-node quadratic rectangular element.

12.2.2 *Quadratic Triangular Element*

We provide below the expressions for the quadratic interpolation functions for the 6-node triangular element shown in 12.2 by employing the expressions for the area coordinates denoted as L_j in eqs. (10.25a - 10.25c)

$$\Omega_1 = L_1(2L_1 - 1) \tag{12.2a}$$

$$\Omega_2 = 4L_1L_2 \tag{12.2b}$$

$$\Omega_3 = L_2(2L_2 - 1) \tag{12.2c}$$

$$\Omega_4 = 4L_2L_3 \tag{12.2d}$$

$$\Omega_5 = L_3(2L_3 - 1) \tag{12.2e}$$

$$\Omega_6 = 4L_1L_3 \tag{12.2f}$$

For the given node numbering shown in Fig. 12.2, it is sufficient, in defining appropriate local coordinates, to consider them with respect to nodes 1 and 2. (Another node numbering will not provide any new information on these coordinates.)

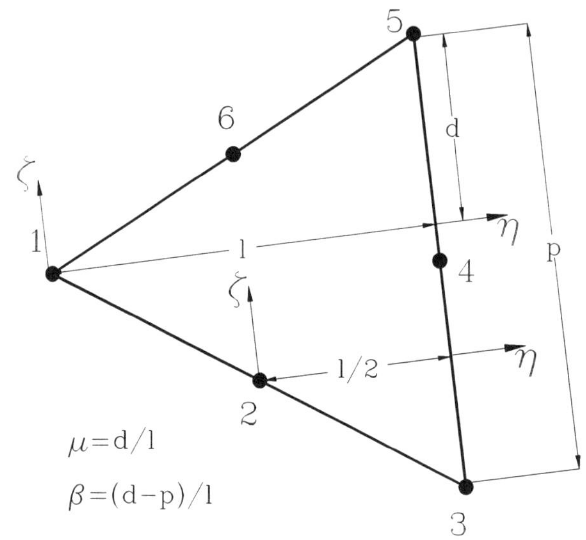

Figure 12.2 6-node triangular element with local coordinates.

12.3 Domain Integrations for Quadratic Elements

It is necessary that the domain integrations associated with quadratic rectangular and triangular elements be accurately done so that the gains of better approximation of the dependent variable are realized. In the case of the quadratic rectangular element, the integrals that deserve consideration arise from the source node located at a corner, a side, and the center of the element. We had earlier in chapter 10 considered the case when the source node is at a corner, and that is provided by eq. (10.32). The remaining integrations to consider are given below:

$$D3_{ij}^{(a)} = \frac{1}{2}\int_{0}^{b}\int_{-a/2}^{a/2}\left(\frac{x}{a}\right)^{i}\left(\frac{y}{b}\right)^{j}\ln(x^2+y^2)\,dx\,dy \tag{12.3a}$$

$$D3_{ij}^{(b)} = \frac{1}{2}\int_{-b/2}^{b/2}\int_{0}^{a}\left(\frac{x}{a}\right)^{i}\left(\frac{y}{b}\right)^{j}\ln(x^2+y^2)\,dx\,dy \tag{12.3b}$$

$$D4_{ij} = \frac{1}{2}\int_{-b/2}^{b/2}\int_{-a/2}^{a/2}\left(\frac{x}{a}\right)^{i}\left(\frac{y}{b}\right)^{j}\ln(x^2+y^2)\,dx\,dy \tag{12.3c}$$

The first integration refers to the source node on the side of the element with length a, the second comes from the source node being on the side with length b, while the third arises when the source node is at the center of the element. The exact expressions for the integrals in eqs. (12.3a) through (12.3c) are given below:

$$D3_{22}^{(a)} = \frac{1}{72}\left[ab[\ln(\frac{a^2}{4}+b^2)-1]+\frac{a^3}{4b}+ 4\frac{b^3}{a}-\frac{a^4}{8b^2}\tan^{-1}(\frac{2b}{a})-\frac{8b^4}{a^2}\tan^{-1}(\frac{a}{2b})\right] \tag{12.4a}$$

$$D3_{21}^{(a)} = \frac{1}{48}\left[ab\ln(\frac{a^2}{4}+b^2)+\frac{3a^3}{20b}\ln(\frac{a^2+4b^2}{a^2})- \frac{19ab}{15}+\frac{16b^3}{5a}-\frac{32b^4}{5a^2}\tan^{-1}(\frac{a}{2b})\right] \tag{12.4b}$$

$$D3_{20}^{(a)} = \frac{1}{24}\left[ab\ln(\frac{a^2}{4}+b^2)-\frac{13ab}{6}+ \frac{2b^3}{a}+\frac{3a^2}{4}\tan^{-1}(\frac{2b}{a})-\frac{4b^4}{a^2}\tan^{-1}(\frac{a}{2b})\right] \tag{12.4c}$$

$$D3_{02}^{(a)} = \frac{1}{6}\left[ab\ln(\frac{a^2}{4}+b^2)+\frac{a^3}{8b}-\frac{13ab}{6} - \frac{a^4}{16b^2}\tan^{-1}(\frac{2b}{a})+3b^2\tan^{-1}(\frac{a}{2b})\right] \tag{12.4d}$$

$$D3_{01}^{(a)} = \frac{1}{2}\left[\frac{ab}{2}\ln(\frac{a^2}{4}+b^2)+\frac{a^3}{24b}\ln(\frac{a^2+4b^2}{a^2})-\frac{7ab}{6}+\frac{4b^2}{3}\tan^{-1}(\frac{a}{2b})\right] \tag{12.4e}$$

$$D3_{00}^{(a)} = \frac{1}{2}\left[ab\ln(\frac{a^2}{4}+b^2)+\frac{a^2}{2}\tan^{-1}(\frac{2b}{a})+2b^2\tan^{-1}(\frac{a}{2b})-3ab\right] \tag{12.4f}$$

The expressions of $D3_{ij}^{(b)}$ can readily be obtained from $D3_{ij}^{(a)}$ by simply interchanging a and b in the above expressions.

$$D4_{22} = \frac{1}{288}\left[ab\ln(\frac{a^2+b^2}{4})-ab + \frac{b^3}{a}+\frac{a^3}{b}-\frac{b^4}{a^2}\tan^{-1}(\frac{a}{b})-\frac{a^4}{b^2}\tan^{-1}(\frac{b}{a})\right] \tag{12.4g}$$

$$D4_{20} = \frac{1}{24}\left[ab\ln(\frac{a^2+b^2}{4})-\frac{13ab}{6}+\frac{b^3}{2a} - \frac{b^4}{2a^2}\tan^{-1}(\frac{a}{b})+\frac{3a^2}{2}\tan^{-1}(\frac{b}{a})\right] \tag{12.4h}$$

$$D4_{02} = \frac{1}{24}\left[ab\ln(\frac{a^2+b^2}{4})-\frac{13ab}{6}+\frac{a^3}{2b} + \frac{3b^2}{2}\tan^{-1}(\frac{a}{b})-\frac{a^4}{2b^2}\tan^{-1}(\frac{b}{a})\right] \tag{12.4i}$$

$$D4_{00} = \frac{1}{2}\left[ab\ln(\frac{a^2+b^2}{4}) - 3ab + b^2\tan^{-1}(\frac{a}{b}) + a^2\tan^{-1}(\frac{b}{a})\right] \tag{12.4j}$$

For the quadratic triangular element, two domain integrations are of interest. One is with reference to the source node located at one of the vertices, while the other is with reference to the source node on a side of the element. The first integration had earlier been considered in chapter 10 (i.e. eq. (10.34)), and the second is given by (see Figure 12.3)

$$D5_{ij} = \frac{1}{2l^{i+j}}\int_{-l/2}^{l/2} x^i dx \int_{\beta x}^{\mu x+(\mu-\beta)l/2} y^j \ln(x^2+y^2)\, dy \tag{12.5}$$

$$\begin{aligned} D5_{00} = \frac{l^2}{2}\Bigg[& \frac{1}{8}\left((3\mu-2\beta)+\frac{\mu(\mu-\beta)^2}{(1+\mu^2)}\right)\ln[1+(2\mu-\beta)^2] \\ & - \frac{1}{8}\left((\mu-2\beta)+\frac{\mu(\mu-\beta)^2}{(1+\mu^2)}\right)\ln(1+\beta^2) + \frac{\mu}{2}\ln(\frac{l}{2}) \quad \frac{3}{2}(\mu-\beta) \\ & + \frac{(\mu-\beta)^2}{4(1+\mu^2)}\tan^{-1}\left(\frac{2(\mu-\beta)}{\beta^2-1-2\mu\beta}\right) + \frac{1}{4}\tan^{-1}\left(\frac{2\mu}{\beta^2+1-2\mu\beta}\right)\Bigg] \end{aligned} \tag{12.6a}$$

$$\begin{aligned} D5_{10} = \frac{l^2}{2}\Bigg[& \frac{(\mu-\beta)}{6}\left(\frac{\mu(\mu-\beta)}{2(1+\mu^2)}-\frac{1}{3}\right) + \frac{1}{48}\left((5\mu-3\beta)+\frac{(1-\mu^2)(\mu-\beta)^3}{(1+\mu^2)^2}\right) \times \\ & - \ln[1+(2\mu-\beta)^2] - \frac{1}{48}\left((\mu+\beta)+\frac{(1-\mu^2)(\mu-\beta)^3}{(1+\mu^2)^2}\right)\ln(1+\beta^2) + \\ & - \frac{(\mu-\beta)}{6}\ln(\frac{l}{2}) + \frac{1}{12}\tan^{-1}\left(\frac{2\mu}{\beta^2+1-2\mu\beta}\right) - \\ & - \frac{(\mu-\beta)^3\mu}{12(1+\mu^2)^2}\tan^{-1}\left(\frac{2(\mu-\beta)}{\beta^2-1-2\mu\beta}\right)\Bigg] \end{aligned} \tag{12.6b)}$$

$$
\begin{aligned}
D5_{20} = \frac{l^2}{24}\Bigg[& -(\mu-\beta)\left(\frac{13}{12}+\frac{(\mu-\beta)^2(\mu^2-1)}{4(1+\mu^2)^2}\right)+\frac{1}{16}\left(7\mu-4\beta+\frac{(\mu-\beta)^4\mu(3-\mu^2)}{(1+\mu^2)^3}\right)\cdot \\
& \ln[1+(2\mu-\beta)^2]-\frac{1}{16}\left(4\beta-\mu+\frac{(\mu-\beta)^4\mu(3-\mu^2)}{(1+\mu^2)^3}\right)\ln(1+\beta^2)\ + \\
& (\mu-\beta)\ln(\frac{l}{2})+\frac{3}{8}\tan^{-1}\left(\frac{2\mu}{\beta^2+1-2\mu\beta}\right)+ \\
& \frac{(\mu-\beta)^4(1-3\mu^2)}{8(1+\mu^2)^3}\tan^{-1}\left(\frac{2(\mu-\beta)}{\beta^2-1-2\mu\beta}\right)\Bigg]
\end{aligned}
\tag{12.6c}
$$

$$
\begin{aligned}
D5_{01} = \frac{l^2}{4}\Bigg[& \frac{(\mu-\beta)}{18}\left(5(\beta-2\mu)-\frac{3(\mu-\beta)}{(1+\mu^2)}\right) \\
& +\frac{1}{24}\left(1+\mu^2+3(2\mu^2-3\mu\beta+\beta^2)+\frac{(\mu-\beta)^3(3\mu+\mu^3)}{(1+\mu^2)^2}\right)\ln[1+(2\mu-\beta)^2] \\
& -\frac{1}{24}\left(3+\mu^2+2\beta+3\beta(\mu-\beta)+\frac{(\mu-\beta)^3(3\mu+\mu^3)}{(1+\mu^2)^2}\right)\ln(1+\beta^2) \\
& +\frac{1}{3}(\mu-\beta)(2\mu-\beta)\ln(\frac{l}{2})+\frac{(\mu-\beta)^3}{6(1+\mu^2)^2}\tan^{-1}\left(\frac{2(\mu-\beta)}{\beta^2-1-2\mu\beta}\right)\Bigg]
\end{aligned}
\tag{12.6d}
$$

$$
\begin{aligned}
D5_{02} = \frac{l^2}{96}\Bigg\{ & -(\mu-\beta)\left(\frac{7\mu^2-3+7(\mu-\beta)^2}{3}+\frac{2(\mu-\beta)^2}{(1+\mu^2)^2}\right)+ \\
& \frac{1}{4}\left[\frac{(\mu-\beta)^4}{(1+\mu^2)^3}(\mu^5+3\mu^3+6\mu)+\mu^2(5\mu-4\beta)+(\mu-\beta)^2(10\mu-4\beta)\right]\times \\
& \ln[1+(2\mu-\beta)^2]-\frac{1}{4}\Bigg[\frac{(\mu-\beta)^4}{(1+\mu^2)^3}(\mu^5+3\mu^3+6\mu)+ \\
& \mu^2(4\beta-3\mu)+(\mu-\beta)^2(2\mu+4\beta)\Bigg]\ln(1+\beta^2)+ \\
& 4(\mu-\beta)[\mu^2+(\mu-\beta)^2]\ln(\frac{l}{2})-\frac{1}{2}\tan^{-1}\left(\frac{2\mu}{\beta^2+1-2\mu\beta}\right)+ \\
& \frac{(\mu-\beta)^4(\mu^3+3)}{2(1+\mu^2)^3}\tan^{-1}\left(\frac{2(\mu-\beta)}{\beta^2-1-2\mu\beta}\right)\Bigg\}
\end{aligned}
\tag{12.6e}
$$

$$
\begin{aligned}
D5_{11} = \frac{l^2}{96}\Bigg[& \mu(\mu-\beta)\left(\frac{(\mu-\beta)^2(\mu^2+5)}{2(1+\mu^2)^2} - \frac{17}{6}\right) + \\
& \frac{1}{8}\left(3(1+\mu^2)+2(\mu-\beta)(7\mu-3\beta) - \frac{(\mu-\beta)^4(\mu^4+6\mu^2-3)}{(1+\mu^2)^3}\right)\ln[1+(2\mu-\beta)^2] - \\
& \frac{1}{8}\left(3(1+\mu^2)-2(\mu-\beta)(\mu+3\beta) - \frac{(\mu-\beta)^4(\mu^4+6\mu^2-3)}{(1+\mu^2)^3}\right)\ln(1+\beta^2) + \\
& 4\mu(\mu-\beta)\ln\left(\frac{l}{2}\right) - \frac{2\mu(\mu-\beta)^4}{(1+\mu^2)^3}\tan^{-1}\left(\frac{2(\mu-\beta)}{\beta^2-1-2\mu\beta}\right)\Bigg]
\end{aligned}
\tag{12.6f}
$$

These exact expressions are utilized in deriving the expressions for the element matrices found in Appendix C.

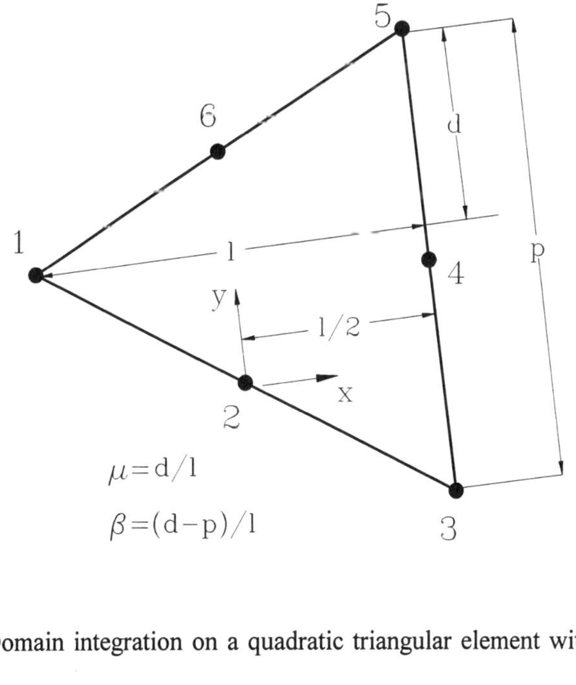

Figure 12.3 Domain integration on a quadratic triangular element with a mid-side source node.

12.4 Simulation of Example with Quadratic Elements

A simple example of flow governed by the Laplace equation is used to demonstrate how the accuracy of **GEM** is enhanced by the incorporation of quadratic elements. This example had been addressed in chapter 10 with linear **GEM**, and here the solutions from the quadratic model are compared with those of the linear model. The problem is governed by

$$\nabla^2 h = 0 \tag{12.7a}$$

with these conditions

$$\begin{aligned} h(0,y) &= 0; \quad \frac{\partial h}{\partial x}(0.5,y) = 0; \quad 0 \le y \le 0.5 \\ h(x,0) &= 0; \quad h(x,0.5) = \sin \pi x; \quad 0 \le x \le 0.5 \end{aligned} \tag{12.7b}$$

In chapter 10, sixteen linear elements were used to discretize the computational domain, whereas four quadratic elements are now used to achieve the same. Computing running time comparisons from the **GEM** simulations indicate the quadratic model offers about 15% savings over the linear model. The solutions from both simulations are shown in Table 12.1 in the form of the relative absolute error. It is observed that **GEM** with quadratic elements consistently has higher accuracy at all nodes than that with linear elements. This result is indicative of the enhancement in accuracy expected from incorporating higher-order elements in **GEM** calculations.

12.5 Isoparametric Elements

The concept of *isoparametric elements* is introduced and considered for **GEM** applications. An in-depth treatment of the subject can be found in these references: Zienkiewicz [1], Reddy [2], Rao [3], and Ottosen and Peterson [4], and Chung [5].

Isoparametric elements are those which use the interpolation polynomials to denote the variation of the dependent variables and the geometry of the element. That is

$$x = \sum_{j=1}^{n} \Omega_j x_j, \quad y = \sum_{j=1}^{n} \Omega_j y_j \text{ and } h = \sum_{j=1}^{n} \Omega_j h_j \tag{12.8}$$

in which $\Omega_j = \Omega_j(\zeta, \eta)$ are the interpolation or shape functions with respect to the local coordinates ζ and η. The concept of isoparametric elements finds profound usefulness in accurate representation of boundaries that are irregular. However, the use of such elements increase the computational effort involved in evaluating the boundary and domain integrations which provide the expressions for the element matrices. Such an effort involves using an invertible transformation between the isoparametric element Λ and a simple and regular element $\hat{\Lambda}$ (*master element*) over which integration can readily be done (see Figure 12.4). The general coordinate transformation can be considered

Table 10.2: Comparison of linear and Quadratic **GEM** solutions.

Position (x,y)	Solution type	Exact solution	Absolute relative error	
			Quadratic **GEM**	Linear **GEM**
(0,0)	$\partial h/\partial x$	1.2530	0.0017	0.0304
(0.125,0)	h	0.1525	0.0105	0.0400
(0.25,0)	h	0.2818	0.0163	0.0397
(0.375,0)	h	0.3590	0.0056	0.0072
(0.5,0)	h	0.3985	0.0196	0.0010
(0,0.125)	$\partial h/\partial x$	1.3498	0.0033	0.0215
(0.125,0.125)	h	0.1644	0.0024	0.0049
(0.25,0.125)	h	0.3038	0.0148	0.0053
(0.375,0.125)	h	0.3970	0.0184	0.0023
(0.5,0.125)	h	0.4297	0.0186	0.0358
(0,0.25)	$\partial h/\partial x$	1.6585	0.0027	0.0266
(0.125,0.25)	h	0.2020	0.0050	0.0094
(0.25,0.25)	h	0.3733	0.0099	0.0067
(0.375,0.25)	h	0.4877	0.0117	0.0133
(0.5,0.25)	h	0.5279	0.0108	0.0457
(0,0.375)	$\partial h/\partial x$	2.2262	0.0135	0.0452
(0.125,0.375)	h	0.2712	0.0077	0.0166
(0.25,0.375)	h	0.5011	0.0006	0.0146
(0.375,0.375)	h	0.6547	0.0020	0.0144
(0.5,0.375)	h	0.7086	0.0071	0.0435
(0,0.5)	$\partial h/\partial y$	3.1416	0.0086	0.0125
(0.125,0.5)	$\partial h/\partial y$	1.1026	0.0524	0.0525
(0.25,0.5)	$\partial h/\partial y$	2.0374	0.0179	0.0532
(0.375,0.5)	$\partial h/\partial y$	2.6620	0.0060	0.1255
(0.5,0.5)	$\partial h/\partial y$	2.8813	0.0031	0.1365

$$x = x(\zeta,\eta) \text{ and } y = y(\zeta,\eta) \tag{12.9}$$

The transformation maps every point (ζ,η) in the master element $\hat{\Lambda}$ unto a point (x,y) on the element Λ. Special cases of eq. (12.9) abound. A mapping, for instance, from a master element that is a square to a quadrilateral in the $x-y$ plane can be achieved by the bilinear interpolation polynomials which were defined in chapter 10.
That is

$$x = \sum_{j=1}^{4} \Omega_j(\zeta,\eta)x_j \text{ and } y = \sum_{j=1}^{4} \Omega_j(\zeta,\eta)y_j \tag{12.10}$$

$0 \le \zeta,\eta \le 1$ and $\Omega_j(\zeta,\eta)$ are the bilinear interpolation polynomials. When Ω_j are the quadratic interpolation polynomials, then the mapping transforms the master element into a curvilinear element. Thus, depending on the nature of Ω_j, an appropriate transformation can map a master element and any arbitrary element in the global coordinates. However, the choice of the transformation is dictated by the ease with which integrations are carried out within the element.

Figure 12.3 Isoparametric and master elements

Element calculations involve evaluation of domain and boundary integrations which give rise to the element coefficient matrices. These integrals are expressed in terms of the global coordinates x and y, and they have to be rewritten in terms of ζ and η (coordinates of the master element) so that they are easier to evaluate. In general, we are interested in evaluating an integral of the form

$$\iint_{\Lambda} F(x,y)\, dx\, dy \tag{12.11}$$

Applying the chain rule of differentiation to eq. (12.9), we obtain

$$dx = \frac{\partial x}{\partial \zeta} d\zeta + \frac{\partial x}{\partial \eta} d\eta \tag{12.12a}$$

and

$$dy = \frac{\partial y}{\partial \zeta} d\zeta + \frac{\partial y}{\partial \eta} d\eta \tag{12.12b}$$

Both equations are expressed in matrix form as

$$\begin{Bmatrix} dx \\ dy \end{Bmatrix} = \begin{bmatrix} \dfrac{\partial x}{\partial \zeta} & \dfrac{\partial x}{\partial \eta} \\ \dfrac{\partial y}{\partial \zeta} & \dfrac{\partial y}{\partial \eta} \end{bmatrix} \begin{Bmatrix} d\zeta \\ d\eta \end{Bmatrix} = [J] \begin{Bmatrix} d\zeta \\ d\eta \end{Bmatrix} \tag{12.13}$$

where $[J]$ is the Jacobian matrix of the transformation from the local coordinates ζ and η to the global coordinates x and y given by eq. (12.9). The inverse of the transformation exist when the determinant of $[J]$, denoted $\boldsymbol{J}$, is nonzero at every point ζ, η in $\hat{\Lambda}$. That is

$$\boldsymbol{J} = \det[J] = \frac{\partial x}{\partial \zeta}\frac{\partial y}{\partial \eta} - \frac{\partial x}{\partial \eta}\frac{\partial y}{\partial \zeta} \neq 0 \tag{12.14}$$

When eq. (12.14) holds, then

$$\begin{Bmatrix} \partial\zeta \\ \partial\eta \end{Bmatrix} = [J]^{-1} \begin{Bmatrix} dx \\ dy \end{Bmatrix} \tag{12.15}$$

and in that case, the following relationship applies

$$\zeta = \zeta(x,y) \text{ and } \eta = \eta(x,y) \tag{12.16}$$

The existence of the Jacobian matrix suggests that ζ and η be continuous and differentiable. When the interpolation polynomial functions are used for the transformation as in eq. (12.8), these conditions of continuity and differentiability of ζ and η are satisfied. In that case, the Jacobian matrix has the expression

$$[J] = \begin{bmatrix} \sum_{j=1}^{n} \frac{\partial\Omega_j}{\partial\zeta} x_j & \sum_{j=1}^{n} \frac{\partial\Omega_j}{\partial\eta} x_j \\ \sum_{j=1}^{n} \frac{\partial\Omega_j}{\partial\zeta} y_j & \sum_{j=1}^{n} \frac{\partial\Omega_j}{\partial\eta} y_j \end{bmatrix} \tag{12.17}$$

The elemental area

$$dA \equiv dx\,dy = \boldsymbol{J}\,d\zeta\,d\eta \tag{12.18}$$

The integral of eq. (12.11) now becomes

$$\iint_{\Lambda} F(x,y)\,dx\,dy \equiv \iint_{\hat{\Lambda}} F(\zeta,\eta)\,\boldsymbol{J}\,d\zeta\,d\eta \tag{12.19}$$

Eq. (12.19) gives the expression for calculating the integration in the local coordinates, which originally was in the global coordinates. It essentially involves computing the Jacobian, and it is for ease of that computation that the transformation from the global to the local coordinates has to be algebraically simple.

12.6 Three-Dimensional Problems

A cursory discussion of the application of **GEM** to three dimensional problems is presented. The additional spatial dimension over 2-D. problems does not alter the theory of **GEM**; it only increases the computational effort involved in evaluating the integrals which involve surface and volume integrals. The computational domain is discretized into three dimensional elements over which is prescribed a distribution for the dependent variables. The flow problem governed by the Poisson equation is used to illustrate **GEM** application in 3-D. It is given by

$$\nabla^2 h = f(x,y,z) \quad \text{on} \quad \Lambda \tag{12.20}$$

in which $\nabla^2 \equiv \partial^2/\partial x^2 + \partial^2/\partial y^2 + \partial^2/\partial z^2$ is, in Cartesian coordinates, the 3-D. Laplacian operator. The complimentary differential equation to eq. (12.20), which applies to an infinitely extensive domain, is

$$\nabla^2 G = \delta(r - r_i) \tag{12.21}$$

in which $r = (x,y,z)$ and $r_i = (x_i,y_i,z_i)$ are the field and source nodes. The solution to eq. (12.21) is

$$G \equiv G(r,r_i) = \frac{1}{|r - r_i|} \tag{12.22}$$

Applying Green's identity to eqs. (12.20) and (12.21) we have

$$\iiint_\Lambda [h\nabla^2 G - G\nabla^2 h]\, dV = \iint_\Gamma [h\nabla G - G\nabla h].\boldsymbol{n}\, dA$$

or

$$-\lambda h(r_i) + \iint_\Gamma (h\frac{\partial G}{\partial n} - G\frac{\partial h}{\partial n})\, dA + \iiint_\Lambda G f dV = 0 \tag{12.23}$$

in which λ is the solid angle at the source point r_i. That is

$\lambda = 4\pi$ if r_i is within Λ,

$\lambda = 2\pi$ if r_i is on Γ and Γ is on a smooth surface at r_i

$\lambda = \theta$ (interior solid angle at r_i) if r_i is on Γ and Γ is not on a smooth surface at r_i

Eq. (12.23) is the integral equation of the differential equation, and it is based on the singular boundary integral theory. The Green element method requires that the computational region be discretized by suitable elements. Some of the simplest elements that can be used are the cube and tetrahedron. Over these elements h and $\partial h/\partial n \equiv \varphi$ are approximated by the expression of the form of eq. (10.13). The reader may refer to Zienkiewicz[1] for detailed treatment of interpolation functions in 3-D. elements. Introducing the interpolation for h and φ into eq. (12.23) yields

$$\boxed{\sum_{e=1}^{M}\left(R_{ij}^{(e)}h_j + L_{ij}^{(e)}\varphi_j + T_{ij}^{(e)}f_j\right) = 0} \tag{12.24}$$

in which the element matrices are given

$$R_{ij}^{(e)} = \iint_{\Gamma^{(e)}} \frac{\partial G(r,r_i)}{\partial n}\Omega_j dA - \delta_{ij}\lambda^{(e)} \tag{12.25a}$$

$$L_{ij}^{(e)} = -\iint_{\Gamma^{(e)}} G(r,r_i)\Omega_j dA \tag{12.25b}$$

$$T_{ij}^{(e)} = \iiint_{\Lambda^{(e)}} G(r,r_i)\Omega_j dV \tag{12.25c}$$

The above element matrices are obtained by numerical evaluation of the surface and volume integrals.

12.7 References

1. Zienkiewicz, O.C., *The Finite Element Method*, 3rd ed., McGraw-Hill, UK, 1977.

2. Reddy, J.N., *An Introduction to the Finite Element Method*, McGraw-Hill, Inc. NY., 1984.

3. Rao, S.S., *The Finite Element Method in Engineering*, 2ed. Pergamon Press plc, Oxford, 1989.

4. Ottosen, N.S. and H. Petersson, *Introduction to the Finite Element Method*, Prentice Hall Int. Ltd. UK., 1992.

5. Chung, T.J., *Finite Element Analysis in Fluid Dynamics*, Mc-Graw-Hill, Inc, New York, 1978.

APPENDIX A

A.1 Capabilities of the Program GEMLN1D

This program solves in 1-D. Spatial dimension steady and transient problems that are governed by:

i. Heat conduction or diffusion equation
ii. Helmholtz equation
iii. Contaminant transport in uniform flow with first-order decay

Essentially, then, it handles the linear problems that are addressed in chapters 2 through 6 of this text.

A.2 User Manual of the Executable Program GEMLN1D

1. Title of the project or the problem or example (Maximum of 80 characters).
2. Is it a steady problem? (y for steady and n for unsteady)
3. Number of segments into which computational domain is divided.
4. X-coordinate of the first node of the computational domain.
5. X-coordinate of the last node of the computational domain.

Implement line 6 for as many number of segments

6. Number of elements generated in each segment.

Implement line 7 for the first and last node of the computational domain

7. Node type, Boundary data (Node type=1 for flux node; =2 for Dirichlet node).
8. Distribution of recharge data (= 0 for uniform data; = 1 for nonuniform data).
9. Strength of distributed recharge (one value for uniform recharge, i.e. when item 8 = 0; nodal values for nonuniform recharge, i.e. when item 8 = 1).
10. No. of point sources/sinks.

Implement line 11 for as many point sources/sinks

11. X-coordinate, Strength (negative value for sink & positive value for source).
12. Diffusivity or conductivity of the medium.
13. Characteristic parameter or eigenvalue (see chapter 4).
14. Uniform velocity of flow in the x-direction (see chapter 6).
15. Rate of first-order decay (see chapter 6).

Implement lines 16 through 26 for transient problems

16. Value of difference time weigthing factor (between 0 and 1).
17. Coefficient of temporal derivative term.
18. Value of initial time.
19. Number of divisions of time dimension (1 or 2 or 3, or ...).

Implement line 20 for as many as number of divisions of the time dimension.

20. Value of time step, Time at end of each division of the time dimension.
21. Distribution of initial data (= 0 for uniform data; = 1 for nonuniform data).
22. Value(s) of initial data (one value for uniform initial data, i.e. when item 21 = 0; nodal values for nonuniform initial data, i.e. when item 21 = 1).
23. Number of time steps to be skipped before output of solutions (Could take 0 or 1 or 2 or ... value, depending on how you want to limit output of results).

Implement line 24 for as many as number of divisions of the time dimension.

24. Should boundary data be updated in current division of time? (y if they are to be update, or n if they are not to be updated).
25. Number of nodes at which data should be updated.

Implement line 26 for as many as number of nodes at which data should be updated

26. Node number, New boundary data value.

A.3 Hint to Solution to some of the Exercises

The data files with the prefix EXER can be run with GEMLN1D to give solutions to some of the exercises. For instance, EXER2-4b.DAT is the data file for exercise 2.4b in the text. Furthermore, data files with the prefix EXMP, when run with GEMLN1D, reproduce the results of some of the examples in chapters 2 through 6. EXMP5-1.DAT, for instance, is the data file for example 1 in chapter 5.

EXER2-4B.DAT

```
Exercise 2.4b of Chapter 2
y  Steady problem=y, and unsteady=n
 1  No. of segments into which the domain is  divided
.0000E+00 x-coordinate of the first node
.1000E+01 x-coordinate of the last node
 4 Number of elements generated in segment  1
 2  .0000E+00 Node type for node  1;  Boundary data
 2  .0000E+00 Node type for node  5;  Boundary data
 0  Recharge data distribution = 0 (uniform); = 1 (nonuniform)
-.1000E+01 Strength of uniform distributed recharge
 0 No. of sources/sinks
.1000E+01 Diffusivity or Conductivity
.0000E+00 Characteristic parameter (chap. 4)
.0000E+00 Uniform velocity
.0000E+00 Rate of first-order decay
```

EXMP5-1.DAT

```
Example 1 of Chapter 5
n  Steady problem=y, and unsteady=n
 1  No. of segments into which the domain is  divided
.0000E+00 x-coordinate of the first node
.1000E+01 x-coordinate of the last node
10 Number of elements generated in segment  1
 2  .1000E+01 Node type for node  1;  Boundary data
 2  .0000E+00 Node type for node  11;  Boundary data
 0  Recharge data distribution = 0 (uniform); = 1 (nonuniform)
.0000E+00 Strength of uniform distributed recharge
 0 No. of sources/sinks
.1000E+01 Diffusivity or Conductivity
.0000E+00 Characteristic parameter (chap. 4)
.0000E+00 Uniform velocity
.0000E+00 Rate of first-order decay
.1000E+01 Difference time weigthing factor
.1000E+01 Coefficient of temporal derivative
.0000E+00 Initial time
 1  No. of divisions of time dimension
.5000E-01  .1000E+01  Value of time step Time at end of subdivision  1
 1 Initial data distribution = 0 (uniform); = 1 (nonuniform)
1. 0. 0. 0. 0. 0. 0. 0. 0. 0. 0.   Value of uniform initial data
 1 No. time steps skipped in time interval  1
n  To update data=y, not to=n
```

APPENDIX B

The expressions for some of the element matrices of the Green element formulation for the steady 2-D. problem encountered in chapter 10 are presented here for linear rectangular and triangular elements. The element matrices are:

$$R_{ij}^{(e)} = \int_{\Gamma^{(e)}} \frac{\partial G(r,r_i)}{\partial n} \Omega_j \, ds - \delta_{ij} \lambda_i^{(e)} \tag{B.1}$$

$$L_{ij}^{(e)} = -\int_{\Gamma^{(e)}} G(r,r_i) \Omega_j \, ds \tag{B.2}$$

$$T_{ij}^{(e)} = \iint_{\Lambda^{(e)}} G(r,r_i) \Omega_j \, dA \tag{B.3}$$

We have utilized the exact expressions of the boundary integrals given by eqs. (10.31a - f) and the domain integrals given by eqs. (10.33a - i) and (10.35a - c) in obtaining the expressions for the element matrices. (Hereafter the superscript e is omitted.)

B.1 Element Matrices for Linear Rectangular Elements

The element matrices for the linear rectangular element shown in Fig. 10.3a are:

$$R_{11} = -\pi/2 \tag{B.4}$$

$$R_{12} = -\frac{a}{2b}\ln\left(\frac{a^2+b^2}{a^2}\right) + \tan^{-1}\left(\frac{b}{a}\right) \tag{B.5}$$

$$R_{13} = \frac{a}{2b}\ln(\frac{a^2+b^2}{a^2}) + \frac{b}{2a}\ln(\frac{a^2+b^2}{b^2}) \tag{B.6}$$

$$R_{14} = -\frac{b}{2a}\ln(\frac{a^2+b^2}{b^2}) + \tan^{-1}(\frac{a}{b}) \tag{B.7}$$

The element matrix L_{ij} that accounts for the coefficients of the flux term needs to be split into two parts: L_{ij}^b and L_{ij}^f to account, respectively, for the flux before and after the node, when the sides of the element are traversed in an anti-clockwise direction.

$$L_{11}^b = -b(\ln b^2 - 3)/4 \tag{B.8}$$

$$L_{12}^b = -a(\ln a^2 - 1)/4 \tag{B.9}$$

$$L_{13}^b = -\frac{1}{4}\left[b\ln(a^2+b^2) + \frac{a^2}{b}\ln(\frac{a^2+b^2}{a^2}) - b\right] \tag{B.10}$$

$$L_{14}^b = -\frac{1}{4}\left[a\ln(a^2+b^2) - \frac{b^2}{a}\ln(\frac{a^2+b^2}{b^2}) - 3a + 4b\tan^{-1}(\frac{a}{b})\right] \tag{B.11}$$

$$L_{11}^f = -a(\ln a^2 - 3)/4 \tag{B.12}$$

$$L_{12}^f = -\frac{1}{4}\left[b\ln(a^2+b^2) - \frac{a^2}{b}\ln(\frac{a^2+b^2}{a^2}) - 3b + 4a\tan^{-1}(\frac{b}{a})\right] \tag{B.13}$$

$$L_{13}^f = -\frac{1}{4}\left[a\ln(a^2+b^2) + \frac{b^2}{a}\ln(\frac{a^2+b^2}{b^2}) - a\right] \tag{B.14}$$

$$L_{14}^f = -b(\ln b^2 - 1)/4 \tag{B.15}$$

$$T_{11} = DI_{11} - DI_{10} - DI_{01} + DI_{00} \tag{B.16}$$

$$T_{12} = DI_{10} - DI_{11} \tag{B.17}$$

$$T_{13} = DI_{11} \tag{B.18}$$

$$T_{14} = DI_{01} - DI_{11} \tag{B.19}$$

B.2 Element Matrices for Linear Triangular Elements

The element matrices for the linear triangular element shown in Fig. 10.4a are:

$$R_{11} = -(\tan^{-1}\mu - \tan^{-1}\beta) \tag{B.20}$$

$$R_{12} = \frac{1}{(\mu-\beta)}\left[\mu(\tan^{-1}\mu - \tan^{-1}\beta) - \frac{\ln(1+\mu^2)\ \ln(1+\beta^2)}{2}\right] \tag{B.21}$$

$$R_{13} = \frac{1}{(\mu-\beta)}\left[-\beta(\tan^{-1}\mu - \tan^{-1}\beta) + \frac{\ln(1+\mu^2) - \ln(1+\beta^2)}{2}\right] \tag{B.22}$$

$$L_{11}^{b} = -\frac{l\sqrt{1+\mu^2}}{4}[2\ln l + \ln(1+\mu^2) - 3] \tag{B.23}$$

$$L_{12}^{b} = -\frac{l\sqrt{1+\beta^2}}{4}[2\ln l + \ln(1+\beta^2) - 1] \tag{B.24}$$

$$L_{13}^{b} = -\frac{l}{(\mu-\beta)}[\frac{(\mu-\beta)^2}{2}\ln l + \frac{(\mu^2+1-2\mu\beta)\ln(1+\mu^2) + (\beta^2-1)\ln(1+\beta^2)}{4} + \frac{(4\mu\beta-3\beta^2-\mu^2)}{4} - \beta(\tan^{-1}\mu - \tan^{-1}\beta)] \tag{B.25}$$

$$L_{11}^{f} = -\frac{l\sqrt{1+\beta^2}}{4}[2\ln l + \ln(1+\beta^2) - 3] \tag{B.26}$$

$$L_{12}^{f} = -\frac{l}{(\mu-\beta)}[\frac{(\mu-\beta)^2}{2}\ln l + \frac{(\mu^2-1)\ln(1+\mu^2) + (1+\beta^2-2\mu\beta)\ln(1+\beta^2)}{4} + \frac{(4\mu\beta-3\mu^2-\beta^2)}{4} + \mu(\tan^{-1}\mu - \tan^{-1}\beta)] \tag{B.27}$$

$$L_{13}^{f} = -\frac{l\sqrt{1+\mu^2}}{2}[2\ln l + \ln(1+\mu^2) - 1] \tag{B.28}$$

$$T_{11} = D2_{00} - D2_{10} \tag{B.29}$$

$$T_{12} = \frac{(\mu D2_{10} - D2_{01})}{\mu-\beta} \tag{B.30}$$

$$T_{13} = D2_{00} - T_{11} - T_{12} \tag{B.31}$$

APPENDIX C

The expressions for some of the element matrices associated with quadratic rectangular and triangular elements encountered in chapter 12 are presented here. The element matrices are:

$$R_{ij}^{(e)} = \int_{\Gamma^{(e)}} \frac{\partial G(r,r_i)}{\partial n} \Omega_j ds - \delta_{ij}\lambda_i^{(e)} \qquad \text{(C.1)}$$

$$L_{ij}^{(e)} = -\int_{\Gamma^{(e)}} G(r,r_i)\Omega_j ds \qquad \text{(C.2)}$$

$$T_{ij}^{(e)} = \iint_{\Lambda^{(e)}} G(r,r_i)\Omega_j dA \qquad \text{(C.3)}$$

C.1 Element Matrices for Quadratic Rectangular Elements

The element matrices for the quadratic rectangular element shown in Fig. 12.1 are:

$$R_{11} = -\pi/2 \qquad \text{(C.4)}$$

$$R_{12} = R_{18} = 0 \qquad \text{(C.5)}$$

$$R_{13} = (1-\frac{2a^2}{b^2})\tan^{-1}(\frac{b}{a})+\frac{2a}{b}-\frac{3a}{2b}\ln(\frac{a^2+b^2}{a^2}) \qquad \text{(C.6)}$$

$$R_{14} = \frac{4a}{b}\left[\frac{1}{2}\ln\left(\frac{a^2+b^2}{a^2}\right) - 1 + \frac{a}{b}\tan^{-1}\left(\frac{b}{a}\right)\right] \tag{C.7}$$

$$\begin{aligned} R_{15} = & -\frac{a}{2b}\ln\left(\frac{a^2+b^2}{a^2}\right) - \frac{b}{2a}\ln\left(\frac{a^2+b^2}{b^2}\right) + 2\left(\frac{a}{b} + \frac{b}{a}\right) - \\ & 2\left(\frac{a^2}{b^2}\tan^{-1}\left(\frac{b}{a}\right) + \frac{b^2}{a^2}\tan^{-1}\left(\frac{a}{b}\right)\right) \end{aligned} \tag{C.8}$$

$$R_{16} = \frac{4b}{a}\left(\frac{1}{2}\ln\left(\frac{a^2+b^2}{b^2}\right) - 1 + \frac{b}{a}\tan^{-1}\left(\frac{a}{b}\right)\right) \tag{C.9}$$

$$R_{17} = \left(1 - \frac{2b^2}{a^2}\right)\tan^{-1}\left(\frac{a}{b}\right) + \frac{2b}{a} - \frac{3b}{2a}\ln\left(\frac{a^2+b^2}{b^2}\right) \tag{C.10}$$

$$R_{21} = R_{23} = \left(1 - \frac{a^2}{2b^2}\right)\tan^{-1}\left(\frac{2b}{a}\right) + \frac{a}{b} - \frac{3a}{4b}\ln\left(\frac{a^2+4b^2}{a^2}\right) \tag{C.11}$$

$$R_{22} = -\pi \tag{C.12}$$

$$R_{24} = R_{28} = \frac{a}{b}\left[\ln\left(\frac{a^2+4b^2}{a^2}\right) - 2 + \frac{a}{b}\tan^{-1}\left(\frac{2b}{a}\right)\right] \tag{C.13}$$

$$\begin{aligned} R_{25} = R_{27} = & -\frac{a}{4b}\ln\left(\frac{a^2+4b^2}{a^2}\right) - \frac{b}{a}\ln\left(\frac{4a^2+b^2}{b^2}\right) + \frac{a}{b} + \frac{2b}{a} \\ & - \frac{a^2}{2b^2}\tan^{-1}\left(\frac{2b}{a}\right) - \frac{4b^2}{a^2}\tan^{-1}\left(\frac{a}{2b}\right) \end{aligned} \tag{C.14}$$

$$R_{26} = -\frac{4b}{a} - \left(\frac{8b^2}{a^2} + 2\right)\tan^{-1}\left(\frac{a}{2b}\right) \tag{C.15}$$

$$R_{91} = R_{93} = R_{95} = R_{97} = \frac{b}{a}\left(1 - \frac{b}{a}\tan^{-1}\left(\frac{a}{b}\right)\right) + \frac{a}{b}\left(1 - \frac{a}{b}\tan^{-1}\left(\frac{b}{a}\right)\right) \tag{C.16}$$

$$R_{92} = R_{96} = -\frac{2b}{a} + 2\left(1 - \frac{b}{a}\tan^{-1}\left(\frac{a}{b}\right)\right) \tag{C.17}$$

$$R_{94} = R_{98} = -\frac{2a}{b} + 2\left(1 - \frac{a}{b}\tan^{-1}\left(\frac{b}{a}\right)\right) \tag{C.18}$$

$$R_{99} = -2\pi \tag{C.19}$$

$$L_{11}^{b} = -\frac{b}{12}\left(lnb^{2} - \frac{17}{3}\right) \tag{C.20}$$

$$L_{12}^{b} = L_{12}^{f} = -\frac{a}{3}\left(\ln a^{2} - \frac{5}{3}\right) \tag{C.21}$$

$$L_{13}^{b} = -\frac{a}{12}\left(\ln a^{2} + \frac{1}{3}\right) \tag{C.22}$$

$$L_{14}^{b} = L_{14}^{f} = -b\left[\frac{1}{3}\ln(a^{2}+b^{2}) + \frac{a^{2}}{b^{2}}\ln(\frac{a^{2}+b^{2}}{a^{2}}) - \frac{5}{9} - \frac{4a^{2}}{3b^{2}} + \frac{4a^{3}}{3b^{3}}\tan^{-1}(\frac{b}{a})\right] \tag{C.23}$$

$$L_{15}^{b} = -b\left[\frac{1}{12}\ln(a^{2}+b^{2}) - \frac{a^{2}}{4b^{2}}\ln(\frac{a^{2}+b^{2}}{b^{2}}) + \frac{1}{36} + \frac{2a^{2}}{b^{2}} - \frac{2a^{3}}{3b^{3}}\tan^{-1}(\frac{b}{a})\right] \tag{C.24}$$

$$L_{16}^{b} = L_{16}^{f} = -a\left[\frac{1}{3}\ln(a^{2}+b^{2}) + \frac{b^{2}}{a^{2}}\ln(\frac{a^{2}+b^{2}}{b^{2}}) - \frac{5}{9} - \frac{4b^{2}}{3a^{2}} + \frac{4b^{3}}{3a^{3}}\tan^{-1}(\frac{a}{b})\right] \tag{C.25}$$

$$L_{17}^{b} = -a\left[\frac{1}{12}\ln(a^{2}+b^{2}) - \frac{3b^{2}}{4a^{2}}\ln(\frac{a^{2}+b^{2}}{b^{2}}) - \frac{17}{36} + \frac{2b^{2}}{3a^{2}} + \frac{b}{a}(1 - \frac{2b^{2}}{3a2})\tan^{-1}(\frac{a}{b})\right] \tag{C.26}$$

$$L_{18}^{b} = L_{18}^{f} = -\frac{b}{3}\left(\ln b^{2} - \frac{5}{3}\right) \tag{C.27}$$

$$L_{11}^{f} = -\frac{a}{12}\left(\ln a^{2} - \frac{17}{3}\right) \tag{C.28}$$

$$L_{13}^{f} = -b\left[\frac{1}{12}\ln(a^{2}+b^{2}) - \frac{3a^{2}}{4b^{2}}\ln(\frac{a^{2}+b^{2}}{a^{2}}) - \frac{17}{36} + \frac{2a^{2}}{3b^{2}} + \frac{a}{b}(1 - \frac{2a^{2}}{3b^{2}})\tan^{-1}(\frac{b}{a})\right] \tag{C.29}$$

$$L_{15}^{f} = -a\left[\frac{1}{12}\ln(a^{2}+b^{2}) - \frac{b^{2}}{4a^{2}}\ln(\frac{a^{2}+b^{2}}{a^{2}}) + \frac{1}{36} + \frac{2b^{2}}{a^{2}} - \frac{2b^{3}}{3a^{3}}\tan^{-1}(\frac{a}{b})\right] \tag{C.30}$$

$$L_{17}^{f} = -\frac{b}{12}\left(\ln b^2 + \frac{1}{3}\right) \tag{C.31}$$

$$L_{21}^{b} = L_{23}^{f} = -b\left[\frac{1}{12}\ln(\frac{a^2}{4}+b^2) - \frac{3a^2}{16b^2}\ln(\frac{a^2+4b^2}{a^2}) - \frac{17}{36} + \frac{a^2}{6b^2} + \frac{a}{2b}(1-\frac{a^2}{6b^2})\tan^{-1}(\frac{2b}{a})\right] \tag{C.32}$$

$$L_{22}^{b} = L_{22}^{f} = -\frac{a}{3}\left(\ln\frac{a^2}{4} - \frac{8}{3}\right) \tag{C.33}$$

$$L_{23}^{b} = L_{21}^{f} = -\frac{a}{12}\left(\ln\frac{a^2}{4} - \frac{2}{3}\right) \tag{C.34}$$

$$L_{24}^{b} = L_{24}^{f} = R_{28}^{b} = R_{28}^{f} = -b\left[\frac{1}{32}\ln(\frac{a^2}{4}+b^2) - \frac{a^2}{4b^2}\ln(\frac{a^2+4b^2}{a^2}) - \frac{5}{9} - \frac{a^2}{3b^2} + \frac{a^3}{6b^3}\tan^{-1}(\frac{2b}{a})\right] \tag{C.35}$$

$$L_{27}^{b} = L_{25}^{f} = -\frac{a}{12}\left(\ln(\frac{a^2}{4}+b^2) - \frac{2}{3} + \frac{8b^2}{a^2}[1 - \frac{2b}{a}\tan^{-1}(\frac{a}{2b})]\right) \tag{C.36}$$

$$L_{25}^{b} = R_{27}^{f} = -b\left[\frac{1}{12}\ln(\frac{a^2}{4}+b^2) - \frac{a^2}{16b^2}\ln(\frac{a^2+4b^2}{a^2}) + \frac{1}{36} + \frac{a^2}{6b^2} - \frac{a^3}{12b^3}\tan^{-1}(\frac{2b}{a})\right] \tag{C.37}$$

$$L_{26}^{b} = L_{26}^{f} = -a\left[\frac{1}{3}\ln(\frac{a^2}{4}+b^2) - \frac{4b^2}{3a^2} - \frac{8}{9} + \frac{2b}{a}(1 + \frac{4b^2}{3a^2})\tan^{-1}(\frac{a}{b})\right] \tag{C.38}$$

$$L_{91}^{b} = L_{95}^{b} = L_{93}^{f} = L_{97}^{f} = -\frac{b}{12}\left(\ln(\frac{a^2+b^2}{4}) - \frac{2}{3} + \frac{2a^2}{b^2}[1 - \frac{a}{b}\tan^{-1}(\frac{b}{a})]\right) \tag{C.39}$$

$$L_{92}^{b} = L_{92}^{f} = L_{96}^{b} = L_{96}^{f} = -a\left(\frac{1}{3}\ln\left(\frac{a^2+b^2}{4}\right)-\frac{8}{9}-\frac{b^2}{3a^2}+\frac{b}{a}\left[1+\frac{b^2}{3a^2}\right]\tan^{-1}\left(\frac{a}{b}\right)\right) \tag{C.40}$$

$$L_{94}^{b} = L_{94}^{f} = L_{98}^{b} = L_{98}^{f} = -b\left(\frac{1}{3}\ln\left(\frac{a^2+b^2}{4}\right)-\frac{8}{9}-\frac{a^2}{3b^2}+\frac{a}{b}\left[1+\frac{a^2}{3b^2}\right]\tan^{-1}\left(\frac{b}{a}\right)\right) \tag{C.41}$$

$$L_{91}^{f} = L_{93}^{b} = L_{95}^{f} = L_{97}^{b} = -\frac{a}{12}\left(\ln\left(\frac{a^2+b^2}{4}\right)-\frac{2}{3}+\frac{2b^2}{a^2}\left[1-\frac{b}{a}\tan^{-1}\left(\frac{a}{b}\right)\right]\right) \tag{C.42}$$

$$T_{11} = 4DI_{22}-6DI_{21}+2DI_{20}+6DI_{12}+9DI_{11}-3DI_{10}+2DI_{02}-3DI_{01}+DI_{00} \tag{C.43}$$

$$T_{12} = 4\left(-2DI_{22}+3DI_{21}-DI_{20}+2DI_{12}-3DI_{11}+DI_{10}\right) \tag{C.44}$$

$$T_{13} = 4DI_{22}-6DI_{21}+2DI_{20}-2DI_{12}+3DI_{11}-DI_{10} \tag{C.45}$$

$$T_{14} = 4\left(-2DI_{22}+2DI_{21}+DI_{12}-DI_{11}\right) \tag{C.46}$$

$$T_{15} = 4DI_{22}-2DI_{21}-2DI_{12}+DI_{11} \tag{C.47}$$

$$T_{16} = 4\left(-2DI_{22}+DI_{21}+2DI_{12}-DI_{11}\right) \tag{C.48}$$

$$T_{17} = 4DI_{22}-2DI_{21}-6DI_{12}+3DI_{11}-DI_{01}+2DI_{02} \tag{C.49}$$

$$T_{18} = 4\left(-2DI_{22}+2DI_{21}-DI_{02}+3DI_{12}-3DI_{11}+DI_{01}\right) \tag{C.50}$$

$$T_{19} = 16\left(DI_{22}-DI_{21}-DI_{12}+DI_{11}\right) \tag{C.51}$$

$$T_{21} = T_{23} = 4D3_{22}^{(a)}-6D3_{21}^{(a)}+2D3_{20}^{(a)} \tag{C.52}$$

$$T_{22} = -8D3_{22}^{(a)}+12D3_{21}^{(a)}-4D3_{20}^{(a)}+2D3_{02}^{(a)}-3D3_{01}^{(a)}+D3_{00}^{(a)} \tag{C.53}$$

$$T_{24} = T_{28} = -8D3_{22}^{(a)}+8D3_{21}^{(a)} \tag{C.54}$$

$$T_{25} = T_{27} = 4D3_{22}^{(a)} - 2D3_{21}^{(a)} \tag{C.55}$$

$$T_{26} = -8D3_{22}^{(a)} + 4D3_{21}^{(a)} + 2D3_{02}^{(a)} - D3_{01}^{(a)} \tag{C.56}$$

$$T_{29} = 4(4D3_{22}^{(a)} - 4D3_{21}^{(a)} - D3_{02}^{(a)} + 3D3_{01}^{(a)}) \tag{C.57}$$

$$T_{91} = T_{93} = T_{95} = T_{97} = 4D4_{22} \tag{C.58}$$

$$T_{92} = T_{96} = -8D4_{22} + 2D4_{02} \tag{C.59}$$

$$T_{94} = T_{98} = -8D4_{22} + 2D4_{20} \tag{C.60}$$

$$T_{99} = 16D4_{22} - 4D4_{20} - 4D4_{02} + D4_{00} \tag{C.61}$$

C.2 Element Matrices for Quadratic Triangular Elements

The element matrices for the quadratic triangular element shown in Fig. 12.2 are:

$$R_{11} = -(\tan^{-1}\mu - \tan^{-1}\beta) \tag{C.62}$$

$$R_{12} = R_{16} = 0 \tag{C.63}$$

$$\begin{aligned} R_{13} = \frac{1}{(\mu-\beta)^2}[&\frac{(5\mu-\beta)}{2}(\ln(1+\beta^2) - \ln(1+\mu^2)) \\ &(3\mu^2 - \mu\beta - 2)(\tan^{-1}\mu - \tan^{-1}\beta) + 2(\mu-\beta)] \end{aligned} \tag{C.64}$$

$$\begin{aligned} R_{14} = \frac{1}{(\mu-\beta)^2}[&\frac{-(\mu+\beta)}{2}(\ln(1+\beta^2) - \ln(1+\mu^2)) + \\ &(1-\mu\beta)(\tan^{-1}\mu - \tan^{-1}\beta) - (\mu-\beta)] \end{aligned} \tag{C.65}$$

$$\begin{aligned} R_{15} = \frac{1}{(\mu-\beta)^2}[&\frac{-(3\mu+\beta)}{2}(\ln(1+\beta^2) - \ln(1+\mu^2)) + \\ &(2-\mu\beta-\mu^2)(\tan^{-1}\mu - \tan^{-1}\beta) - 2(\mu-\beta)] \end{aligned} \tag{C.66}$$

$$L_{11}^{f} = -\frac{l\sqrt{1+\beta^2}}{12}\left(2\ln(l\sqrt{1+\beta^2}) - \frac{17}{3}\right) \tag{C.67}$$

$$L_{12}^{b} = L_{12}^{f} = -\frac{l\sqrt{1+\beta^2}}{3}\left(2\ln(l\sqrt{1+\beta^2}) - \frac{5}{3}\right) \tag{C.68}$$

$$L_{13}^{b} = -\frac{l\sqrt{1+\beta^2}}{12}\left(2\ln(l\sqrt{1+\beta^2}) + \frac{1}{3}\right) \tag{C.69}$$

$$L_{13}^{f} = -\frac{1}{(\mu-\beta)^2}\left[2l^2 B2_2 - (5\mu-\beta)lB2_1 + (3\mu-\beta)B2_0\right] \tag{C.70}$$

$$L_{14}^{b} = L_{14}^{f} = -\frac{4}{(\mu-\beta)^2}\left[(-l^2 B2_2 - (\mu+\beta)lB2_1 - \mu\beta B2_0\right] \tag{C.71}$$

$$L_{15}^{b} = -\frac{1}{(\mu-\beta)^2}\left[-2l^2 B2_2 + (3\mu+\beta)lB2_1 - \mu(\mu+\beta)B2_0\right] \tag{C.72}$$

in which $B2_j$, $j=0,1,2$ are obtained from eqs. (10.31d - f), where $\eta_i = l$, $\zeta_1 = \beta l$ and $\zeta_2 = \mu l$ (see Fig. C.1a).

$$L_{15}^{f} = -\frac{l\sqrt{1+\mu^2}}{12}\left(2\ln(l\sqrt{1+\mu^2}) + \frac{1}{3}\right) \tag{C.73}$$

$$L_{16}^{b} = L_{16}^{f} = -\frac{l\sqrt{1+\mu^2}}{3}\left(2\ln(l\sqrt{1+\mu^2}) - \frac{5}{3}\right) \tag{C.74}$$

$$L_{11}^{b} = -\frac{l\sqrt{1+\alpha^2}}{3}\left(2\ln(l\sqrt{1+\alpha^2}) - \frac{17}{3}\right) \tag{C.75}$$

$$T_{11} = 2D2_{20} - 3D2_{10} + D2_{00} \tag{C.76}$$

$$T_{12} = \frac{4}{(\mu-\beta)}[\mu(D2_{10} + D2_{10}) - D2_{01} + D2_{11}] \tag{C.77}$$

$$T_{13} = \frac{1}{(\mu-\beta)}[-\mu D2_{10} + D2_{01} + \frac{2}{(\mu-\beta)}(\mu^2 D2_{20} - 2\mu D2_{11} + D2_{02})] \tag{C.78}$$

$$T_{14} = -\frac{4}{(\mu-\beta)^2}[\mu\beta D2_{20} - (\mu+\beta)D2_{11} + D2_{02}] \tag{C.79}$$

$$T_{15} = \frac{1}{(\mu-\beta)}[\beta D2_{10} - D2_{01} + \frac{2}{(\mu-\beta)}(\beta^2 D2_{20} + D2_{02} - 2\beta D2_{11})] \tag{C.80}$$

$$T_{16} = -\frac{4}{(\mu-\beta)}[\beta D2_{10} - \beta D2_{20} - D2_{01} + D2_{11}] \tag{C.81}$$

$$R_{22} = -\pi \tag{C.82}$$

$$R_{23} = \frac{1}{p^2}[2B1_2 - (4d+p)B1_1 + d(2d+p)B1_0] \tag{C.83}$$

$$R_{24} = \frac{4}{p^2}[-B1_2 - (2d-p)B1_1 + d(p-d)B1_0] \tag{C.84}$$

$$R_{25}^b = \frac{1}{p^2}[-2B1_2 + (4d-p)B1_1 + d(p-2d)B1_0] \tag{C.85}$$

in which $B1_j$, j=0,1,2 are obtained from eqs. (10.31a - c), where (see Fig. C.1b)

$$\zeta_2 = \frac{(2\alpha-\beta)l}{2}; \quad \zeta_1 = \frac{\beta l}{2}; \quad p = (\alpha-\beta)l; \quad d = \mu l; \quad \eta_i = \frac{l}{2} \tag{C.86}$$

$$R_{25}^f = \frac{1}{p^2}[2B1_2 - (4d+p)B1_1 + d(2d+p)B1_0] \tag{C.87}$$

$$R_{26} = \frac{4}{p^2}[-B1_2 - (2d-p)B1_1 + d(p-d)B1_0] \tag{C.88}$$

$$R_{21} = \frac{1}{p^2}[-2B1_2 + (4d-p)B1_1 + d(p-2d)B1_0] \tag{C.89}$$

$$R_{25} = R_{25}^b + R_{25}^f \tag{C.90}$$

in which $B1_j$, j=0,1,2 are obtained from eqs. (10.31a - c), where (see Fig. C.1c)

$$d = \frac{(1+\mu\beta)l}{2\sqrt{1+\mu^2}}; \quad p = l(\sqrt{1+\mu^2}); \quad \eta_i = \frac{(\mu-\beta)l}{2\sqrt{1+\mu^2}}; \quad \zeta_2 = d; \quad \zeta_1 = \frac{l(\mu\beta-1-2\mu^2)}{2\sqrt{1+\mu^2}} \tag{C.91}$$

$$L_{21}^f = R_{23}^b = -\frac{l\sqrt{1+\beta^2}}{6}[\ln(\frac{l\sqrt{1+\beta^2}}{2}) - \frac{1}{3}] \tag{C.92}$$

$$L_{22}^f = L_{22}^b = -\frac{2l\sqrt{1+\beta^2}}{3}[\ln(\frac{l\sqrt{1+\beta^2}}{2}) - \frac{4}{3}] \tag{C.93}$$

$$L_{23}^f = -\frac{1}{p^2}[2B2_2 - (4d+p)B2_1 + d(2d+p)B2_0] \tag{C.94}$$

$$L_{24}^b = L_{24}^f = -\frac{4}{p^2}[-B2_2 - (2d-p)B2_1 + d(p-d)B2_0] \tag{C.95}$$

$$L_{25}^b = -\frac{1}{p^2}[-2B2_2 + (4d-p)B2_1 + d(p-2d)B2_0] \tag{C.96}$$

in which $B2_j$, j=0,1,2 are obtained from eqs. (10.31d - f), and d, p, η_i, ζ_1 and ζ_2 are given by eq. (C.86) (see Fig. C.1b).

$$L_{25}^f = -\frac{1}{p^2}[2B2_2 - (4d+p)B2_1 + d(2d+p)B2_0] \tag{C.97}$$

$$L_{26}^b = L_{26}^f = -\frac{4}{p^2}[-B2_2 - (2d-p)B2_1 + d(p-d)B2_0] \tag{C.98}$$

$$L_{21}^b = -\frac{1}{p^2}[-2B2_2 + (4d-p)B2_1 + d(p-2d)B2_0] \tag{C.99}$$

in which $B2_j$, j=0,1,2 are obtained from eqs. (10.27d - f), and d, p, η_i, ζ_1 and ζ_2 are given by eq. (C.91) (see Fig. C.1c).

$$T_{21} = -D5_{01} + 2D5_{20} \tag{C.100}$$

$$T_{22} = \frac{4}{(\mu-\beta)}[\frac{\beta}{2}D5_{10} - \frac{D5_{01}}{2} - \mu D5_{20} + D5_{11} + \frac{(\mu-\beta)}{4}D5_{00}] \tag{C.101}$$

$$T_{23} = \frac{2}{(\mu-\beta)^2}[\frac{(\mu-\beta)}{2}(\mu D5_{10} - D5_{01}) + \mu(\mu D5_{20} - 2D5_{11}) + D5_{02}] \tag{C.102}$$

$$T_{24} = \frac{4}{(\mu-\beta)^2}[\frac{(\mu-\beta)}{2}(-\beta D5_{10} + D5_{01}) - \beta\mu D5_{20} + (\mu+\beta)D5_{11} - D5_{02}] \tag{C.103}$$

$$T_{25} = \frac{1}{(\mu-\beta)^2}[(\mu-\beta)(\beta D5_{10} - D5_{01}) + 2\beta(\beta D5_{20} - 2D5_{11}) + 2D5_{02}] \qquad (C.104)$$

$$T_{26} = -\frac{4}{(\mu-\beta)}[\frac{\beta}{2}D5_{10} - \frac{D5_{01}}{2} - \beta D5_{20} + D5_{11}] \qquad (C.105)$$

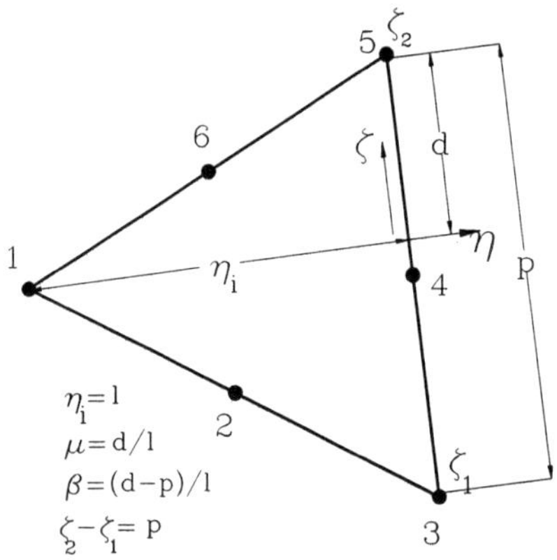

Figure C.1a Source node at 1 and integration on segment with nodes 3,4,5.

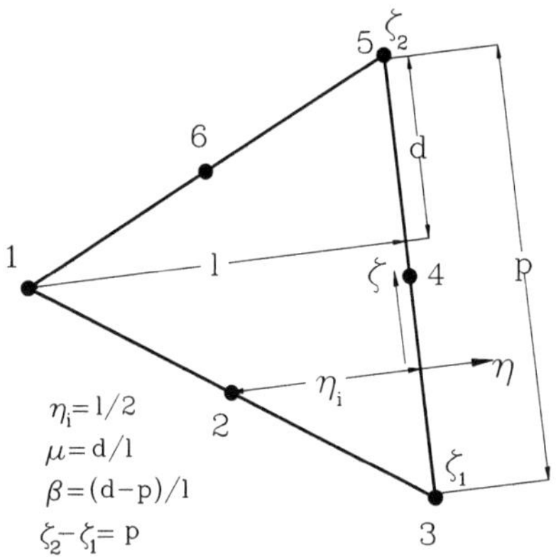

Figure C.1b Source node at 2 and integration on segment with nodes 3,4,5.

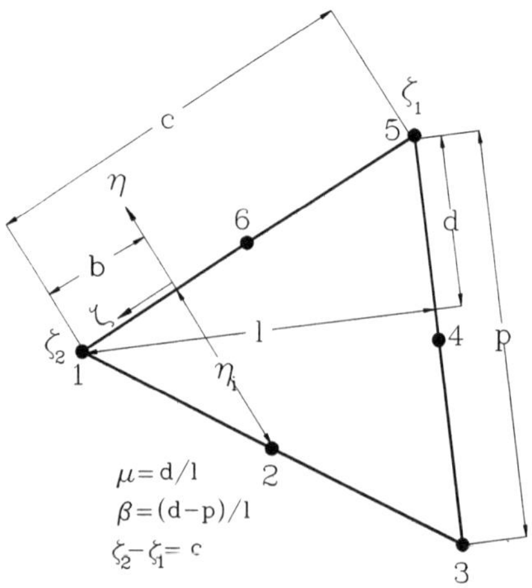

Figure C.1c Source node at 2 and integration on segment with nodes 5,6,1.

AUTHOR INDEX

Abramowitz, M., 36, 50, 299, 313
Alves, W.J., 191, 194
Aral, M.M., 3, 13

Banerjee, P.K., 2, 13, 119, 151, 313
Banks, R.B., 182, 194
Bear, J., 305, 313
Benton, E.R., 196, 215
Bouloutas, E.T., 230, 250
Boussinesq, J., 147, 152
Brebbia, C.A., 2, 3, 13, 21, 49, 96
Brian, P.L.T., 131, 152
Brooks, R.H., 221, 230
Burgers, J.M., 215
Butterfield, R., 2, 13, 119, 151, 218, 230, 313

Celia, M.A., 228-230, 249, 250
Chang, Y.P., 2, 13
Chen, D.J., 13
Chen, C-I., 313
Cheng, A. H-D., 2, 12, 13, 53, 96, 119, 151, 218, 230
Chung, T.J., 324, 331
Cleary, R.W., 306, 313
Clements, D.L., 2, 12, 218, 230
Cole, J.D., 216, 245, 250
Corey, A.T., 221, 230
Cowper, G.R., 270, 271, 288

Davies, T.G., 96

Fairweather, G., 1, 12
Feshbach, H., 296, 313
Finn, W.D.L., 205, 211, 216
Fix, G.J., 58, 96
Fletcher, C.A.J., 196, 215

Gipson, G.S., 50
Gray, W.G., 131, 152, 170, 194, 240, 250
Greenberg, M.D., 288

Haverkamp, R., 221, 228, 230, 249, 250
Hopf, E., 215
Hornbeck, R.W., 80, 96
Huyakorn, P.S., 225, 230, 283, 285, 288

Jacob, C.E., 305, 313
Jaswon, M.A., 1, 2, 12

Kakuda, K., 202, 205, 215
Kang, C.S., 13

Lafe, O.E., 2, 12, 53, 96, 218, 230
Leendertse, J.J., 131, 152
Liggett, J.A., 1-3, 12, 13, 21, 50, 119, 151, 218, 230, 267, 288, 313
Lighthill, M.J., 196, 205, 215, 247, 250
Lomen, D.O., 308, 311, 313

Lynch, D.R., 131, 152, 170, 194, 240, 250

Manolis, G.D., 53, 96
Morse, P.M., 296, 313
Mualem, Y., 221, 230

Nardini, D., 3, 13

Ogata, A., 182, 194
Onyejekwe, O.O., 13, 14, 53, 96, 215, 218, 230, 313
Ottosen, N.S., 61, 96, 324, 331
Ou, K., 119, 151
Ouaza, D., 53, 96

Partridge, P.W., 53, 96
Petersson, H., 96, 331
Philip, J.R., 225, 230
Pinder, G.F., 131, 152, 170, 194, 240, 250
Platzman, G.W., 196, 215
Ponter, A.R., 1, 12
Pullan, A.J., 218, 230

Raats, P.A.C., 218, 230
Rao, S.S., 61, 96, 324, 331
Reddy, J.N., 16, 49, 61, 96, 260, 271, 288, 324, 331
Richards, L.A., 221,230
Rizzo, F.J., 1, 2, 12

Schlichting, H., 215
Shaw, R.P., 2, 13, 313
Shippy, D.J., 2, 12, 271, 288
Skerget, P., 3, 13
Stegun, I.A., 36, 50, 299, 313
Stone, H.L., 131, 152
Strang, G., 58, 96
Symm, G.T., 1, 12

Taigbenu, A.E., 2, 3, 12-14, 53, 96, 215, 218, 230, 240, 250, 258, 288, 313
Tang, Y., 3, 13
Thomas, S.D., 230, 288
Thompson, B.M., 230, 288
Tien, H-C., 313
Tomlin, G.R., 218, 230
Tosaka, N., 215
Touma, J., 230, 250
Tsai, W.F., 310, 311, 313

Vachaud, G., 230, 250
van Genuchten, M.T., 191, 194, 221, 227, 230, 248, 250
Varoglu, E., 205, 211, 216
Vauclin, M., 230, 250

Warrick, A.W., 308, 311, 313
Wierenga, P., 230, 250
Wrobel, L.C., 2, 3, 13, 96
Wu, J.C., 2, 13
Wu, Y.S., 12

Yeh, G.T., 306, 311, 313

Zarba, R.L., 230, 250
Zienkiewicz, O.C., 58, 61, 96, 324, 331

SUBJECT INDEX

advecting velocity, 203
anisotropic (anisotropy), 13, 230, 252
assembling the global matrix, 30

basis
functions of the Lagrange family, 256
functional vectors, 35, 56
Bernoulli's assumption, 87
boundary conditions, 17, 26, 30, 32, 41, 49,56, 67, 69, 85, 87, 88, 91, 95, 114, 120, 140, 141, 151, 154, 185, 191, 193, 196, 205, 211, 244, 245, 247, 252, 253, 282, 290, 300
boundary integrations, 8, 263, 327
boundary-only character, 2, 3, 5, 53, 119
boundary-only method, 1
Boussinesq equation, 6, 218
Burgers equation, 195-198, 200-203, 214, 215, 232, 235, 244-248

Chebyshev polynomials, 206, 211
column vector, 9
complementary differential equation, 21, 53, 125, 144, 154, 158, 161, 222, 290
complementary error function, 131, 170
complete Beta function, 147
conformal elements, 59
conservation of heat, 51, 117, 119
constitutive relations, 15, 144, 218, 221, 225, 227, 228, 231, 237, 247, 248, 283, 285, 286, 310
continuity requirements, 30, 258
control volume, 16, 19, 51, 97, 117-119, 219
convective term, 196, 202
convergence, 2, 58, 77, 78, 80-83, 199, 200, 224, 225, 228, 229, 237, 248, 285, 311
requirement, 58
tolerance, 78, 225, 228, 229, 311
Courant number, 136-139, 150, 173, 176, 177, 179, 182-184, 186, 192, 240, 242

Darcy flux, 219
Darcy law, 16, 98, 217, 219
degrees of freedom, 28, 257
diagonal matrix, 10
difference weighting factor (*see* also temporal discretization), 123, 228, 229, 248, 249, 300
diffusion
equation, (*see* also heat equation) 2, 13, 117, 119, 121-125, 131, 148-150, 153, 154, 158, 161, 170, 196, 199, 200, 289
coefficient, 52, 57, 173
diffusion-advection equation (transport equation), 3, 153, 170, 173, 240
diffusivity, 119, 142, 220, 308
Dirac delta, 21, 22, 24, 34, 102, 125, 154, 162, 268
direct Green function method, 2
distributed recharge, 16, 32, 43, 46, 48, 61, 268, 269
domain integrations, 3, 4, 37, 38, 53, 153, 255, 265, 267, 272, 315, 318, 321, 325
dual reciprocity BEM (DR-BEM), 3
Dupuit-Forcheimmer assumptions, 84

eigenvalue, 98
    problem, 98
eigenvector, 99
Einstein summation convention, 11, 126
element basis functions, 60
element dimensionless diffusion coefficient, 173
element Peclet number, 173, 176, 182, 191, 240, 243
elements
    bilinear rectangular, 260
    linear triangular, 261, 262
    quadratic, 61, 318, 324
    quadratic rectangular, 316-318
    quadratic triangular, 317, 321, 323
element shape functions, 60, 61, 97, 100, 126, 144, 155, 197
elementary functions, 56, 57
element-by-element approach, 4
end support conditions, 87
error function, 131, 170
exponential functions, 57
exponential integral, 299, 302

finite analytic, 311, 313
finite difference method (FDM), 34
finite element method (FEM), 1, 27
finite elements, 13, 57, 216, 230, 288
Fourier law of heat conduction, 52, 118
Fourier-wave components, 132
Fredholm, 19, 20, 24, 48, 97, 154, 197, 255
fundamental solution, 2, 3, 21, 99, 103, 107, 125, 131, 156, 162, 197, 223, 251, 290, 300

Gaussian
    quadrature, 35, 36, 38, 270
    Legendre quadrature, 36, 37
Gauss-hill profile, 185
generalized two-level time scheme, 157
global coefficient matrix, 5, 31-33, 54, 56, 57, 73, 83, 110, 124, 192, 206, 224, 258
global functional approximation, 62
global node numbering, 31
Green's identity, 22, 254, 329

half bandwidth, 10, 31, 56, 124
heat
    capacity, 119
    coefficient, 7, 16, 52, 118
    conduction, 2, 3, 7, 12, 13, 16, 34, 51, 52, 117, 118, 140, 142, 290, 295, 300
    equation, 117, 290, 295
    flux, 18, 51, 52, 117, 118, 120
    source density, 119
Heaviside function, 24, 26, 37
Helmholtz equation, 97-115
Hermitian
    GEM, 233, 235, 237, 240-250
    polynomials, 231-233
heterogeneous, 14-16, 252
homogeneous, 15-16, 251- 252
Hooke's law, 87
hydraulic conductivity, 16, 34, 52, 57, 80, 98, 142, 218, 220, 221, 230, 282, 285, 302, 305, 308
    hydraulic head, 219, 220
hyperbolic partial differential equation, 195

identity matrix or unit matrix, 10
immiscible fluid flows, 217
implementing the BEM, 26-27
incomplete Beta function, 147
increment (vector), 79, 81, 199, 284
inertia, 87, 195, 196
infiltration processes, 219
initial condition, 120, 151
integral of the recharge term, 33, 34, 102
integral representation, 6, 20-22, 24, 27, 53, 120, 125, 144, 149, 158, 159, 196, 222, 237, 257, 288, 290, 294
integration by parts, 22
interpolation, 3, 39, 46, 56-58, 60, 61, 63, 67, 100, 102, 121, 126, 159, 163, 202, 228, 231-233, 235, 237,243, 249, 257, 260-262, 265, 270, 288, 291, 294, 297, 316, 317, 324, 326, 328, 330
    in an element manner, 4, 57
    functions, 60, 61, 121, 222
    in a global manner, 57
    least square sense, 57, 58, 285
    of the Lagrange family, 60, 61, 256, 260, 316
inverse of the matrix, 11
inverse of the transformation, 327
isoparametric elements, 315, 324, 325
    Jacobian matrix, 327-328
iteration number, 77, 82, 202, 204, 293

kernel function, 20
Kronecker delta, 36

Laplace equation, 16, 272, 324
Laplace transform, 2, 119, 140, 141, 151, 289
leakage factor, 98
linear heterogeneous problem, 69
linear interpolating functions, 62, 82, 89, 121, 222
local coordinate, 39, 60, 64, 65, 68, 232
local support, 5
lower triangular matrix, 10

master element, 325-327
medium parameters, 57, 144, 288, 292
mixed formulation, 29
multiply zoned, 55, 148, 282

nodal angle, 6, 254, 255, 296
nonlinear
- acceleration term, 196, 199, 200
- diffusion, 142, 146-150, 153
- solution strategies, 77-81
  - Jacobian matrix, 81, 83-85, 146, 148, 284, 293, 295
  - Newton-Raphson scheme, 77, 80, 81
  - Picard scheme (or linear iteration), 77-79, 201, 236, 239
- unsaturated flow, 218

norm of the error, 112

one-dimensional elements, 58, 260
orthogonal functions (or polynomials), 35, 36

penalty parameter, 102, 103, 105-109, 112, 113
- suitable range of values, 108

perturbation approach, 2
piecewise homogeneous, 94, 148, 283, 293
piezometric head, 15, 16, 98
point or concentrated sources/sinks, 34
Poisson equation, 15-97, 257, 272, 329
polygonal elements, 4, 255, 256, 260, 291
polynomial elementary functions, 57
potential flow, 1, 6
primary or dependent variable, 15, 20

quadratic interpolating functions, 61

recharge term, 33, 34, 39, 53, 62, 82, 94, 102, 121, 268, 272
representation of a complex variable, 133, 171
Reynolds number, 196
Richards equation, 217, 219, 221-223
rotation of the member, 88
row vector, 9

sampling points, 35-37, 43, 270, 271
seepage face, 147
simulation interval (*see* also time step), 299
soil constitutive relations (or data), 218, 221, 225, 227, 228, 231, 237, 247, 248, 283, 285, 286, 310
soil moisture content, 221
soil moisture diffusivity, 220
spatial discretization, 41, 110, 111, 206, 211, 229, 260, 282, 285, 310, 311
spatial frequency, 133, 171
specific discharge, 16
specific heat, 119
specific moisture capacity, 220
spike function, 21
spurious oscillations, 3, 182, 183, 249
square matrices, 9, 11
stability characteristics, 131-140, 170-181, 240, 242, 250
- amplification factor, 138, 140, 172, 173, 176, 177, 180, 181, 240-242
- dispersive effects, 132, 171
- dissipative effects, 132, 138, 170, 176, 177, 179, 183, 189, 241
- numerical amplification (or amplitude modification factor), 134, 137, 139, 172, 175, 176, 240
- numerical phase change, 134, 137, 176, 240
- numerical stability matrix equation, 135, 174
- overshoot, 182
- theoretical amplification or amplitude modification factor, 133, 134, 172
- theoretical phase change, 133, 137, 172
- wiggles, 182, 183, 186

standing (or stationary wave), 133
structural analysis of beams, 86
system of discrete element equations, 29, 71-73, 83, 101, 105, 107, 135, 138, 145, 153,

154, 157, 161, 174, 203, 291
S-profile (or S-shape), 80

Taylor's series expansion, 79, 80
temporal discretization, 176
    Crank-Nicholson scheme, 123, 142, 177, 179, 186, 191, 192, 206, 225, 240, 243, 245
    fully explicit scheme, 123
    fully implicit scheme, 123, 177, 183, 186, 189
temporal frequency, 133, 171
thermal conductivity, 52
time marching scheme, 124
time step, 131-133, 137, 138, 141, 146, 148, 150, 157, 170-172, 176, 177, 179, 189, 199, 200, 225, 227-229, 243-245, 248, 249, 300, 306, 310, 311
translatory wave, 171
transpose of a matrix, 11
transport
    marginal advection-dominant, 176, 177, 179, 186, 240
    of momentum, 195, 231, 244
    strong advection-dominant, 176, 179, 183, 186, 191, 192, 242
tridiagonal matrix, 10
trigonometric elementary functions, 57

undetermined coefficients, 56
unit response function, 1, 6, 21, 162
unsaturated
    flow equation, 13, 96, 218, 219, 230, 237, 250, 313
    soils, 51, 217-219, 230
upper triangular matrix, 9

Volterra, 19, 20
volumetric moisture content, 219

water table, 147
wave length, 133, 138, 171
wave number, 133, 171
well function, 302

Young's modulus of elasticity, 87

zero-order continuity, 231, 232
zone of influence, 4